Geology of the Alps

Geology of the Alps

Revised and updated translation of *Geologie der Alpen*, Second Edition

O. Adrian Pfiffner

WILEY Blackwell

Registered Office
John Wiley & Sons, Ltd, The Atrium, Southern Gate, Chichester, West Sussex, PO19 8SQ, UK

Editorial Offices
9600 Garsington Road, Oxford, OX4 2DQ, UK
The Atrium, Southern Gate, Chichester, West Sussex, PO19 8SQ, UK
111 River Street, Hoboken, NJ 07030-5774, USA

For details of our global editorial offices, for customer services and for information about how
to apply for permission to reuse the copyright material in this book please see our website at
www.wiley.com/wiley-blackwell.

Library of Congress Cataloging-in-Publication Data

Pfiffner, Othmar Adrian, 1947–
    [Geologie der Alpen English]
    Geology of the Alps / O. Adrian Pfiffner. – Revised and translated second edition.
        pages   cm
    Includes bibliographical references and index.
    ISBN 978-1-118-70813-2 (cloth) – ISBN 978-1-118-70812-5 (pbk.)
1. Geology–Alps. I. Title.
    QE285.P4813 2014
    554.947–dc23
                                                        2013050121

A catalogue record for this book is available from the British Library.

Wiley also publishes its books in a variety of electronic formats. Some content that appears in print
may not be available in electronic books.

Cover image: Front Cover: View of Piz Tumpiv (Canton Graubünden, Switzerland) looking
WSW. The steep gully with a string of snow in the foreground marks a folded thrust fault between
crystalline basement (Aar massif) and Mesozoic sediments (Cavistrau nappe). See Fig. 6-10B for
explanation. Photo by O. A. Pfiffner.
Back Cover: Cross-section through the Central Alps of Switzerland showing nappe structures
in Mesozoic-Cenozoic sediments and the involvement of crystalline basement (Aar, Gotthard,
Verampio, Antigorio and Maggia). The deep structure of the crust is based on seismology and
earthquake tomography (Diel et al. 2009), cross-section by O.A. Pfiffner 2014.
Cover design by Steve Thompson
Graphics by A. Baumeler and O.A. Pfiffner

Set in 9.5/12pt Caslon by SPi Publisher Services, Pondicherry, India

Contents

Preface

for Anne-Marie

I have been lecturing on the 'Geology of Switzerland' at the Institute of Geological Sciences, University of Berne, since 1987. The module for students in their second year studying Earth sciences as their main or subsidiary subject forms the basis for excursions, practical work, in-depth courses on regional geology and geodynamics, as well as for Master's and Doctoral research on Alpine geology. When preparing the lectures for this module, it soon became clear to me that an explanatory text was required for the course material, in addition to illustrations. The resultant lecture notes were corrected and amended each year and I undertook a basic revision on two occasions. Over the course of recent years, the desire emerged for the production of a new version in the form of a book. This book was to extend the focus of the text and provide a more detailed illustration of the situation in the neighbouring regions of the Western and Eastern Alps. A sabbatical in the spring semester of 2008 gave me the time to research the literature, write the text and design the many illustrations. This work then continued throughout the following autumn semester, piecemeal and during my 'free' time.

The geology of the Alps is multifaceted. For a start, there are the different types of rocks – sedimentary, igneous and metamorphic – all of different ages and with manifold processes underlying their formation. In addition, plate tectonic processes, the formation of sediment basins, mountain-building or the uplift of the Alps to produce a high-altitude mountain range and its subsequent erosion are of importance. However, rock-forming processes and plate tectonics are intrinsically linked, such that structuring this material is not easy. Division along broadly chronological lines appeared to be the best solution. For this reason, after placing the Alps in a European context, the pre-Triassic basement is discussed first. The subsequent chapters focus on the Mesozoic and then on the Cenozoic building blocks. In each of these three chapters, the rock formations are presented first. This is followed by a discussion of the emergence of these rock formations in the plate tectonic framework. In all cases, the selection of the material posed a real challenge, as each of these chapters is worth a book in its own right. The final chapter on the most recent events in the Alps makes the transition to the current geological situation. The selection of material posed problems here as well and many interesting aspects have been neglected.

I was able to rely on the help of many colleagues during the realization of this book. Above all, I would like to thank Andreas Baumeler. He helped me produce the graphics, provided enormous input for the design of the illustrations, advised me on the use of colours and symbols and improved my drafts. Many of the illustrations moved back and forth between us several times before we were both happy with the end result. A book on the geology of the Alps requires a large amount of illustrative material. The publisher's willingness to accommodate our wishes and print all of the illustrations in colour was very obliging. My thanks also go to the publisher's proof-reader, Claudia Huber, and to Marco Herwegh, who checked the figures and figure legends for consistency and typographic errors. I am grateful to my colleagues at the Institute, in Switzerland and abroad, for numerous discussions and for answering my questions, of whom there are too many to name here.

Finally, I would also like to thank my wife, who has had to go without many things during the writing of the book, but who always showed great understanding for my work. I would like to dedicate this book to her.

Berne, January 2009

Comments on the second edition

After the publication of *'Geologie der Alpen'*, I was sent numerous comments. A commonly expressed wish was for the inclusion of an index. I was happy to comply with this wish in this second edition. The revision also gave me the opportunity to expand the Appendix to include a geological time scale. However, the comments I received also revealed weaknesses that have been addressed in the second edition. I am grateful, in particular, to Hanspeter Funk for his help on the correct usage of stratigraphic terms, to Marino Maggetti for pointing out errors in traces of cross-sections, to Christoph Spötl for his suggestions on ice age stratigraphy, to Wolfgang Frisch for his comment on the position of the Tauern relative to the flysch basins and to Henry Naef for prompting me to think about the deep structure in Vorarlberg. A total of over 50 illustrations have been redesigned to lesser and greater extents.

Berne, January 2010

Comments on the translation

Soon after the publication of the German version of this book, *Geologie der Alpen*, which appeared in the UTB/Haupt series I was urged to consider an English version. Many colleagues considered the book to be a standard work for the Alps that would interest an international audience amongst scientists and students. I was also encouraged to include more photographs and to expand some of the material. In particular I was asked to place more emphasis on Alpine metamorphism. Given that I had more space available I was happy to comply with these demands.

Along with the translation I updated many figures. The palaeogeographical maps showing plate configurations have undergone a major overhaul in the presentation. The cross-sections spanning the entire Alps were harmonized and now include the crustal structure as determined from earthquake tomography. At this point, I wish to thank Edi Kissling for his invaluable help in providing me with geophysical data and keeping an eye on my interpretations. To show the kinematic evolution of the Alps serial cross-sections were overhauled and a new set of cross-sections was produced for the Central Alps of western Switzerland, a classic area for nappe tectonics studied by Argand in the nineteenth century. In the chapter on the most recent history of the Alps, the inclusion of data from Italy and France provided a much more complete image of the present-day uplift pattern of the Alps. Eckhart Villinger pointed out data from the Alpine foreland to me that allowed updating the palaeogeographical maps containing the courses of the ancestral rivers draining the Alps.

The text was translated by Dr Deborah J. Curtis in cooperation with tolingo translations, Hamburg. Deborah quickly grasped what I intended to convey to the reader and spared me innumerable hours of translation that would never have reached the quality she obtained. I express my sincere gratitude to her. For the preparation of the figures, I was again able to count on the assistance of Andreas Baumeler. His eyes quickly caught the weak points that crept in with modifiying the original figures.

Berne, June 2013

1 The Alps in their Plate Tectonic Framework

▶ **Figure 1.1** Tectonic
map of Europe showing
mountain ranges coloured
according to their age of
formation and associated
terranes and continents.

Rocks can be found in the Alps that range in age from one billion years to present times. The rocks themselves – sedimentary, igneous, metamorphic and unconsolidated rock – cover the entire conceivable spectrum. Many of these rocks and their formation can be understood only within the context of the geological structure of Europe and the associated plate tectonic processes. In the following therefore, the plate tectonic framework for Europe, the older mountain chains and the younger Alpine mountain ranges in Europe will be considered briefly.

1.1 Older Mountain Chains in Europe

From a geological perspective, the European continent has a highly chequered history. Although the Alps are an integral component of this continent and are, essentially, a spectacular mountain chain, their origin lies in the recent geological history of the continent. In order to understand the geological structure of Europe, the individual regions need to be classified according to the age of their consolidation. In this case, the term consolidation is taken to mean the welding of continents, following on from the motion of plates. Almost all of the mountain chains in Europe originated as a result of plate movements, where an ancient ocean was swallowed up in a subduction zone and the continental blocks subsequently collided with each other. The density of continental crust is relatively low and, therefore, buoyancy acts against it sinking to greater depths once it has entered a subduction zone. As a result, continental crust remains close to the surface

Baltic Shield

and is compressed. During this process, the uppermost portions of the crust are pushed upwards and gradually build a mountain chain. This process is called orogenesis or mountain-building.

A number of such collisions between continents, or orogenies, have occurred during the geological evolution of Europe. Accordingly, we distinguish between Caledonian, Variscan and Alpine orogens. The continental plates involved in these collisions were North America, Siberia, Baltica/Europe and Africa and are also called terranes. The tectonic map in Fig. 1.1 takes this division into consideration. Europe has also been subdivided into Eo-, Palaeo-, Meso- and Neo-Europe, based on the relative ages of these orogenies. It must be noted that the terranes mentioned above contain rock units that are relics of even older, fully eroded mountain chains.

Eo-Europe is a large geological structure, a welded block that experienced no further orogenies after the Precambrian. Two geological provinces are distinguished within Eo-Europe: the Baltic Shield and the Russian Platform.

The Baltic (or Fennoscandian) Shield is a convex bulge or shield covering a large area, which is composed of a highly metamorphic crystalline basement (Baltica in Fig. 1.1). Multiple, very ancient and fully eroded mountain chains can be distinguished within these series of rock formations. The oldest rocks in the Baltic Shield are three to three and a half billion years old and were encountered in a deep drill core obtained in the region of Kola, to the south of the White Sea, as well as in Lapland.

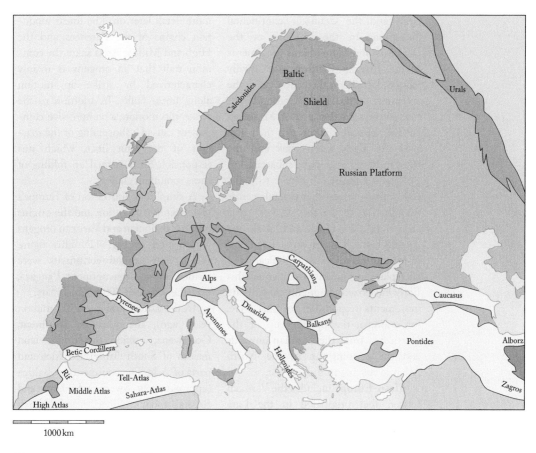

1000 km

Mountain ranges

- Alpine
- Variscan
- Caledonian

Terranes

- North America
- Siberia
- Baltica/Europe

- Africa
- Arabia

- Phanerozoic sediments
- —— Traces of cross-sections

The Russian Platform is the sedimentary cover over the Baltic Shield and is composed of Neoproterozoic non-metamorphosed sediments, overlain by Cambrian rocks as well as a series of rock formations that extend into the Cenozoic. In the southeast, the platform plunges beneath the foreland of the Caucasus, to the north of the Caspian Sea, and in the east and west, beneath the forelands of the Ural and Carpathian Mountains. The internal structure of the plate contains local depressions or basins with thick sedimentary successions as well as zones with a thin sedimentary cover. The sediments of the Russian Platform reflect the later phases of mountain-building that took place at its margins. Examples are the famous Old Red Sandstone, continental fluviatile sediments of the Middle to Late Devonian that are the erosional product from the (Caledonian) mountains in Norway and Scotland, the Permo-Triassic continental lagoon sediments in the foreland of the (Variscan)

Russian Platform

Urals and the Cenozoic continental formations in the foreland of the Caucasus and Carpathians. Sediments of the Russian Platform are usually marine deposits in the centre (with the exception of the Early Carboniferous coal swamps in the area of Moscow), but the sea retreated towards the south after the Early Cretaceous and the Russian Platform became subaerial.

Palaeo-Europe refers to the Caledonian orogen that extends across Scandinavia to Ireland. Other parts are found in Greenland and the Appalachians. This broad geographical distribution is sufficient to indicate that later plate movements fragmented this Early Palaeozoic mountain chain. Plate movements responsible for this were, for example, the opening up of the North Sea from the Permian onwards and the opening up of the North Atlantic starting in the Jurassic.

Meso-Europe includes the Variscan orogen that originated in the Late Palaeozoic. With the exception of the Urals, the Variscan mountain chain can be followed as a continuous range, which in Germany and France is generally completely eroded and covered with younger sediments, as illustrated by the island-like distribution of remnants of these mountains shown in Fig. 1.1.

Finally, **Neo-Europe** comprises a series of mountain chains that originated in the Jurassic (Turkey), in the Cretaceous (parts of the Alps and Pyrenees), but mainly in the Cenozoic. These mountain chains are often winding and arc-shaped. In addition to the Alps, good examples are the Carpathians and the Betic Cordillera–Rif–Tell–Atlas system. This arc shape is essentially due to the geometry of the plate boundaries of the different associated microplates, a point that is discussed in

more detail later on. The linear mountain chains of the Pyrenees and the High and Middle Atlas share the common trait that an orogeny is mainly characterized by strike-slip motion along linear faults. In addition to the strike-slip motion, a compressive component caused a shortening of the margins of the fault lines, which was responsible for the actual 'up-folding' of these mountain chains.

A simplified illustration of Europe's plate tectonic evolution and the origins of the Caledonian and Variscan orogens is provided in Fig. 1.2. This figure shows how several continents were welded into a megacontinent, Pangaea, over the course of 300 million years.

In the Late Cambrian (500 million years ago), the southern continent, Gondwana, unified the extant land masses of South America, Africa and parts of Asia. The continents of Baltica (approximately Sweden, Finland and Russia today), Siberia and North America were surrounded by oceanic basins, in which thick sedimentary deposits accumulated. At the northern continental margin of Baltica, 1400 metres of grey and reddish arkoses, conglomerates, limestones and shales were deposited in the shallow part of the Iapetus Ocean during the Proterozoic (about 600 million years ago). The arkoses also contain tillites, that is, fossilized diamictites (glacial deposits that indicate very ancient glaciations). The Cambrian starts with a basal conglomerate that contains alum slate, that is, a dark pelite rich in iron sulphide. The marine sedimentation continued in the Ordovician–Silurian, with clay, limestone and turbidite deposits. Greenstones with gabbro and peridotite, typical rock associations in a newly developing oceanic crust, originated in the Iapetus Ocean itself. Finally, 6000 metres of Torridonian arkoses, conglomerates, sandstones, greywackes and pelites were deposited at the North American continental margin in the Proterozoic. This was followed by quartzites in the Cambrian and then thick dolostones, which continued to be deposited into the Ordovician.

The Iapetus Ocean was gradually closed through subduction and a large mountain range was formed due to the collision of Baltica with North America: the Appalachians in North America and the Caledonian orogen in Europe (Scandinavia and the Bristish Isles).

Figure 1.3 shows two cross-sections through the Caledonian mountain chain. The cross-section through the Caledonian mountain chain in Scandinavia shows how the Baltic Shield was overthrust in an easterly direction by large thrust sheets containing the Precambrian crystalline basement of the past continental margin of Baltica and its Proterozoic–Palaeozoic sedimentary cover. These crystalline nappes were thrust onto the Baltic shield over hundreds of kilometres, as can be seen from the example of the Jotun Nappe. The thin obducted nappes of Aurdal and Synfjell are mainly composed of Early Palaeozoic sediments. At the extreme east, the Oslo Graben is visible, which is a rift within the Baltic shield that is largely filled with Permian igneous rocks. In the west, towards the North Sea, there are ophiolitic rocks overlying the Jotun Nappe, relics of the Iapetus Ocean. Fragments of this ocean were not subducted during the collision between Baltica and North America, but instead incorporated into the developing mountain chain.

A certain similarity can be seen when comparing the cross-section through Scotland with that through Scandinavia. The cross-section shown in Fig. 1.3 has been adapted from

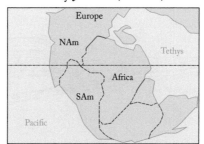
Early Jurassic (200 Ma)

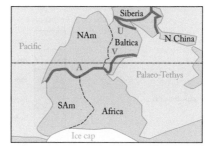
Late Carboniferous (300 Ma)

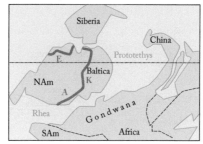
Early Devonian (400 Ma)

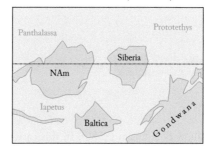
Late Cambrian (500 Ma)

— Rift   ▬ Mountain chain
----- Shape of present-day continents

Figure 1.2 Plate tectonic evolution of Europe shown in four time slices. Positions of plates are based on Blakey (2008) and Scotese & Sager (1988). A, Appalachians; K, Caledonides; E, Ellesmere orogen; V, Variscan orogen; U, Urals; NAm, North America; SAm, South America.

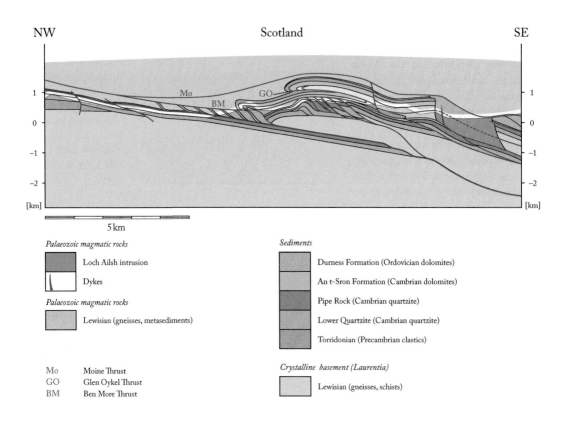

NW   Scotland   SE

Palaeozoic magmatic rocks

Loch Ailsh intrusion

Dykes

Palaeozoic magmatic rocks

Lewisian (gneisses, metasediments)

Mo   Moine Thrust
GO   Glen Oykel Thrust
BM   Ben More Thrust

Sediments

Durness Formation (Ordovician dolomites)

An t-Sron Formation (Cambrian dolomites)

Pipe Rock (Cambrian quartzite)

Lower Quartzite (Cambrian quartzite)

Torridonian (Precambrian clastics)

Crystalline basement (Laurentia)

Lewisian (gneisses, schists)

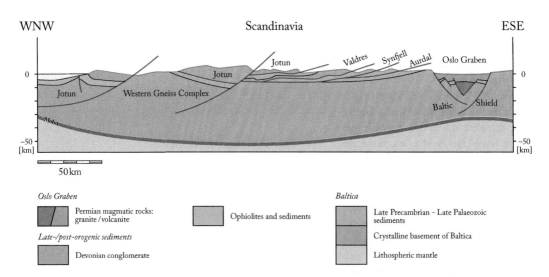

WNW   Scandinavia   ESE

Oslo Graben

Permian magmatic rocks:
granite / volcanite

Late-/post-orogenic sediments

Devonian conglomerate

Ophiolites and sediments

Baltica

Late Precambrian - Late Palaeozoic
sediments

Crystalline basement of Baltica

Lithospheric mantle

Elliott & Johnson (1980). In this case, the Precambrian crystalline basement has also been included in the structure of the nappe. This 'Lewisian' basement is exposed in the Outer Hebrides in the northwest of Scotland. The 'Lewisian' has a greater affinity to the crystalline basement of the North American craton in the foreland of the Appalachians in Canada and Greenland than to the Baltic Shield. From a geological perspective, the Outer Hebrides must thus be regarded as part of North America.

The highest unit, the Moine Nappe Complex, has been almost completely eroded in this cross-section. However, the degree of metamorphism in the rocks below permits the clear conclusion that there was once a thick nappe pile overlying the currently visible rock formations (Strachan et al. 2002), because the Moine Nappe Complex exhibits a higher degree of metamorphism than the rocks below it.

The chronology of the formation of these nappes can be narrowed down based on the example of the Loch Ailsh intrusion: the Glen Oykel Thrust is cross-cut by the intrusion (so, is older), while the Ben More Thrust displaces and transports the intrusion (so, is younger) and the Moine Thrust caps the intrusion and is therefore also younger. The Loch Ailsh intrusion is dated at 434 million years ago (Silurian). It can be regarded as synorogenic, as it is located chronologically between the formation of the Glen Oykel and the Ben More or Moine thrusts.

In great contrast to Scandinavia, the nappes in Scotland (and in the Appalachians) were transported in a northwesterly direction. Overall, the Caledonian mountain chain therefore exhibits a bivergent nappe structure, a structure that is typical for mountain chains that emerge from a continent–continent collision.

The Rheic Ocean remained intact after the collision between Baltica and North America (Rhea in Fig. 1.2). The continent Gondwana, which included the land masses of South America and Africa, lay to the south of the Rheic Ocean. Towards the northeast, the Rheic Ocean met up with the Proto-Tethys Ocean, which separated the continental masses of Siberia and China. The Old Red Sandstone is the product of erosion of the Caledonian mountain chain and the Appalachians, and represents delta deposits at the margin of the continent (i.e., close to the Caledonian mountain chain). Sandy clastic sediments were also deposited at the northern margin of Gondwana (in today's Atlas). Pure limestones and shales were deposited in the Rheic and Proto-Tethys oceans in the Devonian, which are now exposed in the Eastern Alps. A microcontinent can be seen in the centre of the Rheic Ocean. Sedimentation is patchy here, indicating shallow water depth. These sediments have been preserved, for example, in the North Alpine foreland, in the Vosges, the Black Forest and in the Bohemian Massif. This microcontinental zone is also called Moldanubicum. The closure of the Rheic Ocean and collision of the continents North America–Baltica and South America–Africa led to the Variscan orogen (Fig. 1.1). The collision stages mainly occurred around the Devonian–Carboniferous transition, 345 million years ago, and then in the Early Carboniferous, 320 to 300 million years ago. In Europe, the Rhenish Schiefergebirge, the Ardennes, the Cantabrian Mountains on the Iberian Peninsula, as well as the mountainous areas in Brittany and in the Massif Central originated. On the North American side, the Southern Appalachians were uplifted. Therefore, the Appalachians have a more complex history of formation and are the product of more than one collision.

Plate convergence between Siberia and Baltica led to the uplift of the Ural

◀ **Figure 1.3** Geological cross-sections through the Caledonian orogen in Scandinavia and Scotland. In both cases, the crystalline basement is affected by thrusting and is involved in the nappe structure. Transport, however, occurred in opposite directions in Scandinavia and Scotland. Source: Based on Elliott & Johnson (1980).

Caledonides: bivergent orogen, basement involved in thrusting

Mountains (Fig. 1.2). The Proto-Tethys was reduced to an almost closed sea basin, the Palaeo-Tethys. For example, as is shown In Fig. 1.2, Baltica migrated from the Southern Hemisphere, northwards across the Equator and into the Northern Hemisphere during the period from the Late Cambrian to the Early Carboniferous. The southern tip of the welded continent underwent a glaciation in the Early Carboniferous.

The internal structure of the Variscan mountain chain in Europe is illustrated in Fig. 1.4 in a cross-section through Germany (redrawn from Matte 1991). A bivergent nappe pile was formed during the collision, which also includes the crystalline basement, similar to the case for the Caledonian mountain chain. The nappes at the continental margin of Gondwana were transported to the southeast and on the margin of Baltica to the northwest. The thick Palaeozoic sediments of the Rhenish Slate Mountains were pushed over each other like tiles and folded. The basal detachment of the nappes took place in a thick slate horizon. At the core of the orogen there is a vertically oriented fault that can be followed from Portugal to Bohemia, which probably represents a component of strike-slip motion. Even though the exposure of the core of the orogen is only patchy and is often covered by younger sediments, remnants of the Rheic Ocean can be found in a variety of locations in the form of ophiolites.

In the early Jurassic, 200 million years ago, the welded continental mass that became Pangaea had migrated further to the north and experienced a degree of anti-clockwise rotation (Fig. 1.2). The ocean between Africa and Asia became the Tethys. However, movements had now started that slowly led to the break-up of Pangaea. For example, one rift separates the Indian

Variscides: bivergent orogen, basement involved in thrusting

▶ **Figure 1.4** Geological cross-section through the Variscan orogen in central Europe. The crystalline basement is affected by thrusting and involved in the nappe structure. But the transport directions on either side of the orogen are opposite. Source: Matte (1991). Reproduced with permission of Elsevier.

subcontinent from Africa and another rift opened up between Africa and the Variscan mountain chain. This rift formation and the associated ingress of the Tethys Ocean towards the west are crucial to the understanding of the geology of the Alps and are therefore discussed in more detail below.

1.2 Break-up of Pangaea and Opening of the Alpine Tethys

The plate tectonic processes during the break-up of the megacontinent, Pangaea, had a variety of effects on the Alps when they were formed later on. The small ocean basins and microcontinents that originated during this process resulted in a complicated juxtaposition of different sedimentary environments: deep-sea basins, shelf seas and submarine rises. The very different sedimentary facies representing these environments are now visible in the Alps juxtaposed vertically and laterally, in apparent complete disorder. The palaeogeographical shapes of the sedimentary basins affected the architecture

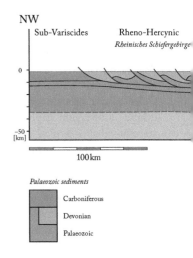

of the Alps when these basins were closed during the formation of the Alps.

The break-up of Pangaea is illustrated in three snapshots in time in Fig. 1.5. A level of uncertainty is associated with all of these plate reconstructions, which is why these palaeogeographic maps also vary greatly from one author to the next. Figure 1.5 was simplified and redrawn from illustrations by Blakey (2008). In the Early Triassic (Keuper), 230 million years ago, Pangaea broke up along a rift that opened up between Gondwana and Laurasia. The rift originated in the Tethys, and extended along an arm of the Tethys between the continental masses of Arabia and Greece–Italy. The Palaeo-Tethys was then closed by subduction, such that the Turkish landmass was welded to Laurasia (Baltica) in the Middle Jurassic (Dogger), 170 million years ago. The eastern part of the rift shifted to the north and now separated the landmass Greece–Italy from Laurasia. This small ocean basin is referred to as the Ligurian or Piemont Ocean in Alpine geology. The rift expanded to the west and separated Africa–South America from North America. This rift was the precursor to the Atlantic and extended as far as Mexico. In the Early Cretaceous, 120 million years ago, North America and Africa drifted further apart and the Central Atlantic was born. In the north, Iberia separated from North America. The movement of Iberia was due to a spreading ridge in the west (mid-oceanic ridge of the Atlantic as it was opening up) and a transform fault to the north and to the south of Iberia. In the north, further spreading systems extended either side of Greenland. These were the forerunners for the opening up of the North Atlantic.

Figure 1.6 shows a simplified plate reconstruction for the Jurassic–Cretaceous transition (about 145 million years ago), adapted from Wortmann et al. (2001). Opinions diverge on the exact geometries of the individual basins, but the solution given in Fig. 1.6 is a combination of their essential characteristics. This shows a continental fragment extending from Iberia in a northeasterly direction and composed of Corsica–Sardinia–Briançon, which became separated from Iberia. The Corsica–Sardinia–Briançon continental

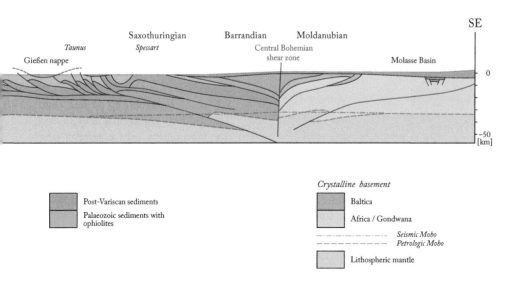

Figure 1.5 The break-up of Pangaea shown in three time slices. Positions of plates are based on Blakey (2008) and Scotese & Sager (1988). Grl, Greenland; It, Italy; Gr, Greece; Tu, Turkey; SAm, South America; Wr, Wrangellia; Mex, Mexico.

Early Cretaceous (120 Ma)

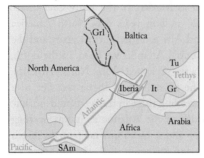

Middle Jurassic/Dogger (170 Ma)

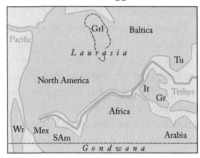

Late Triassic/Keuper (230 Ma)

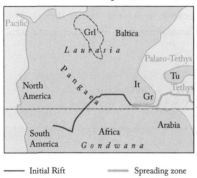

—— Initial Rift   ══ Spreading zone

—— Transform fault

fragment corresponds to the so-called Briançon microcontinent, a submarine rise that can be traced from the Western Alps through to the Swiss Alps. The sea basin to the northwest of the

Briançon microcontinent corresponds to the Valais Basin, and that in the southeast corresponds to the Penninic Ocean. A transform fault separated Iberia from Europe and acted as a local plate boundary during the drift of Iberia away from North America. Another transform fault connects the Piemont with the Penninic Ocean. The opening up of the Atlantic occurred simultaneously with an oblique opening up of the Ligurian–Piemont and Penninic oceans. The Dauphinois–Helvetic realms on the southeastern margins of Europe and the Southalpine Dolomites to the north of the Adriatic and in the Eastern Alps between the Vardar and Penninic Ocean are of particular relevance to today's Alps.

Figure 1.7 illustrates the broad palaeogeographical situation at the Barremian–Aptian transition, 125 million years ago. The reconstruction is based on Wortmann et al. (2001). The Ligurian–Piemont Ocean is characterized by multiple transform faults that indicate progressive oblique opening up of this ocean. A transform fault separates the Adriatic from the microcontinents Bakony, Austroalpine and Tiza. The Valais Basin opened up further due to the thinning of the continental margin of Baltica, or Europe, and evolved oceanic crust only in certain locations, in 'pull-apart basins'. In contrast, the Piemont Ocean had a mid-oceanic ridge that led to the formation of oceanic crust. The Adriatic continental margin and the Austroalpine microcontinent were stretched in an east–west direction by the opening up of the Piemont and the Penninic oceans. The normal faults in the future Austroalpine and Southalpine realms are evidence of this.

The Alps originated as a result of convergent plate movements between

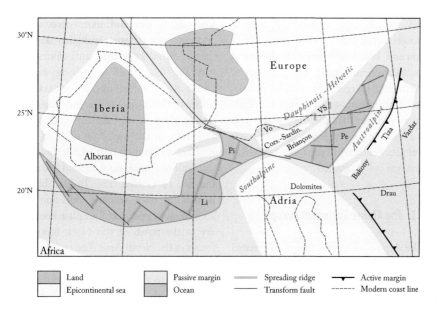

Figure 1.6 Plate reconstruction for the Berriasian (ca. 145 million years ago), simplified after Wortmann et al. (2001). The Ligurian (Li)–Piemont (Pi) Ocean stretches between the microcontinents of Iberia and Adria. It is disrupted by a transform fault and continues as the Penninic (Pe) Ocean between the Briançon and Austroalpine continental fragments. Narrow basins, the Vocontian (Vo) and Valais (VS) straddle the southern margin of the European continent. Cors-Sard, Corsica–Sardinia continental fragment. Source: Wortmann et al. (2001). Reproduced with permission of John Wiley & Sons.

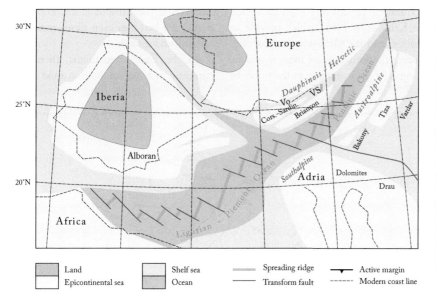

Figure 1.7 Plate reconstruction for the Barremian (ca. 125 million years ago), simplified after Wortmann et al. (2001). The Ligurian–Piemont Ocean is now wider and aligned with the Penninic Ocean. Similarly, the Southalpine and Austroalpine domains are now aligned. Cors-Sard, Corsica–Sardinia continental fragment; Vo, Vocontian basin; VS, Valais basin. Source: Wortmann et al. (2001). Reproduced with permission of John Wiley & Sons.

Baltica/Europe and Africa–Arabia. During this process, the sea basins that lay between, the Piemont Ocean and the Valais Basin, were closed up by subduction. This occurred in two separate stages.

The Piemont–Penninic ocean was closed by subduction proceeding in a westerly direction in the Cretaceous, the Valais Basin by a collision in the Cenozoic between the Briançon microcontinent and the Adriatic continental margin, proceeding in a more north–south direction and, later on, between the Briançon microcontinent and the European continental margin. The complex palaeo-geography illustrated in Fig. 1.7 leads us to surmise that the subduction and the collision process led to an even more complex geometry in the mountain chain that was being uplifted.

1.3 The Alpine System in Europe

The Alpine mountain ranges originated in the Cretaceous and in the Cenozoic. These ranges include, for example, the 'young' European mountain ranges (Betic Cordillera, Pyrenees, Alps, Apennines, Carpathians, Dinarides). Of note is that these mountain ranges are winding and arc-shaped. Figure 1.8 summarizes the continuing present-day motions (based on Kahle et al. 1995), which provide an insight into the plate tectonic processes during the formation of these mountain ranges. Africa is moving to the north by

four millimetres and more each year. Movement in the west is slightly slower, that is, Africa is rotating anti-clockwise very slightly. Arabia is moving much faster, at 25 millimetres per year, in a northern direction. The jump in speed is taking place in a strike-slip fault that originates in the spreading ridge in the Red Sea and extends northwards through the Gulf of Aqaba, via the Dead Sea and the Sea of Galilee. The Turkish block is moving in a westerly direction by 25 millimetres per year. The plate boundary in the north of this block is to be found in the North Anatolian fault line, a seismically active dextral strike-slip fault. This drift to the west changes its direction to south-southwest in the Aegean. Its speed increases, as the Aegean is expanding in the same direction. The plate movements are revealed to be quite complicated, even just between Africa, Arabia and the Turkish block. Further north, this becomes even more complicated to understand.

In the East Carpathians, there is currently an active subduction zone that

Figure 1.8 The tectonic plates in the present-day Alpine system. Open arrows with velocities (mm/a) show the direction of plate motions, simple arrows indicate directions of thrusting within the Alpine orogens. Double arrows denote areas of extension and opening of ocean basins. Source: Kahle et al. (1995). Reproduced with permission of Elsevier.

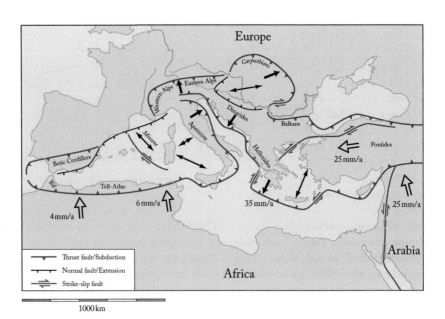

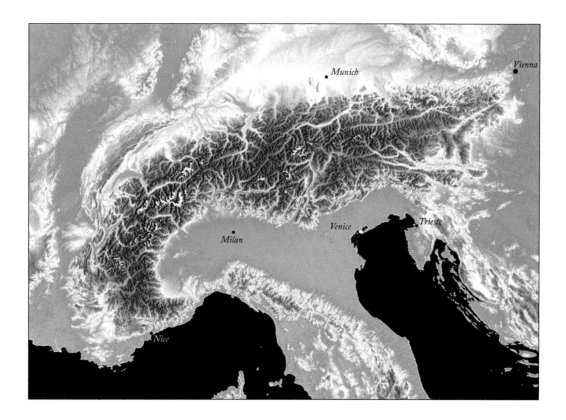

▲ **Figure 1.9** Digital elevation model of the Alps and neighbouring areas. Within the Alps, major valleys running parallel and across the orogen are clearly discernable. Large lowland areas lacking relief stretch across the foreland of the Alps. They correspond to the Rhine Graben in the north, the Bresse–Rhone Graben in the west, the Po Basin in the south and the Pannonian Basin in the east. Source: US Geological Survey.

is plunging towards the west. The Eurasian plate is sinking and simultaneously exhibiting slab retreat or roll back, that is, the plate boundary is moving towards the east in this subduction zone. As a result of this, the Pannonian Basin on the plate above is being stretched in an east-west direction. The Tiza block at the base of the Pannonian Basin is being squeezed out laterally towards the east by the pincer movement between Europe and Africa (or Apulia). However, at the same time, the Apennines and the Dinarides are also moving towards each other. Stretching and new formation of oceanic crust has been detected in the Tyrrhenian Sea in the hinterland of the active Apennines (Facenna et al. 2002). This process started about five million years ago. A little earlier, in the Miocene, the Ligurian Basin opened up under similar circumstances. At that point, the Corsica–Sardinia microcontinent separated from Europe and, during the subsequent rotation away from Europe, new oceanic crust developed in the Ligurian Sea. Therefore, movements of the smallest blocks between the two colliding continental plates also occurred in this case.

Even today, active horizontal motion can be measured in the Alps (Tesauro et al. 2005). The southern margin of the body of the Alps is moving slowly towards the north-northwest, at about 1.2 millimetres per year, and the northern margin is moving at only about 0.7 millimetres per year, but in the same direction. This indicates that the Alps are contracting by about 0.5 millimetres per year in a NNW–SSE direction.

The complicated present-day pattern of movement gives an impression of how the movements that occurred during the formation of the Alps must be envisaged. The sizes of the ocean basins and continents or microcontinents that were involved were modest in comparison with the dimensions in the classic subduction orogens of the Andes or the North American Cordillera, or in the collision mountain chains of the Himalayas or Appalachians. However, the convergent movements were qualitatively comparable and made the Alps into a mountain chain with such a highly heterogeneous structure.

1.4 Structure of the Alps

The mountainous body of the Alps extends in a wide arc from Nice to Vienna. The Po Basin lies within the arc. It is morphologically distinct due to its low altitude and minimal relief, as is made clear in the digital elevation model in Fig. 1.9. Long, narrow basins with no relief are visible outside the Alpine arc: the Rhone–Bresse Graben in the extreme southwest, the Rhine Graben in the north. In the far east of the region, the Alps disappear under the Vienna Basin.

Along this mountain chain, the Alps are subdivided into the Western Alps, Central Alps and Eastern Alps. The Eastern Alps run more or less east–west and their western boundary is located on an approximate line through St Margrethen–Chur–Sondrio. In the Central Alps, the course of the mountain chain changes from east–west to almost north–south. The Western Alps run from north to south, but form a tight arc round the western end of the Po Basin. The boundary between the Central and Western Alps is diffuse.

Western Alps
Central Alps
Eastern Alps

Some authors therefore only subdivide the Alps into the Western Alps and the Eastern Alps. While the three-part division is preferred here, this is simply due to the internal structures, which are hereby easier to classify in a comprehensible manner.

Across this mountain chain, the Alps are subdivided into tectonic units that belong to specific palaeogeographical domains. The palaeogeographical affiliation is defined by the Mesozoic sedimentary environment in these units. Based on this structure, we can distinguish a belt of sedimentary rocks that belong to the European continental margin and are exposed in the extreme external regions of the Alps, that is, extreme west and north. These rock formations are referred to as 'Dauphinois' and 'Helvetic'. A second belt of sedimentary rocks, summarized using the term 'Penninic', is located in a more central position, that is, it lies further to the east or south. The associated Mesozoic sediments were deposited in marine basins between the European and Adriatic continental margins. A third belt of sedimentary rocks is to be found mainly in the most internal location, towards the Po Basin. These units are referred to as 'Austroalpine' and 'Southalpine' and are allocated to the Adriatic continental margin. In general, the Penninic units lie on top of the Helvetic and the Austroalpine unit on top of the Penninic. All these units are actual nappe complexes that were transported hundreds of kilometres from their substratum in the form of relatively thin sheets of rock. Figure 1.10 shows the distribution of these nappe complexes throughout the Alps.

The Austroalpine nappe complex makes up almost the entire Eastern Alps. It is only at the outer margins in the north and east that Penninic and Helvetic nappes can still be recognized in the footwall of the Austroalpine nappes. In the centre of the Eastern Alps, in the Tauern Window, the Austroalpine nappe complex is eroded, such that a spectacular view of the Penninic and Helvetic nappes lying below is revealed. A smaller, but otherwise equivalent, window is found slightly to the west, in the Lower Engadin. Further to the west, the Austroalpine nappe complex is almost fully eroded in the Central Alps. However, small erosional remnants, called klippen or outliers, remain as evidence for the original distribution. The largest of these klippen is to be found in the region of the Dent Blanche.

The Southalpine nappe complex and the adjacent Dolomites to the east are separated from the Austroalpine nappe complex by a major fault, the peri-Adriatic fault system. This continues eastwards into the Karawanks, where it separates the Dinarides from the Eastern Alps. The Southalpine nappe complex, the Dolomites and the Dinarides, were tectonically independent of the Austroalpine nappe complex. Only the affinity of the Mesozoic sediments to the Adriatic continental margin constitutes a common element.

The distribution of the Penninic nappe complex also reveals a large klippen in the region of the Central Alps, on the northern margin of the Alps. This is to be found in the French–Swiss Prealpine and in the Chablais region in France. Further smaller klippen are to be found in central Switzerland. These klippen also provide evidence that the Penninic nappes once covered large parts of the Alps.

The Jura Mountains are visible in Fig. 1.10 at the outermost margin of the Central Alps. At the end of nappe formation in the Alps, this banana-shaped mountain range was compressed, folded and pushed to the northwest.

Younger, Cenozoic basins demarcate the edge of the Alps. In the north of the Alps, the Molasse Basin extends from Vienna, via Munich into the Swiss Central Plateau and peters out in a westerly direction. The Molasse Basin is a foreland basin that developed in the Oligocene–Miocene after Alpine nappe formation and was then filled with clastic deposits from the uplifting Alps. In the course of the most recent nappe movements, the Molasse Basin was mainly compressed at its southern margin and even pushed to the northwest in the region of the Jura Mountains.

The rift system with the Rhine Graben and the Rhone–Bresse Graben is visible outside the Jura. The two rift basins are connected to each other by a transform fault system. Basement uplifts exposing crystalline rocks flank the rift basins on both sides: the Black Forest and Vosges, and the Massif Central and Massif de la Serre.

Finally, the Po Basin is visible in the south of the Alps, a foreland basin shared by the Alps and the Apennines. Clastic sediments up to ten kilometres thick were deposited in this basin in the Cenozoic. The basin fill was partially affected by the Alpine nappe movements, resulting in fold and thrust structures.

Figure 1.11 shows three schematic, simplified cross-sections through the Alps. These show commonalities and differences between the West, Central and Eastern Alps. All three cross-sections are based on insights gained from reflection seismic investigations conducted within the framework of three large national and international research programmes. The cross-section through the Western Alps is

▶ **Figure 1.10** Simplified tectonic map of the Alps and their foreland. The Jura Mountains, and Dauphinois–Helvetic nappe system are part of the European continental margin, the Austroalpine and Southalpine nappe systems represent the Adriatic margin. The Penninic nappe system in between is derived from the Valais basin, the Briançon microcontinent and the Penninic Ocean. Two tectonic windows (inliers) in the Engadin and Tauern prove that the Penninic and Helvetic nappes have a subsurface continuation towards the east. On the other hand, klippen (outliers) of Austroalpine units at the transition between the Central and Western Alps indicate the former extension of the Austroalpine nappes towards the west. A, B and C indicate the locations of the cross-sections shown in Fig. 1.11.

European margin
Piemont Ocean
Adriatic margin

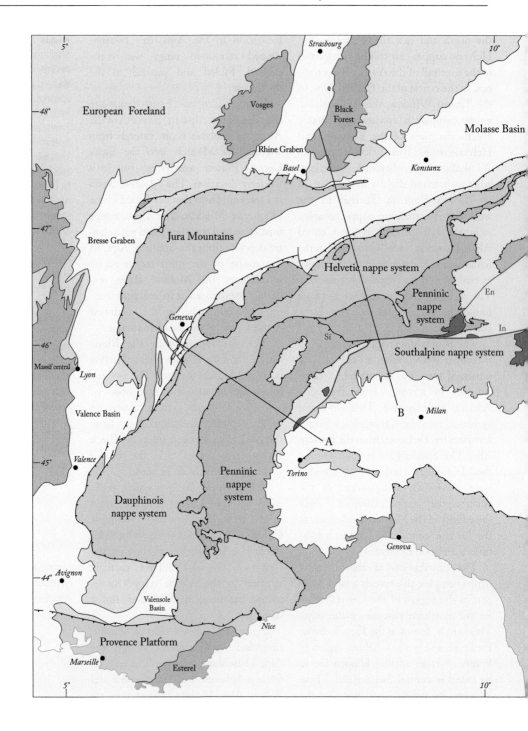

based on the Franco-Italian project ECORS-CROP (Nicolas et al. 1990, Roure et al. 1996, Schmid & Kissling 2000), that through the Central Alps on the Swiss National Research Project, NFP 20 (Pfiffner et al. 1997) and, finally, the cross-section of the Eastern Alps is based on the Germano-Austro-Italian

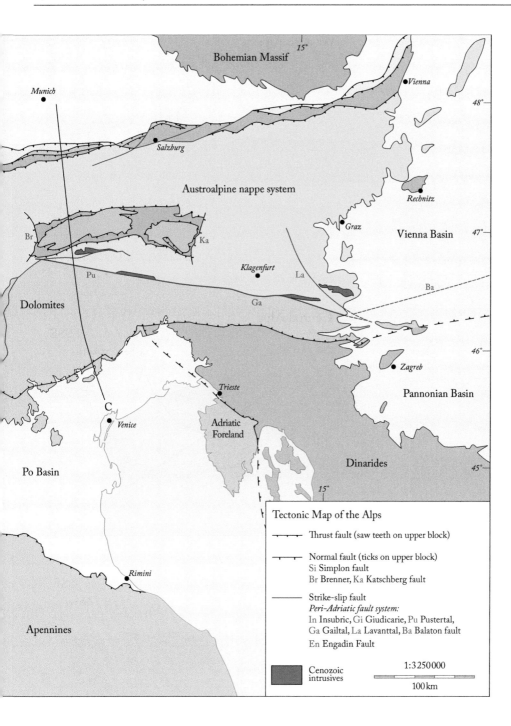

Tectonic Map of the Alps

+—+ Thrust fault (saw teeth on upper block)

—⊥— Normal fault (ticks on upper block)
Si Simplon fault
Br Brenner, Ka Katschberg fault

——— Strike-slip fault
Peri-Adriatic fault system:
In Insubric, Gi Giudicarie, Pu Pustertal,
Ga Gailtal, La Lavanttal, Ba Balaton fault
En Engadin Fault

Cenozoic
intrusives

1:3250000

100 km

project TRANSALP (TRANSALP working group 2002, Lüschen et al. 2004). The structure of the lower crust has been determined from controlled-source seis-mology, as well as local earthquake tomography studies by Waldhauser et al. (2002), Diel et al. (2009) and Wagner et al. (2012).

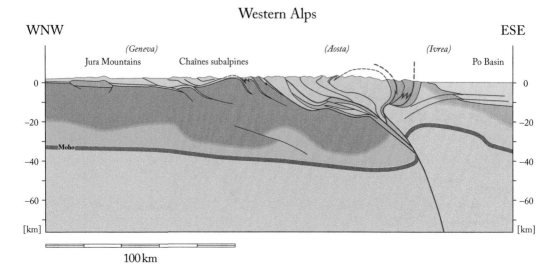

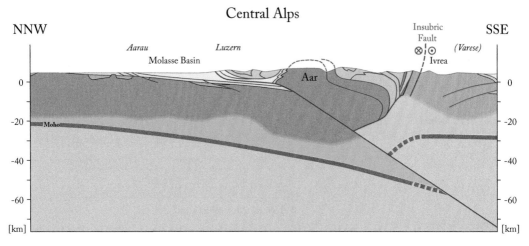

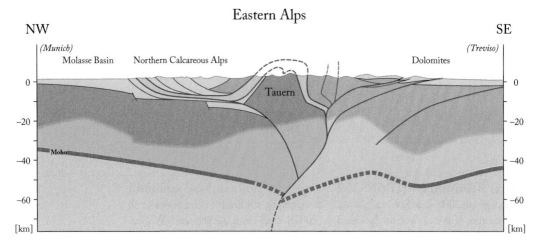

European continental margin

Jura Mts & Helvetic nappe system

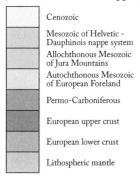

Cenozoic

Mesozoic of Helvetic - Dauphinois nappe system

Allochthonous Mesozoic of Jura Mountains

Autochthonous Mesozoic of European Foreland

Permo-Carboniferous

European upper crust

European lower crust

Lithospheric mantle

Penninic nappe system

Mesozoic oceanic crust

Mesozoic sediments

Crystalline basement/ European upper crust

Strike-slip along Insubric Fault:

⊗   motion away from observer

⊙   motion towards observer

Adriatic continental margin

Austroalpine & Southalpine nappe systems

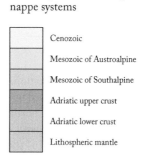

Cenozoic

Mesozoic of Austroalpine

Mesozoic of Southalpine

Adriatic upper crust

Adriatic lower crust

Lithospheric mantle

The cross-section through the Western Alps shows how the crust of the European continental margin plunges below the Alps in an east-southeast direction, beneath both the mantle and the crust of the Adriatic continental margin. The steep contact contains a strike-slip component (the eastern portion moved towards the north). Thrust faults in the European crust indicate substantial compression in an east–west direction, which at least doubled the thickness of the crust. Larger basement uplifts in the Chaînes subalpines and the Penninic nappe complex lead to the conclusion that isolated crystalline basement blocks were overthrust in a westerly direction on to the foreland over a distance of more than 100 kilometres. The Mesozoic sediments of the Jura Mountains were also affected by this and were actually thrust on to the Cenozoic sediments of the Bresse Graben. On the Adriatic side, the crustal blocks were thrust on top of each other in an easterly direction. These structures are covered by the Po Basin fill and are known only from seismic investigations. Lower crust reached the surface of Earth in the Ivrea Zone, and mantle rocks almost did. This elevated location of the boundary

between crust and mantle is unique and is due to an inherited geometry from the time of the formation of the Piemont Ocean. Remnants of this ocean can be found in the thin band of ophiolites interlayered with detached pieces of upper crustal crystalline basement nappes.

The European crust also plunges below the Alps in a south-southeast direction in the cross-section of the Central Alps. The upper crust has been peeled off the lower crust and piled up into a nappe pile composed of crystalline basement nappes. The lower crust runs underneath the compressed margin of the Adriatic plate. Similar to the cross-section of the Western Alps, the Adriatic mantle and the lower crust are in a shallow position. They were pushed northward along a south-dipping thrust fault cutting across the entire lithosphere, a style that is referred to as thick-skinned tectonics. A steep fault, the Insubric fault, separates the rocks of the Adriatic and European plates. The fault exhibits two movement components. The packages of upper crust from the European margin that had been peeled off moved southwards and up along a steeply dipping thrust fault and, in some instances simultaneously, the

▲ **Figure 1.11** Three schematic cross-sections through the Western, Central and Eastern Alps based on geological and geophysical data. The upper crust in these cross-sections is shortened considerably by thrusting and folding, whereas the lower crust and lithospheric mantle show a much simpler structure. Locations of the cross-sections are shown in Fig 1.10.

Adriatic block moved horizontally westwards in a dextral fashion. Klippen of Helvetic and Penninic nappes made of sedimentary rocks lie to the north of the Aar massif. These were sheared off their crystalline substratum and pushed over 100 kilometres in a northerly direction. During this process, they also ended up on top of the internal, southern portions of the Cenozoic fill of the Molasse Basin.

In the cross-section through the Eastern Alps, although the European crust also sinks below the Adriatic continental margin in a southerly direction, the extent of shortening appears slightly lower. Similarly, the Adriatic lithosphere plunges toward the European plate in a northerly direction. In both cases, the lower crust is thickened considerably at their interface. A prominent basement uplift of crystalline rocks is present and emerges in the Tauern Window. A steep fault on the south side of the Tauern Window, the Pustertal fault, separates the Eastern Alps from the Dolomites. In the Dolomites, multiple thrust faults are visible. They have a south-directed transport and extend deep down into the crystalline rocks of the crust. The continuation of these thrust faults deeper down and their merger with the large thrust fault at the base of the Tauern massif is speculation. The structure of the lower crust, at least, is fairly certain based on the teleseismic investigations. There are Mesozoic sediments overlying the Tauern crystalline basement that are comparable to those in the Helvetic nappe complex in the Central Alps. In turn, Penninic nappes overlie these sediments and are themselves overlain by Austroalpine nappes. A larger complex of Austroalpine nappes, called the Northern Calcareous Alps, lies to the north of the Tauern Window on a cushion of Penninic nappes. The Northern Calcareous Alps were telescoped together into a nappe complex as early as the Cretaceous, when some of the thrusting occurred in a westerly direction.

References

Blakey, R., 2008, Regional Paleogeographic Views of Earth History, Northern Arizona University, http://jan.ucc.nau.edu/~rcb7/RCB.html

Diel, T., Husen, S., Kissling, E. & Deichmann, N., 2009, High-resolution 3-D P-wave model of the Alpine crust. Geophysical Journal International, 179, 1133–1147.

Elliott, D. & Johnson, M. R. W., 1980, Structural evolution in the northern part of the Moine thrust belt, NW Scotland. Edinburgh Geological Society Transactions, Earth Sciences, 71, 69–96.

Facenna, C., Speranza, F., D'Ajello Caracciolo, F., Mattei, M. & Oggiano, G., 2002, Extensional tectonics on Sardinia (Italy): insights into the arc-back-arc transitional regime. Tectonophysics, 356/4, 213–232.

Kahle, H.-G., Müller, M. V., Geiger, A., Danuser, G., Mueller, St., Veis, G., Billiris, H. & Paradissis, D., 1995, The strain field in northwestern Greece and the Ionian Islands: results inferred from GPS measurements. Tectonophysics, 249, 41–52.

Lüschen, E., Lammerer, B., Gebrande, H., Millahn, K., Nicolich, R. & TRANSALP Working Group, 2004, Orogenic structure of the Eastern Alps, Europe, from TRANSALP deep seismic reflection profiling. Tectonophysics, 388, 85–102.

Matte, Ph., 1991, Accretionary history and crustal evolution of the Variscan belt in Western Europe. Tectonophysics, 196, 309–337.

Nicolas, A., Polino, R., Hirn, A., Nicolich, R. & Ecors-Crop working group, 1990, Ecors-Crop traverse and deep structure of the western Alps: a synthesis. In: Roure, F., Heitzmann, P. & Polino, R. (eds), Deep Structure of the Alps, Mémoires de la Société Géologique de France, Paris, 156, 15–28.

Pfiffner, O. A., Lehner, P., Heitzmann, P., Mueller, S. & Steck, A. (eds), 1997, Deep Structure of the Swiss Alps: Results of NRP 20, Birkhäuser, 380 pp.

Roure, F., Bergerat, F., Damotte, B., Mugnier, J.-L. & Polino, R. (1996), The Ecors-Crop Alpine seismic traverse, Mémoires de la Société Géologique de France, 170, 113 pp.

Schmid, S. M. & Kissling, E., 2000, The arc of the Western Alps in the light of geophysical data on deep crustal structure. Tectonics, 19/1, 62–85.

Scotese, C. R. & Sager, W. W., 1988, Mesozoic and Cenozoic plate reconstructions. Tectonophysics, 155, 27–48.

Strachan, R. A., Smith, M., Harris, A. L., & Fettes, D. J., 2002, The Northern Highland and Grampian terranes. In: Trewin, N. H. (eds), The Geology of Scotland, Geological Society of London, 81–147.

Tesauro, M., Hollenstein, C., Egli, R., Geiger, A. & Kahle, H. G., 2005, Continuous GPS and broad-scale deformation across the Rhine Graben and the Alps. International Journal of Earth Sciences, 94, 525–537.

TRANSALP working group, 2002, First deep seismic reflection images of the Eastern Alps reveal giant crustal wedges. Geophysical Research Letters, 29/10, 10.1029/2002GL014911, 92-1-92-4.

Wagner, M., Kissling, E. & Husen, S., 2012, Combining controlled-source seismology and local earthquake tomography to derive a 3-D crustal model of the western Alpine region. Geophysical Journal International, 191, 789–802.

Waldhauser, F., Lippitsch, R., Kissling, E. & Ansorge, J., 2002, High-resolution teleseismic tomography of upper-mantle structure using an a priori three-dimensional crustal model. Geophysical Journal International, 150, 403–414.

Wortmann, U. G., Weissert, H., Funk, Hp. & Hauck, J., 2001, Alpine plate kinematics revisited: The Adria Problem. Tectonics 20/1, 134–147.

2 The pre-Triassic Basement of the Alps

This chapter deals with the pre-Triassic rock formations that form the direct substratum for the Mesozoic sedimentary rock suites in many parts of the Alps. Numerous different rock formations can be distinguished in this basement. Some of these are similar in formation and can be seen in different locations. This applies, in particular, to the following three groups of rocks:

Basement:
Polymetamorphic
rocks,
Igneous rocks,
Sediments

- Polymetamorphic crystalline rocks that are also referred to as 'Altkristallin'. The age of these rocks extends far back into the Precambrian. Originally, these were usually clastic sediments and basalts.
- Igneous rocks, in particular, granites and volcanics. Granite intrusions dating from the Ordovician, which are now present in the form of orthogneiss, granites from the Late Carboniferous and a group of Permian magmatites are particularly widespread.
- Palaeozoic sediments, with ages extending from the Ordovician and into the Carboniferous, as well as Late Palaeozoic sediments from the Permo-Carboniferous.

With regard to their history of formation, the pre-Triassic basement rocks experienced multiple phases of mountain-building and basin formation. Three major Palaeozoic events had a sustainable effect on the overall picture: mountain-building on the continent of Gondwana during the Pan-African cycle (870–550 million years ago), the Caledonian orogeny in north-western Europe and the Variscan orogeny in Central Europe. Evidence for orogenic activity is provided by metamorphic overprinting – which produced high-pressure paragenesis (to ecologite grade) and also caused differential melting (anatexis) at high temperatures and the formation of migmatites – and by igneous activity, including widespread production of granitic rocks and volcanics. A variety of sedimentary rocks were deposited in both large and small basins before, during and after these orogenies. Even though these sediments are only sparsely preserved, they fit into a general picture. The pre-Triassic basement was fragmented into small blocks by the most recent, Alpine orogeny and by the preceding basin formation during the break-up of Pangaea. In order to understand the overall relationships we need to try and piece together a picture from these geographically isolated blocks. Such an undertaking is fairly difficult. However, as we repeatedly find a similar local evolutionary history in very different locations in the Alps, the resultant synthesis can be viewed as highly credible. The pre-Mesozoic geology of the entire Alps has been discussed in depth recently in an overview by von Raumer & Neubauer (1993a); this provides numerous local, regional and supra-regional compilations that are referred to many times in this chapter. A more recent publication by von Raumer et al. (2013) discusses the pre-Mesozoic basement in a larger, European framework.

The three-part division of the rock formations in the pre-Triassic basement listed above can be traced from the North Alpine foreland (Black Forest–Vosges, Bohemian Massif) into the actual Alps. Figure 2.1 shows where the basement in the Alps is encountered in isolated outcrops or blocks at the surface of the Earth. Examples of such outcrops are the external massifs in the Helvetic nappe complex and the nappes composed of crystalline rocks in the Penninic nappe complex, the Austroalpine nappe complex and the Southalpine nappe complex, all units that were moved over great distances relative to each other over the course of Alpine mountain-building. These isolated and patchily exposed outcrops or blocks make it difficult to understand the relationships between the individual deposits. In the following, some of these basement blocks are discussed in more detail. The selected examples are unique with regard to the spectrum of rocks and their origins and have been particularly well documented by more recent investigations.

2.1 The pre-Triassic Basement in the Black Forest and Vosges

The pre-Triassic basement is exposed in two geological windows on the shoulders of both sides of the Rhine Graben to the north of Basel. In the northern part of these windows, rocks from the Variscan orogen that are were moderately overprinted by metamorphism lie exposed at the surface. In the middle section, these are in the form of polymetamorphic gneisses belonging to what is called the Central Gneiss Complex. To the south of this, a narrow belt of Palaeozoic sediments traverses the Black Forest. The adjacent southern polymetamorphic gneiss formations form the 'Southern Gneiss Complex'. Figure 2.2 is based on Eisbacher et al. (1989), with additions according to Huber & Huber (1984), and shows the structural relationships in the southern Black Forest.

The Central Gneiss Complex is composed of banded polymetamorphic paragneisses, orthogneiss, metamafic and meta-ultramafic rocks, that all dip to the north. The composition of the paragneiss formations leads to the conclusion that the protolith was composed of greywacke and shale that originated in a source area containing Proterozoic rocks. Geochemical data indicate that the rocks of the Central Gneiss Complex experienced at least one metamorphic event in the Early Palaeozoic, with a peak temperature in the Middle Ordovician, 480–460 million years ago (cf. Eisbacher et al. 1989). Differential melting can also be observed at many locations in the Black Forest, but there is uncertainty associated with the dating of these anatectic processes. According to Huber & Huber (1984), a first anatectic event is thought to have taken place in the Early Cambrian (550–520 million years ago) and a second, the 'main anatectic event', in the earliest Ordovician (490–480 million years ago). Eisbacher et al. (1989) state that petrological data obtained from ultramafic rocks, eclogitic amphibolite and granulitic paragneisses indicate that eclogitic–granulitic high-pressure metamorphism occurred, which took place before regional metamorphism under intermediate pressure. An even younger high-pressure metamorphic overprinting event took place subsequently during the Variscan orogeny. All these processes provide evidence that the Central Gneiss Complex experienced multiple orogeneses in the Palaeozoic. The high degree of metamorphic transformation is such that only few and imprecise statements can be made on the early evolutionary history of these rocks.

A thrust that dips to the north, the Todtnau thrust, forms the boundary

▶ **Figure 2.1**
Tectonic map of the Alps. Also highlighted are the occurrences of pre-Triassic crystalline basement and Palaeozoic sediments within each unit. A-R, Aiguilles Rouges massif; M-B, Mont Blanc massif.

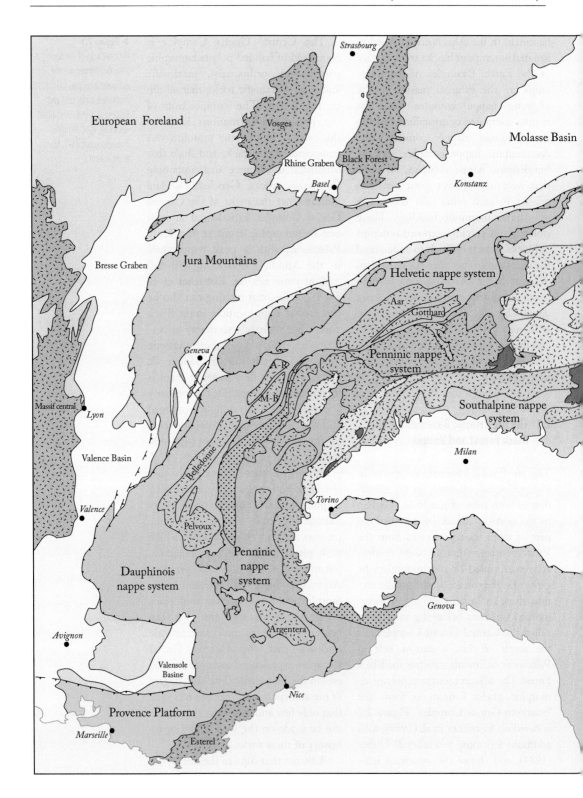

European Foreland

Molasse Basin

Strasbourg

Vosges

Rhine Graben   Black Forest

Basel   Konstanz

Bresse Graben   Jura Mountains

Helvetic nappe system

Aar
Gotthard

Geneva

A-R

Penninic nappe
system

M-B

Massif central
Lyon

Southalpine nappe
system

Valence Basin

Belledonne

Milan

Valence

Torino

Pelvoux

Penninic
nappe
system

Dauphinois
nappe system

Genova

Avignon

Argentera

Valensole
Basine

Nice

Provence Platform

Marseille   Esterel

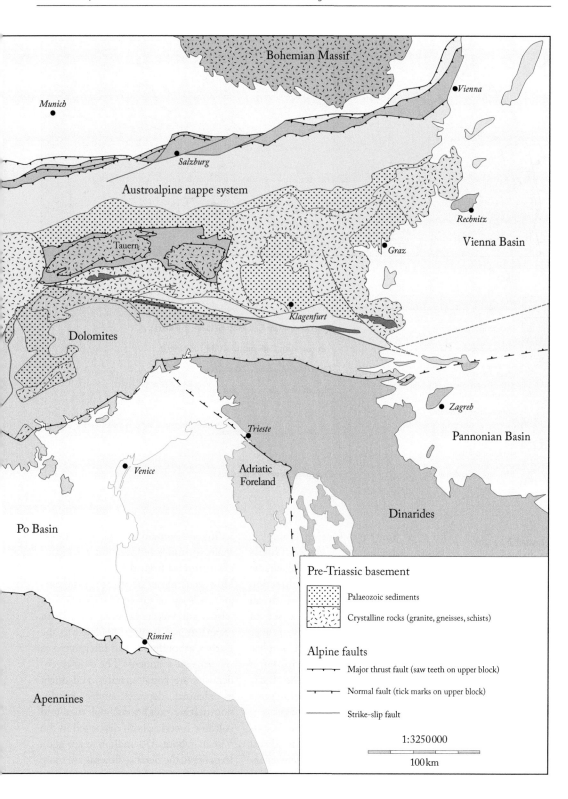

Pre-Triassic basement

- Palaeozoic sediments
- Crystalline rocks (granite, gneisses, schists)

Alpine faults

- Major thrust fault (saw teeth on upper block)
- Normal fault (tick marks on upper block)
- Strike-slip fault

1:3250000

100km

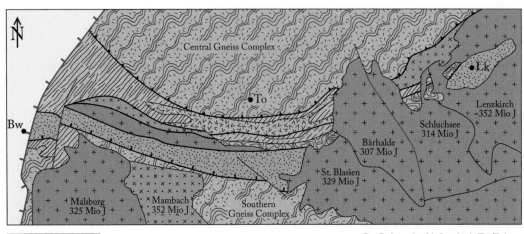

Bw: Badenweiler; Lk: Lenzkirch; To: Todtnau

10 km

Cenozoic fill of the Rhine Graben		
Undeformed granite	Post-Variscan; Late Carboniferous (330 – 310 Mio J)	
Deformed granite	Variscan; Early Carbonifereous (358 – 352 Mio J)	
Greywackes, slates, volcanics, conglomerates	Late Devonian – Early Carbonifereous	
Mylonitic gneisses, amphibolite, metagranitoids	Monometamorphic metasediments and magmatics	
bi-mu-cor-gr-sill Schist		
bi-mu-cor-gr-and Schist	Polymetamorphic metasediments	
mu-chlorite Schist		
Mylonitic banded gneisses	Proterozoic greywackes and slates	

▲ **Figure 2.2** Geological map of the southern Black Forest, based on Eisbacher et al. (1989) and Huber & Huber (1984). The structural relations illustrate the Late Palaeozoic evolution of the area: and, andalusite; bi, biotite; cor, cordierite; gr, garnet; mu, muscovite; sill, sillimanite.

to the Southern Gneiss Complex, along which the Central Gneiss Complex is thrust on to Palaeozoic sediments in a southeasterly direction (Eisbacher et al. 1989). The Todtnau thrust is associated with a set of parallel, retrograde shear zones within the Central Gneiss Complex. Thrusting and uplifting of this latter formation took place in the Early Carboniferous (340–330 million years ago), based on K/Ar cooling ages.

The belt of Palaeozoic sediments to the south of the Todtnau thrust is referred to as the 'Badenweiler–Lenzkirch Zone'. At their base, these sediments contain marine limestones, shale, siltstones and sporadic calcareous olistostromes from the Late Devonian. However, most of these sediments are made up of turbidite deposits with felspar and volcanic components. With regard to age, they are assigned to the Early Carboniferous and interpreted as synorogenic sediments. The most recent deposits are conglomerates, laid down as channel deposits under paralic, non-marine conditions, and manifold volcanic breccias, both deposited in the Visean (about 330 million years ago). In general, the older sediments are more likely to be found in the northern part of the belt and the youngest conglomerates in the south, at shallow depth. The cross-section thus reveals a series of rock formations that is overturned in the south-southeast direction.

Figure 2.2 shows how a normal fault forms the boundary between the Badenweiler–Lenzkirch Zone and the Southern Gneiss Complex. This normal fault brings the gneiss rock formations in the south (Hotzenwald Group) into direct contact with the unmetamorphosed Palaeozoic sediments. The most important rocks in the Southern Gneiss Complex are the polymetamorphic and monometamorphic shales, gneissic metavolcanics, ultramafic rocks, rare gabbro and lens-shaped occurrences of mylonitic leucogranite (Eisbacher et al. 1989). Many of these rocks are preserved only as large-scale schollen within the younger granites. Shear-sense indicators in the mylonitic leucogranite indicate thrusting in a southeasterly direction. However, multiple dextral lateral motions (some with a normal fault component) are also revealed.

In the Late Carboniferous, a series of granites intruded the old gneiss formations and the Palaeozoic sediments. The age of these granites varies from 330 to 310 million years, with a peak around 330 million years. They cut across the Variscan structures such as the Badenweiler–Lenzkirch Zone or the Todtnau thrust and therefore must be considered as post-Variscan. Isolated older granites intruded into the Early Carboniferous (Randgranit, 358 million years old; Mambach Granite 352 million years old; Huber & Huber 1984) and were deformed in the course of the Variscan orogeny.

A series of Late Carboniferous and Permian clastic sediments and volcanics discordantly overlie the rock units discussed above. These Permo-Carboniferous basins started to subside in the Westphalian (ca. 310 million years ago), culminating in the Rotliegend 307–286 million years ago. Between about 307 and 286 million years ago (see Eisbacher et al. 1989), volcanic activity led to the production of felsic magmas (rhyolite). Basins were created in the form of local graben, in which a sequence, two to three kilometres thick, of fluviatile and limnic sandstones, clay and marl, as well as crystalline breccia were deposited in alluvial fans (Matter 1987). Similar basins are also found in all of the Alpine regions, as will be discussed later.

2.2 The pre-Triassic Basement of the External Massifs

The pre-Triassic basement is directly visible in several crystalline uplifts in the external massifs of the Alps. The interpretation of the rock formations in these isolated exposed areas is even more difficult in the individual cases due to the addition of Alpine overprinting, which is absent in the Black Forest. Even so, some commonalities can be determined. For example, to cite the most important elements, Proterozoic metasediments, Ordovician orthogneisses, Variscan migmatites and late or post-Variscan granites from the Early Carboniferous, as well as Permian granites, volcanics and continental clastics occur in several locations. Other rock formations, such as unmetamorphosed Palaeozoic sediments from the Ordovician, Silurian and Devonian, are limited in occurrence to selected regions, more specifically, to the Austroalpine nappe complex. The discussion of the pre-Triassic basement is conducted on a regional basis and is

limited to some particularly well-docu-
mented examples, in order to exemplify
the regional differences.

External Massifs in the Western Alps

The crystalline uplifts in Argentera,
Pelvoux, Belledonne and Aiguilles
Rouges/Mont Blanc are strung together
in the external part of the Alps, as
is made clear in Fig. 2.1. According to
von Raumer et al. (1993c), a formation
of **polymetamorphic metasediments**,
which was overprinted by high-pressure
eclogite-grade metamorphism in the
Silurian or Early Devonian and, later
on, in the Late Devonian or Early
Carboniferous during regional metamorphism associated with the Variscan
orogeny, is common to all these external
massifs. Finally, the most recent overprinting occurred during the Alpine
orogeny. These metasediments are also
referred to using the collective term
'Altkristallin' ('old crystalline rocks').
With regard to their age, they probably
can be assigned to the Late Proterozoic
and Early Palaeozoic (von Raumer et al.
1993c). A formation of metagreywacke,
intercalated with quartzite and metapelites, and rare carbonates can possibly
be interpreted as platform sediments
that are associated with a rift. Another
formation, composed of mica schists
with layers of amphibolite and diopside marble, associated with banded
metagreywacke and felsic gneisses and
amphibolite, is more likely to be attributed to an oceanic environment.

Felsic volcanics are assumed as the
protolith for the granitic banded gneisses
that are widespread in the Argentera
massif, while the augen gneisses and fine-
grained granitic gneisses in the Aiguilles
Rouges massif are interpreted as metavolcanics from a past island arc. Finally,
the augen gneisses in the Mont Blanc
massif, which can be dated to 460 million years ago, are to be interpreted as
Ordovician granite, whereas the age of
the metapelites (called Série satiné) in
the Belledonne massif is unclear.

In addition to these polymetamorphic metasediments, **monometamorphic formations** are to be found mainly
in the Belledonne massif (von Raumer
et al. 1993c). One of these formations,
the Chamrousse ophiolite, is composed
of ultramafic rocks and gabbro with a
crystallization age of 497–496 million
years: it is assumed to have formed at
the transition from thinned oceanic
continental crust to oceanic crust.
The Rioupéroux–Livet plutonic–volcanic
complex containing amphibolite and
trondhjemite is much younger (365–350
million years old). Finally, the Taillefer
formation contains metapelites, meta-
arenites, metaconglomerates, as well as
metaspilites and metakeratophyres. The
latter formation could be metamorphosed sediments and volcanics that
should be viewed as occurring in the
context of an intracontinental pull-
apart basin.

Variscan migmatites, some of which
contain cordierite, are also widespread.
Exhumation through nappe stacking
and erosion in the Late Devonian and
Early Carboniferous led to decompression and an increase in the geothermal
gradient in the developing Variscan
orogen. As a result, differential melting
(anatexis) of crustal rocks occurred.

The deformation that occurred
during the Variscan orogeny started in
the Late Devonian. A main cleavage
overprinted early, pre-existing folds, and
stretching lineations are visible on this
main cleavage that indicate north–south
shearing (von Raumer et al. 1993c).
Later on, in the Early Carboniferous,
large folds developed that overprinted all
early structures.

Argentera,
Pelvoux,
Belledonne,
Aiguilles Rouges,
Mont Blanc

In addition, the external massifs contain a number of **late** and **post-Variscan granites**. The granitoid melts intruded into the polymetamorphic crystalline rocks ('Altkristallin'), into the monometamorphic formations and the migmatites. According to Bonin et al. (1993), the older intrusions that occurred in the Early Carboniferous (350–330 million years ago) are K-rich and can be assigned to the calc-alkaline series, and their porphyritic structure points to a shallow intrusion level. These granitoids intruded into an extant crustal stack after the collision between Gondwana and Baltica, and are therefore late Variscan. In the Early Carboniferous (320–290 million years ago), volcanic–plutonic complexes developed that can be allocated to the alkaline to calc-alkaline series. Their emplacement must be viewed within the context of post-Variscan extensional tectonics.

Finally, even younger volcanic and plutonic activities are noted in the Permian. These are to be interpreted in the context of a western Mediterranean province and the break-up of Pangaea, or the opening of an arm of the Tethys in a westerly direction (see Fig. 1.5). In the Argentera massif (Fig. 2.3), we can

see how a post-Variscan granite, the age of which is the transition from the latest Carboniferous to earliest Permian (293–285 million years), cuts through a body of orthogneiss and the boundary between the Variscan migmatite and the polymetamorphic crystalline rocks of the 'Altkristallin'.

In case of the southwestern Belledonne massif (Fig. 2.4), major Variscan thrust faults transport the Chamrousse Ophiolite and the polymetamorphic crystalline rocks of the 'Altkristallin' on to the Rioupéroux–Livet intrusion complex from opposite directions. The Sept Laux Granite, which extends as a relatively narrow band over 100 kilometres in a north-westerly direction, can be considered as late Variscan based on its age of 330 million years, that is, Early Carboniferous. According to Bonin et al. (1993), the melts were produced in the lower crust by continuous anatexis.

Of note in the Aiguilles Rouges massif (Fig. 2.5), is the fact that the Devonian and Carboniferous sediments that are folded into the old crystalline rock run from north to south in the southwest, that is, diagonally to the general Alpine strike, which is indicated by the elongate shape of the two massifs.

▼ **Figure 2.3**
Geological map of the Argenterea massif, an external massif in the Western Alps. A Variscan thrust fault places Variscan migmatites onto mica schists of the Valetta Formation. Source: von Raumer & Neubauer (Hrsg) (1983). Reproduced with permission of Springer Science+Business Media.

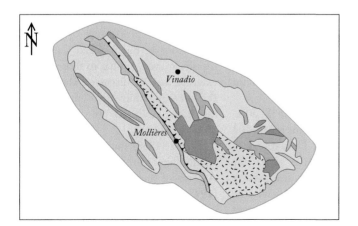

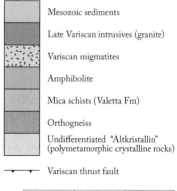

Mesozoic sediments

Late Variscan intrusives (granite)

Variscan migmatites

Amphibolite

Mica schists (Valetta Fm)

Orthogneiss

Undifferentiated "Altkristallin" (polymetamorphic crystalline rocks)

Variscan thrust fault

30km

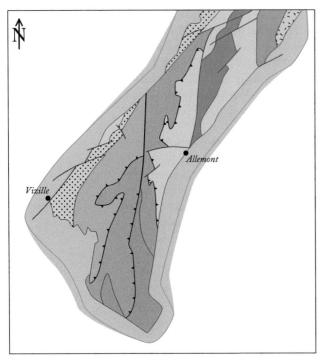

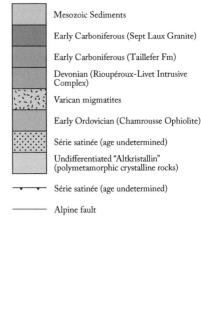

Mesozoic Sediments

Early Carboniferous (Sept Laux Granite)

Early Carboniferous (Taillefer Fm)

Devonian (Rioupéroux-Livet Intrusive Complex)

Varican migmatites

Early Ordovician (Chamrousse Ophiolite)

Série satinée (age undetermined)

Undifferentiated "Altkristallin" (polymetamorphic crystalline rocks)

Série satinée (age undetermined)

Alpine fault

20 km

▲ **Figure 2.4**
Geological map of the southwestern Belldonne massif, an external massif in the Western Alps. Variscan thrust faults place older rocks ('Altkristallin' and the Early Ordovician Chamrousse Ophiolite) onto younger rocks (Devonian sediments). Source: von Raumer & Neubauer (Hrsg) (1983). Reproduced with permission of Springer Science+Business Media.

In contrast, the Early Palaeozoic sediments in the northwest of the massif run parallel to the Alpine strike and parallel to the narrow band of the Vallorcine Granite. According to Bonin et al. (1993), this granite intruded in the form of a slab-shaped pluton 320 million years ago, namely, in a transtensive regime. The melts point to anatexis within the crust (differential melting of metapelites), but the presence of specific mafic components means that a contribution from mantle melts cannot be excluded.

The Mont Blanc Granite in the adjacent massif of the same name is 316–304 million years old (Late Carboniferous), and thus slightly younger. A spectacular view of this granite is obtained looking from Chamonix up to the Aiguilles de Chamonix (see Fig. 2.5). Emplacement of this granite apparently also occurred in a transtensive shear zone. However, due

to subsequent rapid exhumation, a large body of rhyolite was deposited on the granite. Based on more recent dating (Capuzzo & Bussy, in von Raumer 1998), this is 295 million years old (earliest Permian).

External Massifs in the Central Alps

Overall, the pre-Triassic geological successions in the Aar and Gotthard massifs are very similar to those in the external massifs discussed above. For this reason, only two particularly instructive examples are described in more detail.

Franks (1968) studied the structural relationships at the eastern end of the Aar massif and Schaltegger & Corfu (1995) provide an overview of magmatic events. Figure 2.6 provides a schematic illustration of the structural relationships in the basement and the overlying

(A)

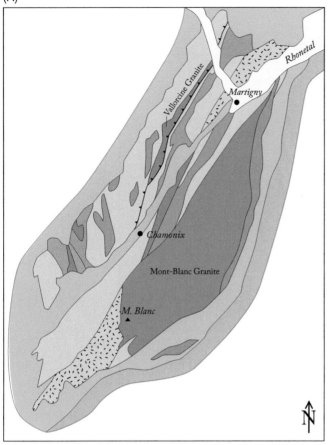

Quaternary

Penninic nappe system

Penninic nappes, undifferentiated

Helvetic nappe system

Mesozoic sediments

Late Carboniferous

Devonian and Early Carboniferous

Post Variscan rhyolite

Post Variscan intrusives (granite)

Variscan migmatites

Metagranite

Undifferentiated "Altkristallin" (polymetamorphic crystalline rocks)

⊢⊸⊸ Variscan thrust fault

20 km

Figure 2.5 (A) Geological map of the Aiguilles Rouges and Mont Blanc massifs. Folded Devonian–Early Carboniferous sediments strike in a NNW–SSE direction, whereas folded Late Carboniferous strata strike in a NNE–SSW direction. The post-Variscan Mont Blanc Granite cuts Variscan migmatites, metagranites and polymetamorphic crystalline rocks of the 'Altkristallin'. Source: von Raumer & Neubauer (Hrsg) (1983). Reproduced with permission of Springer Science+Business Media.

(B) A view of the Aiguilles de Chamonix looking south-southeast (Haute Savoie, France). The pyramids of the high peaks Blaitière and Grand Charmoz are composed of the post-Variscan Mont Blanc Granite, the Petit Charmoz (dark) comprises polymetamorphic gneisses.

(B)

(A)

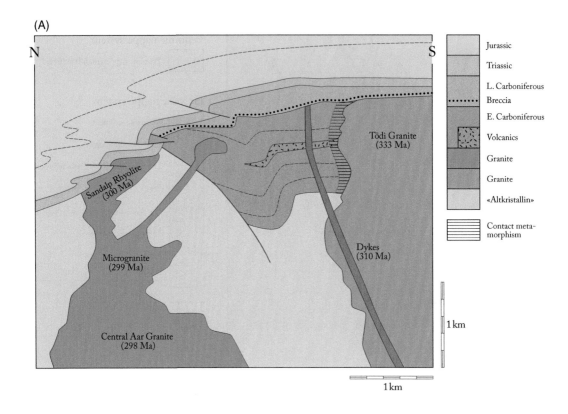

(B)

Mesozoic cover (based on the author's research), with additional information on the age of intrusions. The following sequence of geological events can be separated out in this area: in the Early Carboniferous, probably between 350 and 340 million years ago, a graben developed in the crystalline basement of the eastern Aar massif, which was filled with clastic sediments and volcaniclastics, which are now present in the form of metaconglomerates and hornfels (called the Bifertenfirn metasediments by Franks 1968). The metasediments were overprinted by contact metamorphism during the intrusion of the Tödi granite, the age of which is 333 million years (Early Carboniferous). However, the metasediments were also folded and overprinted by regional metamorphism, but the age of these processes cannot be determined precisely. Dykes cross-cut the metasediments and the Tödi granite 310 million years ago. Somewhat later, just under 300 million years ago, the Central Aar granite intruded, forming a large pluton that can be followed across a distance of well over 100 kilometres, in the direction of the western end of the Aar massif. Microgranites (aged 299 million years) intruded into the metasediments in the roof of the intrusion and, in the case of the Sandalp rhylolite (aged 300 million years), the melt even reached the surface of the Earth. Around this time (300 million years ago), the Tödi granite and all surrounding rocks were exhumed and then immediately covered by volcaniclastics in the Late Carboniferous (called the Bifertengrätli Formation by Franks 1968). Based on plant fossils, these rocks were dated to between Westphalian D and Stephanian, that is, Late Carboniferous (310 to 300 million years ago). The Bifertengrätli Formation is therefore almost the same age as the Sandalp Rhyolite.

The Bifertengrätli Formation was deposited in a tectonic graben, the northern margin of which corresponds approximately to the continuation of the older graben. The formation starts with a basal conglomerate, followed by a volcanic member; the next, estuarine, member contains cross-stratified sandstone, arkose and coal seams containing plant remains that were used for dating, indicating when the graben was filled; the sequence is completed by the youngest, lacustrine member. The younger deposits in the graben were folded, such that the Triassic sediments were deposited discordantly on the sediments and volcanics after erosion and peneplanation. Clearly, compression, or possibly transpression, took place in this part of the Aar massif as late as at the very end of the Carboniferous (or in the Permian?).

The series of rock formations discussed above can easily be traced further west into the Maderaner valley (Franks 1968, Schaltegger & Corfu 1995). However, volcanic rocks are dominant in this area, covering clastic sediments. The Central Aar Granite extends over a distance of 100 kilometres towards the west-southwest. The contact to the encasing polymetamorphic gneisses is, for example, exposed in the north-north-west flank of the Bietschhorn (Fig. 2.6).

In the case of the Gotthard massif, the discussion focuses on the region between the Gotthard and Lukmanier passes. The geological map in Fig. 2.7 is based on the publication by Mercolli et al. (1994). In the east of the map, this figure shows how the Medel and Cristallina intrusives sharply and discordantly cut the contacts between Ordovician orthogneiss and Precambrian migmatite, metagabbro and amphibolite. The same can be seen in the western part in the Fibbia Granite. Furthermore, large-scale schlingen

◄ **Figure 2.6** (A) Synoptic cross-section through the eastern termination of the Aar massif, based on Franks (1968), Schaltegger & Corfu (1995) and own work. The structural relationships demonstrate the polyphase evolution of the rock units. The Early Carboniferous sediments and volcanic rocks show a metamorphic overprint at the contact with the Tödi Granite, which can be attributed to the latter's intrusion. Source: Adapted from Franks (1968) and Schaltegger & Corfu (1995).

(B) A view of the Bietschhorn looking south-southeast across the Lötschen valley (Canton Valais, Switzerland). The contact between the post-Variscan Central Aar Granite (light colour of summit pyramid of Bietschhorn) and the polymetamorphic gneisses (darker colour in the lower part of the mountain flank) is highlighted by the colour difference.

Aar,
Gotthard

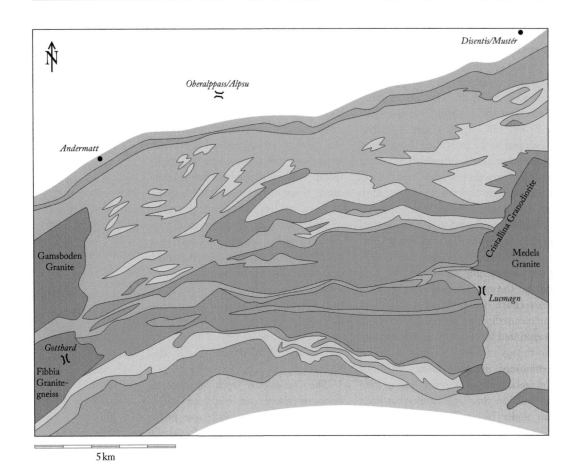

N

Disentis/Mustér

Oberalppass/Alpsu

Andermatt

Cristallina Granodiorite

Gamsboden
Granite

Medels
Granite

Lucmagn

Gotthard

Fibbia
Granite-
gneiss

5 km

Age:	Protolith:	Present-day rock types:
Triassic	Evaporites and carbonates	Cargneule and marble
Permian	Clastic sediments, volcanics	Slate, quartzite, breccia, prasinite
Late Carboniferous	Granite (303–300 Ma)	Mildly deformed to undeformed granite
Silurian–Devonian	Clastic sediments	Quartzite, hbl garben schists, ser phyllite, hbl-gr-mica schist
Late Ordovician	Granitoids (439–446 Ma)	Orthogneiss
Precambrian–Cambrian	Clastic sediments	Migmatite, banded gneisses
Precambrian–Cambrian	Gabbro and ophiolites	Metagabbro, amphibolite, serpentinite

structures are visible in the Ordovician orthogneiss, the intrusion age of which is 439 to 436 years, that is, at the very end of the Ordovician. According to Mercolli et al. (1994), these tectonic schlingen structures can be attributed to the Variscan orogeny. The legend to Fig. 2.7 also provides information on the ages and types of rocks in the different sections of the crystalline basement. Overall, there are certain similarities with the sequences in the external massifs in these Alps and the Black Forest. Precambrian gabbro, ophiolite and clastic sediments are now present in the form of polymetamorphic metagabbro, amphibolite, serpentinite, migmatite and banded gneisses, which are intruded by Ordovician granites that were overprinted by metamorphism and folded in the Variscan orogeny. Silurian and Devonian(?) clastic sediments were transformed into quartzite, different types of slate and phyllite in the Variscan orogeny. The Late Carboniferous granites are substantially less deformed than the older rock formations and the deformation that is present can be attributed to the Alpine orogeny. The age of the intrusives is between 303 and 300 million years (Schaltegger 1994), that is, the intrusives are the same age as the Central Aar Granite in the Aar massif.

External Massifs in the Eastern Alps

A pre-Triassic basement is exposed in the core of the Tauern Window in the Eastern Alps that exhibits great similarity to the external massifs discussed above. In the literature, these rocks are usually allocated to the Penninic nappe system. The outer shell of the Tauern Window is, indeed, composed of sediments and ophiolites that are attributable to the Penninic realm, but the frame of the window is formed of Austroalpine nappes. A more detailed analysis of the sedimentary cover of the basement in the Tauern Window (Lammerer 1986), however, reveals autochthonous Mesozoic sediments with affinity to the European margin. For example, the thick Hochstegen limestones of the Late Jurassic are directly comparable to the Quinten Limestone of the same age in the Helvetic nappe system of Vorarlberg and Switzerland. The core of the Tauern Window therefore can be assigned to the Helvetic realm, bearing in mind that the outer shell is composed of Penninic nappes.

Three important rock formations can be distinguished in the basement itself: polymetamorphic crystalline rocks of the 'Altkristallin', the rocks of the Habach–Storz Group and the Central Gneiss. In some cases, the nomenclature and classification of individual rock units is not entirely clear (Höck 1993).

The **polymetamorphic crystalline rocks of the 'Altkristallin'** are characterized by an amphibolite-facies metamorphic overprint that most probably can be attributed to the Variscan orogeny. Common rock types include orthogneisses, paragneisses, mica schist, migmatites and lenses of amphibolite (Höck 1993).

The younger **Habach–Storz Group** exhibits less overprinting by metamorphism and the protoliths are thought to be Early Palaeozoic sediments and migmatites. The group can be divided into three units: ophiolites with an amphibolite at their base, an island-arc sequence composed of metamorphic mafic rocks, intermediate and felsic lavas and tuffs and, uppermost, the Eiser Formation composed of biotite shale, greywacke, quartzite, garnet mica schists and felsic and mafic volcanics. Only some of the ages of these rocks are known. The ophiolites are 540–500 million years old (Cambrian), parts of the island-arc sequence could have originated in the Late Proterozoic, while other parts are substantially younger, at 600–330 million years old. The Eiser Formation contains plant fossils from the Permo-Carboniferous; however, it might be that the plant fossils pertain to the island-arc series. Either way, the Eiser Formation must be older than the Central Gneiss that intrudes into it. The Central Gneiss is no older than 320 million years

◀ **Figure 2.7** Geological map of the central part of the Gotthard massif, an external massif in the Central Alps. Easily recognizable are the large-scale schlingen structures that are cut by the late-Variscan intrusives: gr, garnet; hbl, hornblende; ser, sericite. Source: Mercolli et al. (1994). Reproduced with permission of the author.

Tauern

▶ **Figure 2.8** Geological map of the Tauern massif, an external basement uplift in the Eastern Alps. The pre-Triassic crystalline basement appears in two windows beneath the Glockner nappe. A Variscan thrust fault puts the polymetamorphic crystalline rocks of the 'Altkristallin' and the Zillertal and Venediger intrusives on top of the Habach–Storz Group. Source: Adapted from Frisch et al. (1993), Höck (1993) and Finger et al. (1993).

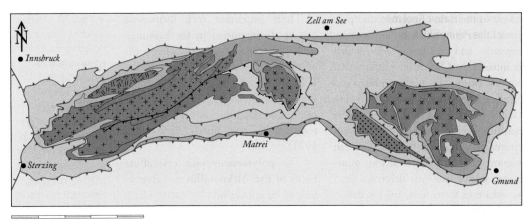

50km

Austroalpine nappe system

Austroalpine nappes undifferentiated

Penninic nappe system

Matrei nappe (ophiolites)

Glockner nappe (sediments and prasinites)
Modereck nappe (sediments)
Wolfendorn nappe (sediments)

  Alpine thrust fault (sawteeth on upper block)

Alpine normal fault (tick marks on upper block)

Riffl thrust (? Variscan; sawteeth on upper block)

Helvetic Venediger nappe system

Mesozoic cover sediments

Late Carboniferous – Permian sediments

Central Gneiss

Ahorn Augen gneiss (porphyric)

Tux Metagranite and metagranodiorite

Zillertal-Venediger metatonalite and metagranodiorite

Granatspitz Metagranite

Badgastein-Hochalm metagranitoids

Sonnblick Metagranite (porphyric)

Habach-Storz Group: Late Proterozoic – Early Carboniferous
metasediments and metavolcanics

Stubach Group (metaophiolites)

Polymetamorphic crystalline rocks («Altkristallin»)

(Höck 1993), that is, the Eiser Formation can be no younger than the Early Carboniferous.

Figure 2.8 has been compiled based on Frisch et al. (1993), Höck (1993) and Finger et al. (1993) and shows how a large proportion of the basement block is made up of the Variscan intrusives known as the **Central Gneiss**. A more detailed description of the Central Gneiss can be found in Finger et al. (1993). In contrast to the situation in the Black Forest, these Variscan intrusives have been substantially deformed by the Alpine orogeny. Based on structural geological analyses, Lammerer & Weger (1998) postulate that the Tauern massif was shortened horizontally to about half its original width. The age of the intrusives determined using the U–Pb zircon method is between 310 and 333 million years (314–313 million years for the Badgastein–Hochalm Metagranitoid, 320–333 million years for the Granatspitz Metagranite). A younger overall age (Permian) is obtained with the Rb–Sr method and this is most probably due to Alpine overprinting (see Finger et al. 1993).

2.3 The pre-Triassic Basement of the Penninic Nappes

Compared with the external massifs, the pre-Triassic basement units in the Penninic nappes are overprinted to a far greater extent and are split into smaller fragments. A relatively complete sequence of pre-Triassic basement is present in the Briançonnais in the Ligurian Alps and will be used as an example and discussed in more detail below: the discussion that follows is based on the publication by Cortesogno et al. (1993). According to these authors, two Alpine units can be distinguished in the Briançonnais: a lower nappe complex, which is located external to the Briançonnais, and an allochthonous upper nappe complex that corresponds to the internal Briançonnais and is composed of numerous smaller elements.

Similar to the case in the Black Forest and the external massifs, within the pre-Late Carboniferous basement, a group of older, polymetamorphic rocks can be distinguished from a younger, monometamorphic sequence, that exhibits only Variscan overprinting.

The polymetamorphic rocks (see Cortesogno et al. 1993) include **parag-neisses**, which were produced from the original greywackes and pelites and, to a lesser extent, from quartzite. Carbonates are absent. These paragneisses are intruded by felsic magmatic rocks, the **Older Orthogneiss** and **Older Migmatite**. The Older Orthogneiss is present in the form of plutons that are several hundred metres thick and in the form of dykes. Lenses of augen gneisses are interpreted as volcanic or subvolcanic rocks (meta-rhyolite). The Older Migmatite is composed of bands of orthogneiss, 10 to 100 metres thick, within the paragneisses and contains almost monomineral biotite lenses. The **Older Metamafics** are the third type of rock that must be mentioned here. This is present in the form of bands of amphibolite and contains local lenses of ultramafic and eclogitic rocks. The amphibolites and eclogites exhibit tholeiitic affinity, but the secondary alterations are such that an unambiguous geodynamic interpretation is not possible.

According to Cortesogno et al. (1993), with the exception of paragneisses, similar rock types to those found in the polymetamorphic sequence can also be recognized in the younger, monometa-morphic rock sequence. The **Younger Orthogneiss** builds up larger bodies within the nappe complexes. These granites intruded more than 327 million years ago (in the Silurian or Devonian). In contrast, the **Younger Metamafics** occur in the form of lenses of metagabbro, in which locally magmatic cumulus textures are still preserved. The **Younger Migmatite** is rare and usually linked to the Younger Metamafics and ecologites within the Older Metamafics.

The **late** and **post-Variscan intru-sives** that are common in the areas discussed above (Black Forest, external massifs) are, according to Cortesogno et al. (1993), only rarely encountered in the Ligurian Alps. Examples (as yet undated) of such intrusives are the small body of Borda Granodiorite and the Rio-Castorello Granophyre dykes.

Above these crystalline rocks there follows a sequence of **Permo-Carboniferous sediments** that were deposited in tectonic grabens. These grabens were formed during continental extensional tectonics in the Late Palaeozoic that affected the entire southern portion of the Variscan orogen. The sedimentary sequence starts with arkoses (erosional products of the exposed Younger Orthogneiss) that then makes a transition into a conglomeratic sandy-argillaceous fluvio-lacustrine succession, which also contains coal seams (or graphite lenses). Fine clastic sediments, in the form of graphitic phyllite in some locations, complete the sedimentary sequence. The thicknesses of the individual strata reflect the extensional tectonic regime that initiated synsedimentary normal faults, and thus the individual strata vary enormously laterally.

However, **volcanics** also appear sporadically in these Permo-Carboniferous sediments, which erupted in three pulses of volcanic activity: volcanic activity was mainly explosive. In the Early Carboniferous, ignimbritic and trachyandesitic pyroclastics erupted, the latter sometimes with shoshonitic composition. Eruptions during the third pulse of volcanic activity in the Early Permian, once again, produced mainly ignimbrites.

Cortesogno et al. (1993) produced a sketch of the evolutionary history of the pre-Triassic basement in the form of a series of snapshots in time, based on a regional synthesis and supported by an in-depth analysis of the debris in the clastic sediments. Figure 2.9 is based on this research.

In a pre-Silurian rifting phase (?Proterozoic to Ordovician), arkoses and clays were deposited in a basin. This rifting phase also led to the intrusion of granites and granodioritic–leucogranitic dykes and, to a lesser extent, synsedimentary felsic volcanism. Basaltic eruptions were of greater importance and were embedded in parallel layers (Fig. 2.9). A first orogeny affected these rocks in the Ordovician–Silurian time period and caused differential melting, as indicated by the Older Migmatite, leading to the formation of eclogites in the Older Metamafics and to the deformation and transformation

of granite into the Older Orthogneiss. The metamorphic rocks were then subsequently exhumed.

Thick, fine clastic sediments were then deposited in the Silurian and Devonian (Fig. 2.9). Monzogranitic plutons and dykes intruded at depth and were accompanied by basaltic and gabbroic dykes locally. A second, the Variscan, orogeny led to an amphibolite-facies metamorphic overprint of the rocks. The monzogranite was transformed into the Younger Orthogneiss, and the basaltic and gabbroic dykes into the Younger Metamafics. The latter was subjected to local differential melting (Younger Migmatite). The Silurian and Devonian sediments were folded. However, most of these sediments were subsequently eroded, and Fig. 2.9 shows the situation as it was in the Early Carboniferous.

Subsequently, the Late Carboniferous and Permian (Fig. 2.9) were then characterized by the deposition of volcano-sedimentary sequences in subsiding continental grabens. First, arkoses were deposited in depressions. This was followed by graben or rift formation, accompanied by explosive volcanism. The filling of the continental basins with, sometimes, coarse clastic sediments in the Early Carboniferous occurred simultaneously with the subsidence of the floors of these basins. In the Permian, thick layers of rhyolite, ignimbrite and dacite covered the graben fill and graben shoulders.

2.4 The pre-Triassic Basement of the Austroalpine Nappes

There are several large nappes in the Austroalpine nappe complex that are composed essentially of pre-Triassic basement rocks. The crystalline basement is particularly well exposed in the Silvretta nappe and has been studied

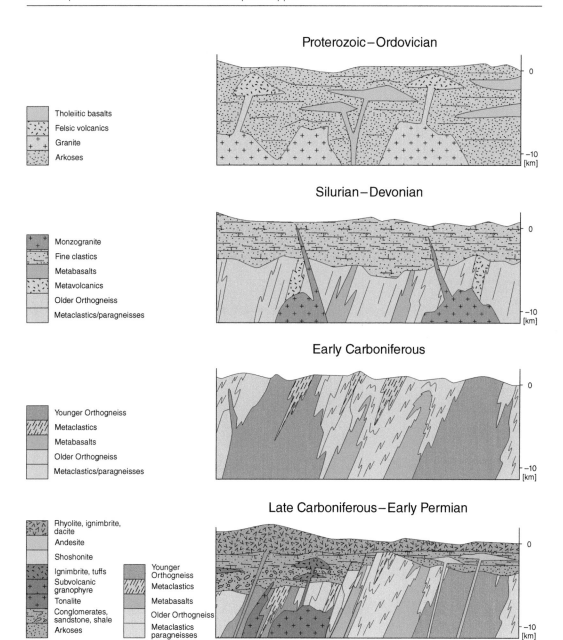

Proterozoic–Ordovician

Tholeiitic basalts
Felsic volcanics
Granite
Arkoses

Silurian–Devonian

Monzogranite
Fine clastics
Metabasalts
Metavolcanics
Older Orthogneiss
Metaclastics/paragneisses

Early Carboniferous

Younger Orthogneiss
Metaclastics
Metabasalts
Older Orthogneiss
Metaclastics/paragneisses

Late Carboniferous–Early Permian

Rhyolite, ignimbrite, dacite
Andesite
Shoshonite
Ignimbrite, tuffs
Subvolcanic granophyre
Tonalite
Conglomerates, sandstone, shale
Arkoses

Younger Orthogneiss
Metaclastics
Metabasalts
Older Orthogneiss
Metaclastics paragneisses

▲ **Figure 2.9** The geological evolution of the crystalline basement of the Penninic Briançon unit in the Ligurian Alps. Source: von Raumer & Neubauer (Hrsg) (1983). Reproduced with permission of Springer Science+Business Media.

intensively. As an example, its development will be discussed below. To this end, the compilation by Maggetti & Flisch (1993) is drawn upon.

The oldest rocks in the basement of the Silvretta nappe, **paragneisses** and **mica schists**, are interpreted to have clastic sediments, mainly greywacke and clay, as protoliths (Maggetti & Flisch 1993). Secondarily, limestone and marl were deposited and are now present as marble and skarn. Tholeiitic basalts (later transformed to **amphibolite**) – possibly of the same age – were

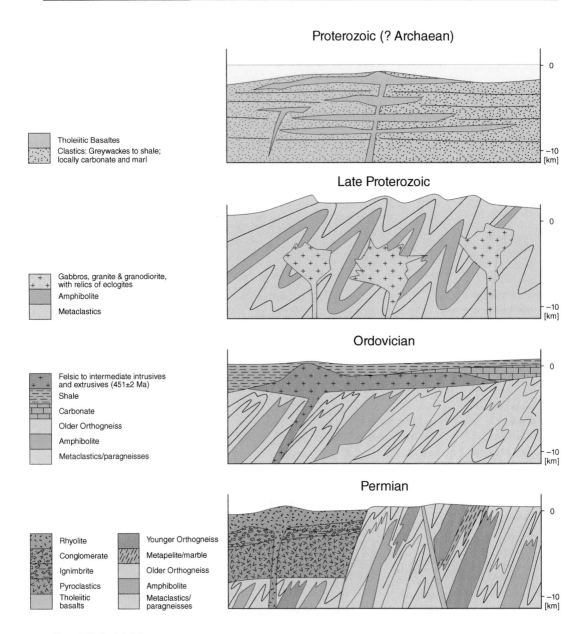

Proterozoic (? Archaean)

Tholeiitic Basaltes
Clastics: Greywackes to shale; locally carbonate and marl

Late Proterozoic

Gabbros, granite & granodiorite, with relics of eclogites
Amphibolite
Metaclastics

Ordovician

Felsic to intermediate intrusives and extrusives (451±2 Ma)
Shale
Carbonate
Older Orthogneiss
Amphibolite
Metaclastics/paragneisses

Permian

Rhyolite
Conglomerate
Ignimbrite
Pyroclastics
Tholeiitic basalts

Younger Orthogneiss
Metapelite/marble
Older Orthogneiss
Amphibolite
Metaclastics/paragneisses

▲ **Figure 2.10** The geological evolution of the crystalline basement of the Austroalpine Silvretta nappe. Source: von Raumer & Neubauer (Hrsg) (1983). Reproduced with permission of Springer Science+Business Media.

intruded in these metasediments and may indicate the formation of a back-arc basin (Fig. 2.10). The age of these rocks is Proterozoic, possibly even Archaean. Subduction processes deformed these rocks (Fig. 2.10) and led to an amphibolite-facies metamorphic overprint, even to eclogitization later. At the end of this episode, in the

Late Proterozoic, granitic, granodioritic and gabbroic melts intruded the now metamorphosed sediments. According to Maggetti & Flisch (1993), the crystallization age is 895 million years.

Another orogeny that is linked to a second high-pressure metamorphism, transformed the Early Proterozoic intrusives into orthogneiss, known as the **Older Orthogneiss**. It is difficult to determine the age of this metamorphism; however, it is probably younger than the intrusion of the Older Orthogneiss and older than the intrusion of the Younger Othogneiss and may therefore be correlated with the Pan-African orogenic cycle. In this cycle, a series of orogens took place on the past continent Gondwana, about 870 to 550 million years ago.

In the Late Ordovician, 451 million years ago, felsic to intermediate melts intruded the older crustal rocks: parts of these probably reached the surface of the Earth at the time (see Maggetti & Flisch 1993). Such Ordovician magmatics are quite prevalent in the Austroalpine nappes and have been explained in the framework of continental rifting, in spite of being the same age as the Caledonian synorogenic intrusives. This would imply that sedimentation also took place in the rifts at that point in time. The carbonates and clays mentioned in Fig. 2.10 are intended to represent such Ordovician basin fill. The Ordovician magmatics were transformed into the **Younger Orthogneiss** over the course of the Variscan orogeny. However, the other rock formations were also deformed and metamorphosed. The metamorphism reached amphibolite facies in the rocks of the Silvretta nappe. According to Maggetti & Flisch (1993), peak temperatures (600–650 °C at pressures of 5.5–7.5 kilobars) were reached 370 million years ago (in the Late Devonian). A subsequent decompression is revealed in the growth of andalusite at 550–600 °C and only 2–3 kilobars; this was followed by compressional tectonics that led to the formation of schlingen structures that extended over kilometres. The age of this youngest Variscan deformation phase is estimated at 340–310 million years (Early Carboniferous).

In the Late Carboniferous and in the Permian, continental extensional tectonics have been determined in the area of the Silvretta nappe (Fig. 2.10). Crustal rocks were exhumed and exposed on the graben shoulders and **pyroclastic**, felsic **volcaniclastic** and **ignimbritic rocks** were deposited in the graben. A large number of tholeiitic **diabase dykes** that can be related to extensional tectonics, cross-cut the crystalline basement.

2.5 The pre-Triassic Basement of the Southern Alps

The basement in the northwestern part of the Southern Alps is of particular interest, as it exhibits an almost complete exposed cross-section through the lower and upper crust. Two large units can be distinguished: the Ivrea Zone and the Strona–Ceneri Zone. The latter zone is also called the 'Serie dei Laghi'. The Insubric Line separates this basement from the Austroalpine nappes (in the west) and the Penninic nappes (in the north). The Pogallo Line forms the boundary between the Ivrea Zone located in the northwest and the Strona–Ceneri Zone. The geological map in Fig. 2.11 shows the general and internal structure of this basement complex. The Ivrea Zone contains a geological succession from the lower crust that reached the surface through a series of tectonic processes. From bottom to top, there are ultramafic rocks, a thick sequence of mafic rocks (referred to as the 'Main Mafic Formation') and paragneisses (Fig. 2.12).

The **ultramafic rocks** comprise peridotite and pyroxenite and, as can be seen in Fig. 2.11, crop out along the Insubric Line mainly as small bodies

▶ **Figure 2.11** Geological map of the Southalpine Ivrea Zone and the Strona–Cenreri Zone. Source: Adapted from the Geological Map 1:500'000 of Switzerland with kind permission of Swiss Topo.

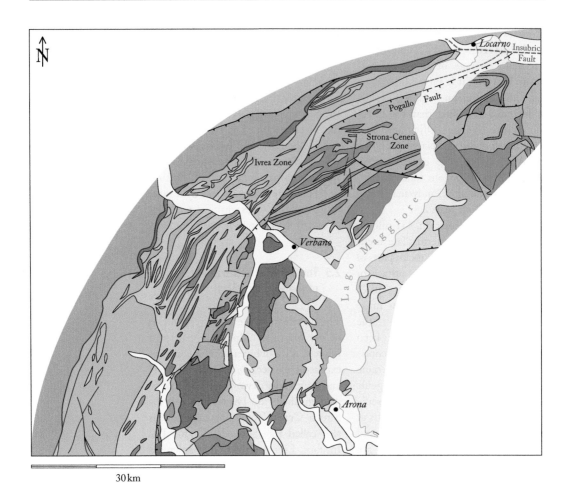

30 km

Quaternary

☐ Alluvium

☐ Glacial deposits

Penninic nappe system

☐ Penninic nappes, undifferentiated

Austroalpine nappe system

☐ Sesia Zone, undifferentiated

☐ Mylonites of the Ivrea Zone and the Canavese Zone

Southalpine nappe system

☐ Mesozoic sediments

☐ Rhyolite, dacite

☐ Permian granites

☐ Diorite, gabbros

☐ Ultramafic rocks

☐ Early Palaeozoic metagranite

☐ Metagabbros

☐ Amphibolite

☐ Gneisses and mica schists metasediments

☐ Kinzingite gneiss

☐ Marble, calcsilicate fels

and lenses. Among the peridotites, spinel peridotite and lherzolite in the Baldissero and Balmuccia bodies can be interpreted as being derived from the mantle (intrusion of melts from the upper mantle into the lower crust). The actual mantle lies slightly beneath the Earth's surface and is called the Ivrea

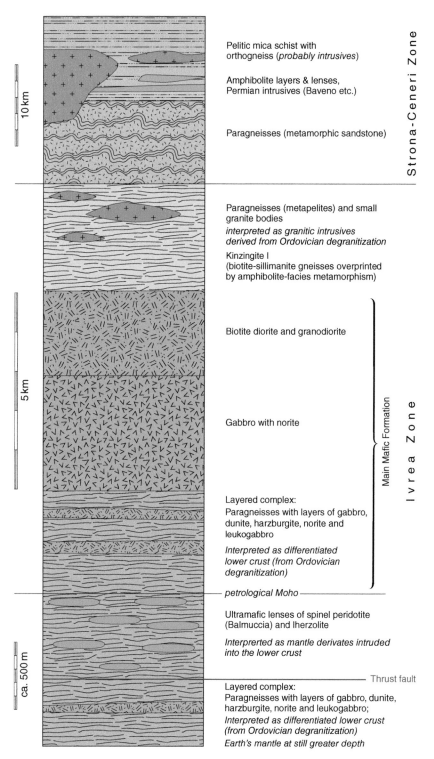

Strona-Ceneri Zone

Pelitic mica schist with orthogneiss (*probably intrusives*)

Amphibolite layers & lenses, Permian intrusives (Baveno etc.)

Paragneisses (metamorphic sandstone)

Paragneisses (metapelites) and small granite bodies
interpreted as granitic intrusives derived from Ordovician degranitization

Kinzingite I
(biotite-sillimanite gneisses overprinted by amphibolite-facies metamorphism)

Ivrea Zone — Main Mafic Formation

Biotite diorite and granodiorite

Gabbro with norite

Layered complex:
Paragneisses with layers of gabbro, dunite, harzburgite, norite and leukogabbro

Interpreted as differentiated lower crust (from Ordovician degranitization)

petrological Moho

Ultramafic lenses of spinel peridotite (Balmuccia) and lherzolite

Interperted as mantle derivates intruded into the lower crust

Thrust fault

Layered complex:
Paragneisses with layers of gabbro, dunite, harzburgite, norite and leukogabbro;
Interpreted as differentiated lower crust (from Ordovician degranitization)
Earth's mantle at still greater depth

Figure 2.12 The sequence of rock types and units in the crustal profile of the Ivrea Zone (lower crust) and the Strona–Ceneri Zone (upper crust).

▶ **Figure 2.13** The
geological evolution of the
crystalline basement of the
Southalpine Ivrea Zone and
Strona–Ceneri Zone. Source:
von Raumer & Neubauer
(Hrsg) (1983). Reproduced
with permission of Springer
Science+Business Media.

Body. This body of dense mantle rocks is responsible for the extreme positive gravity anomaly that extends along the eastern margin of the Western Alps (see below). The nomenclature of ultramafic rocks:

peridotite:	>40% olivine, besides also pyroxene
lherzolite:	pyroxenes are orthopyroxene and clinopyroxene
harzburgite:	pyroxenes are only orthopyroxene
dunite:	>90% olivine
gabbro:	olivine + pyroxene + Ca-plagioclase
norite:	olivine-gabbro

The **Main Mafic Formation** includes two layered complexes containing dunite, harzburgite, norite and gabbro, which may represent the same tectonically repeated horizon. Their formation through magmatic differentiation in the lower crust could be linked to an Ordovician episode of 'degranitization'. Ultramafic lenses (e.g. Balmuccia spinel peridotite) are interspersed in between the two layers.

The **paragneisses** contain kinzigite, that is, biotite–sillimanite paragneiss with an amphibolite-facies metamorphic overprint. The paragneisses probably can be interpreted as metapelites that were chemically transformed through 'degranitization'. Ordovician 'degranitization' would also explain the presence of the smaller granitic bodies within the paragneisses, such as the aplite and pegmatite dykes. Finally, amphibolites and amphibolitic gneisses are also present in the form of thin layers in the paragneisses.

The Strona–Ceneri Zone (or the 'Serie dei Laghi') comprises rocks from the intermediate crust, principally polymetamorphic metasediments. They are mainly psammitic and granitic **gneisses** in the northwest, which are interpreted as metamorphic sandstones, while metapelites (metamorphosed shales) predominate in the southeast. The metasediments were intruded by Ordovician granites (466 million years old) that are now present in the form of **orthogneiss**. In addition, **amphibolites** occur in layers and lenses between the gneisses and schists. The Strona–Ceneri Zone also contains **post-Variscan intrusives**. Of these, the Baveno (276 million years old, i.e., Early Permian) and Mount Orfano granites (283 million years old, earliest Permian) are well-known, as they are quarried for building (pale pink granite).

There is a small deposit of **Late Palaeozoic sediments** near Manno (north of Lugano); these contain isolated coal seams that can be dated to the Late Carboniferous (Westphalian A and B). Younger **Permian volcanics** are exposed to the south of Lugano: the Lugano Quartz porphyry (rhyolite), which is quarried for use as cobbles, and porphyrite (andesite) dated to the (?Early) Permian.

The rocks in the crustal sequences in the Ivrea and Strona-Ceneri zones reveal an evolution that ranges from Proterozoic sedimentation, through Ordovician magmatism to Variscan orogeny and, finally, to post-Variscan magmatism and sedimentation. The discussion below is based on a synthesis produced by Schmid (1993) and is illustrated in Fig. 2.13. Schmid (1993) proposes the following scenario to explain the formation of metapelite, metapsammite and pelitic mica schist in the Strona-Ceneri and the Ivrea zones. In the Late Proterozoic or Early Palaeozoic, an accretionary prism built up through frontal accretion of sediments

Late Proterozoic–Early Palaeozoic

Accretionary wedge (clastics)
Hypothetical continental crust
Oceanic crust with sediments
Lithospheric mantle

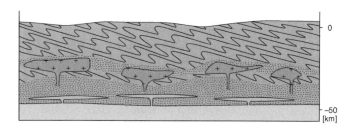

Ordovician

Granite
Mafic intrusives
Metaclastics
Anatexis
Lithospheric mantle

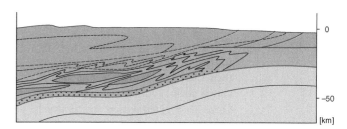

Late Devonian–Early Carboniferous

Carbonates and clastics
Orthogneiss
Metaclastics
High-temperature mylonite
Lower crust
Lithospheric mantle

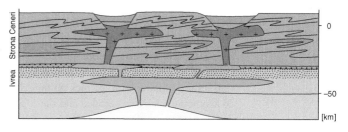

Late Carboniferous–Early Permian

Permian volcaniclastics
Granite
«Apinnite»
Main mafic formation

Orthogneiss
Metaclastics
High-temperature mylonite
Anatexis
Lower crust
Lithospheric mantle
Asthenosphere

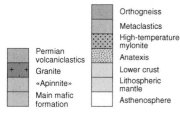

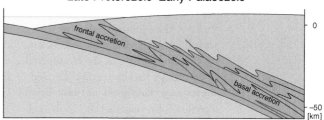

at a subduction zone where an oceanic plate plunged under a continental plate. Some of the sediment was also transported down to great depth and became attached to the base of the continental plate through basal accretion. The sediments were sandstone and clay. Basal accretion also resulted in individual fragments of oceanic crust being scraped off the descending plate and inserted into the sedimentary rock suites of the upper plate (Fig. 2.13). The sediments advected by basal accretion were overprinted by high-pressure metamorphism.

In the Ordovician, a high geothermal gradient was achieved in the upper plate, as is indicated by an initial differential melting of gneisses, dated to 478 million years ago (see Schmid and references therein), and the simultaneous intrusion of granites 466 million years ago into the now metamorphosed sediments. It is possible that mafic intrusions also occurred in the lower crust at this time (Anzola Gabbro type).

In the Late Devonian to Early Carboniferous, the Variscan orogeny then caused intensive shearing of the crustal rocks. During this process, the Ordovician granites were transformed into orthogneiss, which can now be seen in the Strona–Ceneri Zone in the form of bands with large-scale, narrow folds, known as schlingen structures. The end of Variscan mountain-building, as indicated by cooling of the rock formations due to exhumation, is placed by Schmid (1993) in the time interval between 325 and 310 million years before present (Early Carboniferous). This exhumation resulted in a column of rock up to 30 kilometres thick being eroded and the deeper portions of the crust, the Strona–Ceneri Zone, were then exposed at the Earth's surface. Interestingly, there are

marked differences between the metamorphic evolution of the Ivrea and the Strona–Ceneri zones. Due to rapid exhumation the high-pressure paragenesis of the Variscan orogeny remained preserved in the rocks of the Strona–Ceneri Zone, whereas the rocks in the Ivrea Zone were transformed by the prevailing high temperatures, such that the new paragenesis indicates moderately high pressures. It appears that the two blocks of crust were already uncoupled from each other during the Variscan orogeny. A zone of high-temperature mylonites, that are now visible along the Pogallo Line, separated the Strona–Ceneri Zone from the Ivrea Zone. The former moved upwards towards the surface of the Earth (exhumation) and cooled down, while the latter remained at great depth and was heated.

A further phase of magmatism affected both the Ivrea and the Strona–Ceneri zones in the Late Carboniferous and Early Permian (Fig. 2.12). The intrusion of mafic melts into the lower crust, now to be found in the form of the Main Mafic Formation in the Ivrea Zone, caused differential melting of metasediments (a second anatexis) in the lower crust. These melts rose upwards and intruded into the upper crust, the Strona–Ceneri Zone, or even reached the Earth's surface. Examples of these magmatic rocks are the Baveno and Mont Orfano granites mentioned above, as well as the Permian Lugano Quartz porphyry. Deep-seated deformation that involved the entire crust and the uppermost mantle can explain this magmatic activity. Transtensive strike-slip ('wrench tectonics' in Schmid 1993 and references therein) would be a good explanation for this and would fit perfectly into the concept of the post-Variscan continental extensional tectonics that has been mentioned previously in connection with the Black Forest, the external massifs and the Penninic and Austroalpine nappe complexes.

2.6 Palaeozoic Sediments in the Eastern and Southern Alps

Palaeozoic sediments and volcanics that were not, or only weakly, overprinted by the Variscan (and earlier) orogenies are exposed at several locations in the Eastern and Southern Alps. Some of these sequences are up to several kilometres thick, but are subject to substantial local variation, as is also the case for the individual strata. The Palaeozoic in the Carnic Alps, the Greywacke Zone and the Innsbruck Quartz Phyllite have been selected as typical examples.

The Palaeozoic in the Carnic Alps

The most complete sequence, dated based on fossils, can be found in the Carnic Alps, that lie south of the peri-Adriatic fault system. The stratigraphic cross-section of the column shown in Fig. 2.14 is based on the compilation by Schönlaub & Heinisch (1993). Thick layers of Ordovician greywacke and shale are overlain by laterally intercollated volcanics, sandstones and shales, which are in turn overlain by Silurian limestone. The global regression at the end of the Ordovician caused erosion and interrupted sedimentation, such that the Silurian sediments lie discordantly on top of the Ordovician sediments. Several sedimentary facies zones that correspond to varying conditions during deposition can be distinguished in the Silurian deposits: limestone was deposited under shallow marine to pelagic conditions, while black clay was deposited in deeper, more

stagnant parts of the basin. A carbonate platform then became established in the Devonian, with the rate of subsidence increasing, but varying locally: there was widespread development of reefs, which varied both temporally and spatially. These depositional conditions continued into the Early Carboniferous, when an important change took place at the transition to the Hochwipfel Formation, which records the sudden appearance of flysch-like clastics over (palaeo)-karstified carbonates, still in the Early Carboniferous, that is associated with the Variscan orogeny (see Schönlaub & Heinisch 1993 and references therein). Sandy and pelitic turbidites were deposited in a flysch basin at an active continental margin associated with the developing orogen. Sedimentation in this flysch basin continued up into the Serpukhovian/Namurian (326–318 million years ago).

The Ordovician to Early Carboniferous sediments are capped discordantly by a post-Variscan sedimentary sequence with a transgressive conglomerate at its base, the Waidegg Formation. This conglomerate is assigned to the Moscovian/Westphalian (310–307 million years old). In this region, the Variscan orogeny can thus be placed in the time interval of between 318 and 310 million years ago.

The sediments above the transgressive conglomerate (Auernig Formation) contain coastal conglomerates, cross-bedded sandstone, siltstone and clay with bioturbation, as well as limestone. The entire sequence is also interpreted as a molasse (in the sense of 'post-orogenic' clastic sediments). However, according to Krainer (1993), these are to be interpreted as clastic-carbonate transgressive and regressive cycles or cyclothems. The individual cycles lasted about 100 000 years each and

▶ **Figure 2.14**
The Palaeozoic sedimentary sequence in the Austroalpine Carnic Alps. The simplified stratigraphic succession is a compilation based on Schönlaub & Heinisch (1993). Source: von Raumer & Neubauer (Hrsg) (1983). Reproduced with permission of Springer Science+Business Media.

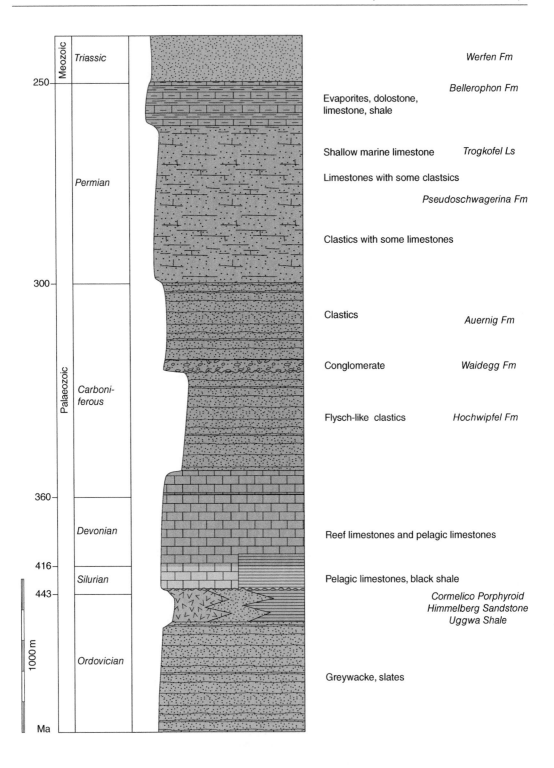

probably resulted from eustatic sea-level fluctuations after glaciations on Gondwana. An intramontane basin that was filled with sediments and volcanics of different provenance can be interpreted as the possible depositional environment.

The subsequent Permian clastics (Pseudoschwagerina Formation) are partially composed of sediments that had been reworked from the footwalls (the Auernig Formation) of faults: tectonic movements (the so-called Saalian Phase) appear to have caused significant uplift and erosion at the margins of the basin (see Krainer 1993). The deposition of these Permian sediments thus also took place in intramontane basins or grabens, and they are discussed in section 2.7 in the context of the Permo-Carboniferous troughs.

The Palaeozoic of the Greywacke Zone

The Eastern (Styria) and Western Greywacke zones (Salzburg–Tyrol) are distinguished geographically based on their spatial distribution: Fig. 2.15 shows the succession in the Eastern Greywacke Zone, summarized based on Schönlaub & Heinisch (1993).

A fossil-free sequence of slates, phyllites and metapyroclastics is found above a conglomerate that lies on the crystalline basement. This conglomerate contains pebbles of orthogneiss in some locations, the crystallization of which was dated to 500 million years ago (the very end of the Cambrian), which suggests an Ordovician age for the overlying metasediments. The next sequence contains volcanics (ignimbrite, tuffs, pyroclastics), the so-called Blasseneck Porphyroid. The Blasseneck Porphyroid is locally interbedded with limestone, the age of which has been determined unequivocally as Late Ordovician. A thick sequence of shale (now present in

the form of metapelite) completes the Ordovician and extends into the Silurian. According to Schönlaub & Heinisch (1993), the continuous Ordovician sequence exhibits no influence or indications of Caledonian orogeny. The Blasseneck Porphyroid volcanics could be correlated with the Pan-African orogenic cycle.

The Early Silurian mafic volcanics can be interpreted as intraplate volcanics based on geochemical data. The overlying black slate and *Orthoceras* limestones were deposited in a tectonically quiet environment.

There is an abrupt change with the calcareous breccias that follow discordantly, which are interpreted as a product of karstification in the Early Carboniferous (Schönlaub & Heinisch 1993 and references therein). This karstification could be explained by emersion of the foreland of the Variscan mountain chain: due to the weight of the nappes in the mountain chain itself, the subducting plate, in this case Gondwana, is flexed downward. However, the plate has a certain elastic stiffness, such that the downflexing and bending produced a bulge in the foreland. Such a bulge would adequately explain the emersion and karstification. Thus the sedimentary succession in the Greywacke Zone shows clear signs of the Variscan orogeny.

The Palaeozoic of the Innsbruck Quartz Phyllite

At a variety of locations in the Eastern and Southern Alps, the Palaeozoic is dominated by phyllitic, in particular, quartz-phyllitic sedimentary sequences. As an example of this, the Innsbruck Quartz Phyllite will be briefly discussed below. The illustration in Fig. 2.16 is based on the work by Neubauer & Sassi (1993).

▶ **Figure 2.15** The Palaeozoic sedimentary sequence in the Austroalpine Eastern Greywacke Zone. Summarized after Schönlaub & Heinisch (1993). Source: von Raumer & Neubauer (Hrsg) (1983). Reproduced with permission of Springer Science+Business Media.

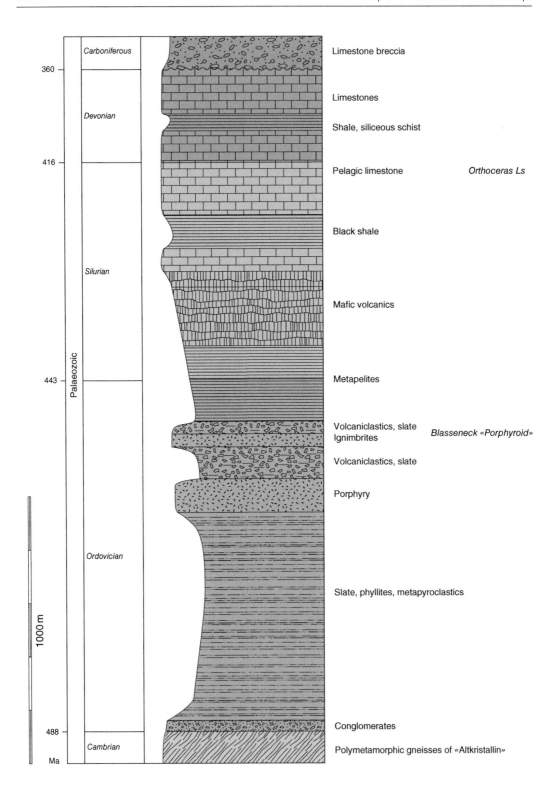

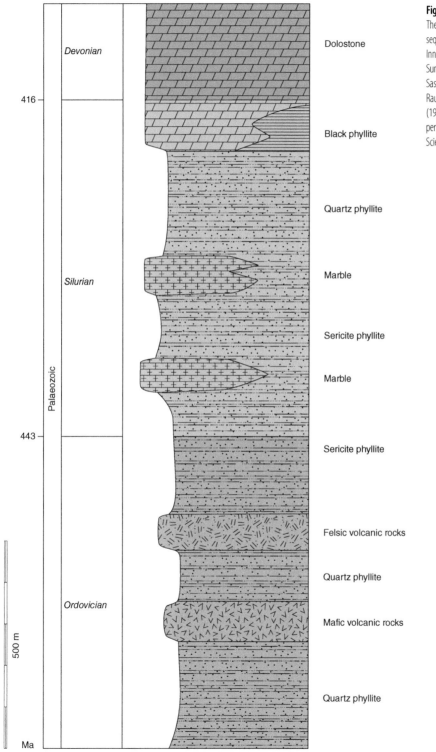

Figure 2.16
The Palaeozoic sedimentary sequence in the Austroalpine Innsbruck Quartz Phyllite. Summarized after Neubauer & Sassi (1993). Source: von Raumer & Neubauer (Hrsg) (1983). Reproduced with permission of Springer Science+Business Media.

Ordovician quartz phyllites and sericite phyllites contain layers of mafic and felsic volcanics. This sequence can be interpreted as the fill of a back-arc basin in the Pan-African orogenic cycle. The younger portions (Silurian?) of the quartz and sericite phyllites exhibit layers of marble, followed by a thick layer of dolostone at the top, which is dated to span the time interval from the Late Silurian to the Middle Devonian. Neubauer & Sassi (1993) interpret this sedimentary sequence as the product of Silurian rifting and the development of a carbonate platform on a submarine rise.

Based on a correlation between the many quartz phyllite series, the conclusions of Neubauer & Sassi (1993) can be summarized as follows: Aa thick clastic sequence was deposited in a back-arc basin at the Pan-African plate margin in the Ordovician; the onset of rifting in the Silurian caused lithological differences in the younger quartz phyllites (conglomeratic sandy in the south, argillaceous in the north); rifting is indicated by the type of volcanic deposits and signs of synsedimentary faulting; from the Late Silurian to the Early Devonian, the basin was filled with clastics from alluvial fans and carbonate tempestites; the maturity of the clastics indicates a long transport distance and a continental source area; from the Middle Devonian onwards and into the Early Carboniferous a carbonate platform became established, possibly on a rise within the basin; the basin was subsequently closed by the Variscan orogeny, starting in the Early Carboniferous.

2.7 The Variscan Orogen at the Close of the Palaeozoic

As has been mentioned in the previously, in the Alps and their foreland late- and post-Variscan intrusives intruded the older rock formations of the crystalline basement and/or Late Palaeozoic volcaniclastics were deposited on top of this basement. Particularly impressive examples of this can be observed in the external massifs of the Alps (see Figs 2.3–2.8). A variety of sediments were deposited at other locations over the same time period. In the Alps, such sedimentary rock suites are exposed, namely, in the Austroalpine and Southalpine nappe complexes (see Figs 2.14–2.16). However, when comparing and contrasting these rock formations, we are inevitably confronted with the problem that the majority of these deposits are covered with younger rock and are thus only exposed in isolated cases in the form of individual outcrops. In addition, over the course of the Mesozoic during the break-up of Pangaea and during the later Alpine orogeny, these outcrops were displaced over large distances relative to each other and thrust on top of each other. So, how can we now piece together these outcrops with their different rock formations to produce a picture of the Variscan orogeny at the close of the Palaeozoic?

A palaeogeographical map was constructed for this moment in time in order to obtain an overview of the distribution of the various rock suites and the local structures. To this end, the different blocks and fragments of crystalline basement first needed to be moved back into their original relative positions before the Alpine orogenesis and before the opening of the Valais Trough and the Penninic–Piemont–Ligurian Ocean. However, as our knowledge of directions and transport distances is only patchy for the building of the Alpine nappe stack, this reconstruction is associated with numerous uncertainties. Once the Alpine nappe movements had been reversed, the same process had to be applied to the plate movements during the break-up of Pangaea, which made the scope for uncertainty even greater. The resultant palaeogeographical map in Fig. 2.17 is therefore associated with some uncertainty with reference to the detail, but fairly secure statements can be made on the large-scale relationships. The illustration is based partially on the compilations by Pfiffner (1993a), de Graciansky (1993) and Ratschbacher & Frisch (1993), but with additions.

The outlines of the outcrops of pre-Triassic basement are depicted on the palaeogeographical map as they appear on the current geological map. Their 'future' allocation to the Alpine nappe complexes (Helvetic, Penninic, Austroalpine and Southalpine) is also indicated. This shows quite unambiguously that deposition of Palaeozoic sediments (Ordovician to Devonian) is clearly restricted to the southeast of the Austroalpine nappe complex and to the Southalpine nappe complex, while the central portions of the future Alps (northwestern Austroalpine, Penninic and Helvetic nappe complexes) are composed of crystalline rocks. In many locations, Devonian to Early Carboniferous metamorphism can be observed in these crystalline portions (see compilation by von Raumer & Neubauer 1993b). These portions are apparently the core of the Variscan mountain chain, where upper crustal crystalline rocks of the orogen were exhumed and exposed at the surface. A limiting factor in all of this is that Variscan metamorphic features in the central portion, in particular in the area of the future Penninic nappe complex, were overprinted by Alpine metamorphism to such an extent that statements on Variscan metamorphic conditions are very difficult to make. Even so, it is possible to reach the conclusion that a large overthrust is present between this crystalline core of the orogen and the Palaeozoic sediments at the same level to the southeast, as is outlined in Fig. 2.17. Analogous overthrusts can be assumed between the Saualpe–Koralpe crystalline basement block and the Palaeozoic Gurktal block, and between the Ötztal block and the Southalpine nappe complex. No conclusions, however, should be drawn concerning the more detailed history of these overthrusts, given the inherent uncertainty associated with the reconstruction of the palaeogeographical map. Smaller Late Devonian to Early Carboniferous sedimentary deposits are also exposed in the Northalpine foreland (Black Forest and Vosges) (e.g. in the Badenweiler–Lenzkirch Zone, to the south of the Todtnau thrust in Fig. 2.2). In the case of the Todtnau thrust, southeast-vergent nappe movement can be assumed based on structural geological findings (Eisbacher et al. 1989). The Ordovician–Devonian portion of the Palaeozoic Graz and Gurktal blocks within the Austroalpine and Southalpine nappe complexes are assumed to represent the passive continental margin of the Gondwana continent, whereas the Greywacke and Quartz Phyllite Zones are more representative of a more external depositional area characterized by extensional tectonics. Volcanic activity accompanied the basin-filling process at different points in time. From the Middle Devonian onwards, a carbonate platform also became established in the quartz phyllite zone. The character of the basin changed in the Early Carboniferous. As a result of Variscan orogenesis, this passive continental margin was transformed into a syn-orogenic foreland basin with local emersion and karstification (calcareous

▶ **Figure 2.17**
Palaeogeographical map of the future Alpine realm at the close of the Palaeozoic. The present-day outcrop shapes of the various basement units are given as a frame of reference. These units have been shifted back into their relative position by accounting for displacements due to Alpine thrusting as well as the preceding opening of oceanic basins. Also indicated are the boundaries of the future Northalpine foreland, Dauphinois–Helvetic, Penninic and Southalpine–Austroalpine realms. Source: von Raumer & Neubauer (Hrsg) (1983). Reproduced with permission of Springer Science+Business Media.

Traces of Variscan thrust faults in the Alps

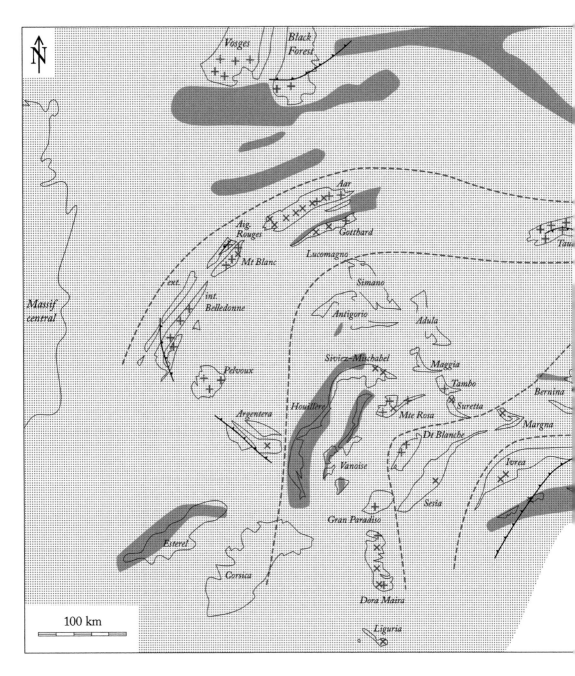

breccia in the Greywacke Zone) and flysch-like deposits (Hochwipfel Formation in the Carnic Alps). In the Early Carboniferous (between 350 and 320 million years ago), the quartz phyllites were overprinted by Variscan metamorphism in many locations. The basin was clearly closed and the basin fill was incorporated into the nappe pile of the Variscan mountain chain.

Carboniferous and Permian granitoid intrusives within the crystalline

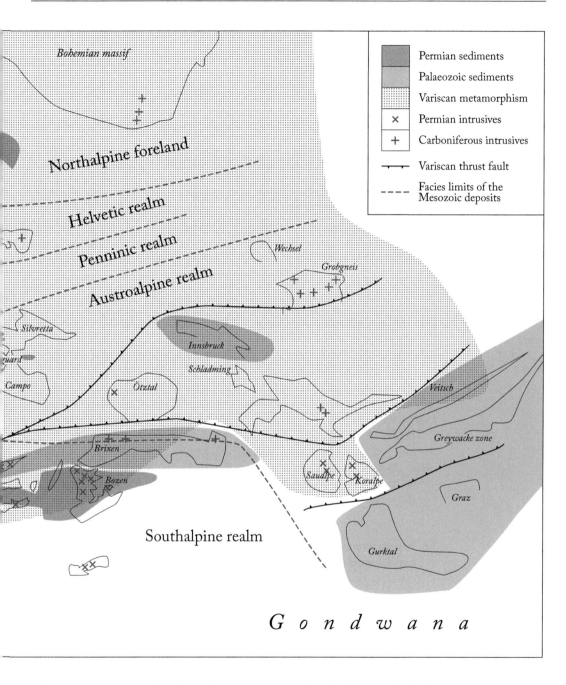

Bohemian massif

Northalpine foreland

Helvetic realm

Penninic realm

Austroalpine realm

Wechsel

Grobgneis

Silvretta

Innsbruck

Campo

Ötztal

Schladming

Veitsch

Greywacke zone

Brixen

Bozen

Saualpe

Koralpe

Graz

Southalpine realm

Gurktal

G o n d w a n a

Legend:
- Permian sediments
- Palaeozoic sediments
- Variscan metamorphism
- × Permian intrusives
- + Carboniferous intrusives
- Variscan thrust fault
- Facies limits of the Mesozoic deposits

rock units are also indicated in the palaeogeographical map. The granitoid intrusives in the crystalline core of the Variscan orogen exhibit a tendency to be older in the northwest. A granite belt with a Late Carboniferous intrusion age can be identified in the foreland (Black Forest) and in the external massifs. Most of these intrusions are post-Variscan, as they clearly cut the Variscan structures. There are exceptions, for example, in the Black Forest (see Fig. 2.2). A younger Permian group of intrusives is mainly concentrated in a southern belt (southwestern and south-eastern Austroalpine and Southalpine nappe complexes). There are mixed intrusions that date from the Early Carboniferous and Permian in the central region, and also in the Aar and Gotthard massifs. The post-Variscan intrusions are commonly linked to the presence of Carboniferous graben structures, as is illustrated based on the example of the eastern Aar massif (see Fig. 2.6). The interplay between intrusion and the formation of graben with clastic and volcaniclastic fills indicates a fundamental reorganization of the tectonic processes, with a transition from collision to extensional tectonics. The extension was coupled with large-scale strike-slip motion, caused by the easterly drift of Eurasia relative to Gondwana. This possibly facilitated the rise of magma in sections undergoing transtension.

Extensional tectonics after collision

The formation of graben structures continued in the Permian and is discussed in greater detail in the next section, although the known Permian troughs are also indicated in Fig. 2.17.

2.8 Post-Variscan Sediments and Volcanics of the Permian

Permian sediments and associated volcanics are known from many locations in the Alps and in their foreland. In terms of area covered, there are three dominant trough systems (Fig. 2.17): one in the Northalpine foreland, just south of the Black Forest and the Vosges, one in the future Penninic nappe complex (Zone Houillère) and one near Bozen in the Southalpine nappe complex. In addition, there is an entire series of smaller troughs.

Some Permian troughs continued development within a trough that had formed previously in the Carboniferous: others did not start development until the Permian and, conversely, not all Carboniferous troughs continued their activity into the Permian. Some of these troughs were partially or fully inverted during Alpine orogenesis, that is, the graben fill was squeezed out between the margins of the graben being pushed together. The examples given in the following have been selected to cover the entire spectrum of the processes involved, ranging from the development of a graben through to its inversion.

The North Swiss Permo-Carboniferous Trough

In the base of the eastern Jura Mountains and the northern Molasse Basin in northern Switzerland, a large east–west trending Permo-Carboniferous trough was identified during a search for locations to deposit radioactive waste. Figure 2.18 shows a cross-section that is based on the interpretation of seismic reflection lines and deep boreholes. The younger, Permian trough fill is delimited by relatively shallow, gently dipping synsedimentary normal faults. In contrast, the older, Carboniferous trough fill is accompanied by steep synsedimentary normal faults, but also by flatter, gently dipping normal faults. The older trough was apparently a narrow graben, the fill of which was deformed before the graben was widened. This deformation could correspond to local transpression consistent with large-scale dextral shearing.

The graben fill itself is composed of fluviatile, lacustrine and playa sediments (Matter 1987), with the older portion, dating to the Late Carboniferous (Stephanian), being encountered in a deep borehole at Weiach

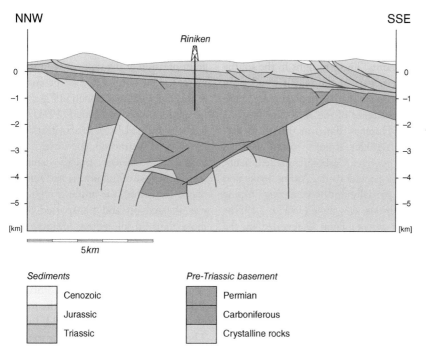

NNW   SSE

Riniken

Figure 2.18 Geological cross-section trough the North Swiss Permo-Carboniferous Trough located beneath the Molasse Basin and the easternmost Jura Mountains. Source: Adapted from Diebold et al. (1991) and Bitterli-dreher et al. (2007).

5km

Sediments

☐ Cenozoic
☐ Jurassic
☐ Triassic

Pre-Triassic basement

☐ Permian
☐ Carboniferous
☐ Crystalline rocks

(north Switzerland, ca. 20 kilometres east-northeast of Riniken). Small cyclic sequences of sandstone–shale formations within the graben can be interpreted as floodplain deposits and fine conglomeratic to course sandy formations as river bed deposits. Thin seams of coal are interspersed in these formations, interpreted as swamp deposits formed on floodplains. The whole formation represents an anastomosing fluviatile system, which continued into the Permian until it was replaced by a lake that gradually silted up due to the progradation of an alluvial fan. The base of this fan in the deep borehole at Weiach is characterized by polymict breccias containing pebbles of crystalline rocks, which rapidly change to fine-grained red strata that were deposited in a playa. The location of the Carboniferous–Permian boundary in Fig. 2.18 is uncertain, as the deeper portion of the trough was not reached by the Riniken borehole,

but is known only from seismic data. There are no recognizable discordant features in the Weiach borehole, which passed through the entire Permo-Carboniferous rock suite.

The Mesozoic sediments overlying the Permo-Carboniferous trough were sheared off along evaporites of the Triassic Anhydrite Group and piled up in a northerly direction as schuppen (or imbricates) during the formation of the Jura Mountains. The hanging wall of the Permo-Carboniferous trough was apparently not substantially affected by this deformation, that is, the basement was resistant to the deformation.

The Permo-Carboniferous in the Helvetic Nappe Complex

In the Helvetic nappe complex, Permo-Carboniferous rock suites are known from one large trough and from several smaller troughs in the external massifs.

A number of older troughs are found in the Aiguilles Rouges massif that are infilled with Devonian(?) and Carboniferous sediments. These troughs in the south-western part of the massif trend from north to south, that is, diagonally to the general strike of the Alps (see Fig. 2.2). The Dorénaz Basin in the northeast-ern Aiguilles Rouges massif is filled with younger, Carboniferous sediments, trends parallel to the Alpine strike and was partially inverted during Alpine orogenesis, as is shown in Fig. 2.19: the latter phenomenon is revealed by the bulge in the stratigraphic contact between the Palaeozoic and Mesozoic sediments, as well as in the bulge in the basal over-thrust of the Morcles nappe above it. At the southeast margin of the trough, fluviatile Carboniferous sandstones and greywackes were rotated into a vertical position when they were squeezed out (Fig. 2.10A).

The trough fill itself is composed of alluvial and fluvio-lacustrine sediments. Niklaus & Wetzel (1996) demonstrated that the oldest portion of the trough fill is made up of dominantly sandy depos-its from a braided riverine system that flowed towards the southeast, tran-versely across the graben. With increas-ing subsidence, an axial basin drainage system then developed, represented mainly by pelitic sediments. Sandstone from the graben margins was deposited in alluvial fans. The sandstone was then oxidized and now has a remarkable wine-red coloration. As the gradient of the river channels decreased, palustrine sediments containing seams of coal were accumulated. Finally, further subsidence of the graben resulted once again in an increase of the gradient of the river channels and the sedimentary deposits indicate the presence of a meandering river system flowing towards the southwest. The sedimentary

fill apparently reflects the tectonic subsidence history of the graben: the photographs in Fig. 2.20A and B give an impression of the types of sediments present. The retro-deformed geometry of the Carboniferous trough in Fig. 2.19 is based on work by Pilloud (1991) but has been updated. In eastern Switzerland, the Glarus Verrucano deposited in a Permo-Carboniferous trough demon-strates the case of full inversion. During nappe formation, the entire trough-fill was sheared off and transported over more than 30 kilometres towards the north. Frontal portions of the crystal-line basement (paragneisses and mica schists) of the Tavetsch massif were also affected by this movement (see Pfiffner 1980) and are now present in the form of a large-scale inverted fold limb. The dislocated trough fill came to rest on a far younger substratum made up of Mesozoic–Cenozoic sediments. The cross-section in Fig. 2.21 shows a reconstruction of the trough in the Mesozoic. The synsedimentary normal faults at the margins of the trough are no longer recognizable in the field, as these were used as lateral ramps when the fill was squeezed out during Alpine nappe formation and were then signifi-cantly sheared by internal deformation. They would now need to be recognized in portions of the basal overthrust of the Helvetic nappes (the Glarus Thrust). Figure 2.21B illustrates the process of squeezing out the Permo-Carboniferous trough fill. The trough fill lies on a smoothed thrust surface. The former margins of the trough are present in the form of recumbent folds, the axes of which form an arc of 180° around the trough fill that has been squeezed out. The Mesozoic sediments in the hanging wall of the trough fill were raised to higher elevation than those to the east and west of the trough.

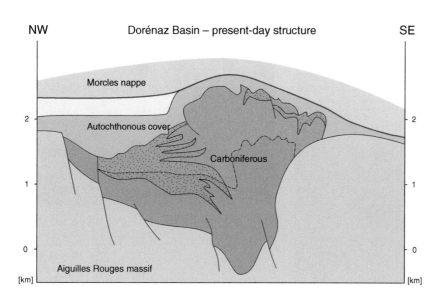

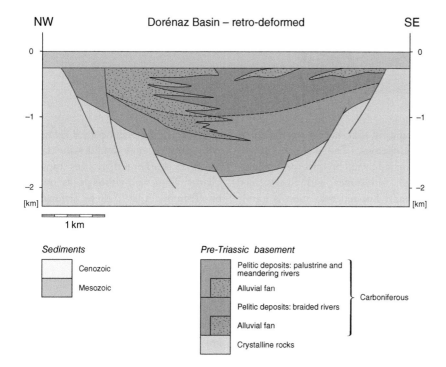

Figure 2.19 Geological cross-section and retro-deformed cross-section through the Carboniferous Dorénaz Basin (Aiguilles Rouges massif). Source: Adapted from Pilloud (1991).

As a consequence, the axes of the folds in these Mesozoic sediments exhibit a notable axial plunge towards the east and west.

The trough fill (Fig. 2.21A) contains very thin layers of black phyllite and greywacke at the base, which are not dissimilar from the formations in the

(A)

(B)

▲ **Figure 2.20**
Carboniferous sediments of
the Dorénaz Basin.

(A) Vertical fluvial sequence
with cross-bedded
sandstones and strings of
conglomerates.

(B) Detailed view of
conglomerate with rounded
and subrounded clasts of
quartz, feldspar and
anthracite.

Aar massif that have been dated as Carboniferous. The Permian fill is dominated by red breccias (or fanglomerates), sandstones and fine-grained red beds. The red coloration is locally completely overprinted by greenschist-facies metamorphism. The coarse clastic sediments (called sernifite locally) are preferentially located at the margins of the trough and represent proximal alluvial fans, while the finer grained clastics are more prevalent in the centre of the trough. The fine-grained red beds (called Schönbüel Shale locally) represent the youngest portion and probably can be equated to those in the North Swiss Permo-Carboniferous Trough. Layers of felsic and mafic volcanics (rhyolite, dacite and/or basalt, now present as spilite) are interbedded

with the calstic sediments. The term 'Verrucano' is now traditionally used to refer to the red clastics of the Glarus Verrucano, although, as noted by Trümpy (1980), these sediments have little in common with the Triassic conglomerates, sandstones and shales of Verruca near Pisa.

The Permo-Carboniferous in the Penninic Nappe Complex

Permian sediments of modest thickness are found in many locations in the Penninic nappe complex, but a larger complex is identified in the Briançonnais of the Western and Central Alps (the Zone Houillère Basin in France and the Mont Fort Basin of the Bernhard nappe complex in Switzerland). According to

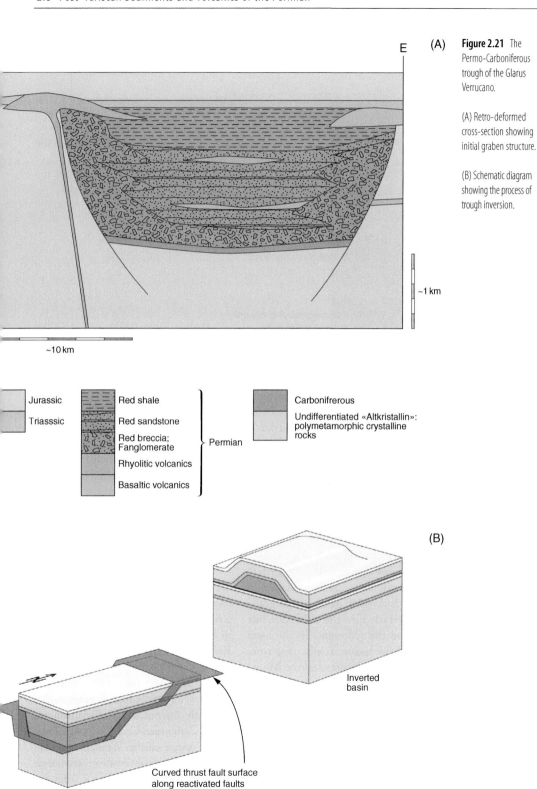

E (A)

Figure 2.21 The Permo-Carboniferous trough of the Glarus Verrucano.

(A) Retro-deformed cross-section showing initial graben structure.

(B) Schematic diagram showing the process of trough inversion.

~1 km

~10 km

Jurassic	
Triasssic	

Red shale
Red sandstone
Red breccia; Fanglomerate   ⎫ Permian
Rhyolitic volcanics
Basaltic volcanics   ⎭

Carbonifrerous

Undifferentiated «Altkristallin»: polymetamorphic crystalline rocks

(B)

Inverted basin

Curved thrust fault surface along reactivated faults

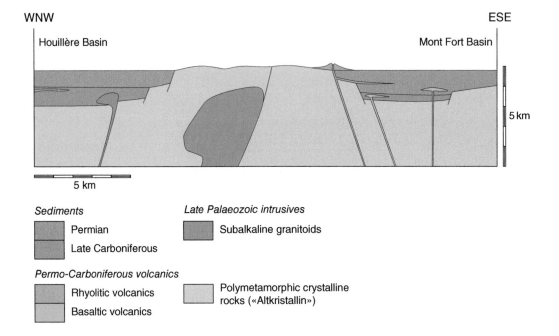

WNW   ESE

Houillère Basin   Mont Fort Basin

5 km

5 km

Sediments

▢ Permian

▢ Late Carboniferous

Permo-Carboniferous volcanics

▢ Rhyolitic volcanics

▢ Basaltic volcanics

Late Palaeozoic intrusives

▢ Subalkaline granitoids

▢ Polymetamorphic crystalline rocks («Altkristallin»)

▲ **Figure 2.22** The Permo-Carbonifereous of the Zone Houillère and Mont Fort basins (Briançon basement). Magmatic rocks associated with the Late Carboniferous to Permian sediments include subalkaline granitoids and rhyolitic and basaltic volcanics.

Thélin et al. (1993), this is an example of two (possibly more) troughs running parallel to each other that are filled with Late Carboniferous to Permo-Triassic fluviatile sediments. Metapelites containing graphite and greywackes occur in the older formations, suggesting they date from the Carboniferous. Quartzites complete the younger Permian trough fill and there is no visible transition to the quartzites assigned to the Triassic. The trough fills are also distinguished by different volcanic rocks, as is indicated in Fig. 2.22. Permian gabbros and granitoids intrude into the old crystalline rocks and the sediments in the form of laccoliths and/or as sills. Volcanics are far more widespread in the Mont-Fort Basin and Thélin et al. (1993) interpret this as indicating a pull-apart basin (transtensive tectonic regime). The orientation of the basins would indicate large-scale dextral shearing, similar to the case of the North Swiss Permo-Carboniferous Trough.

The Permo-Carboniferous in the Austroalpine Nappe Complex

Permian sediments are also widespread in the Austroalpine nappe complex in the Eastern Alps. According to Krainer (1993), these were deposited in troughs trending east–west to northeast–southwest, the opening up of which is explained by large-scale dextral shearing linked to the easterly drift of Eurasia relative to Gondwana at this time.

The Late Palaeozoic of the Gurktal nappe will be briefly discussed below as an example of these many deposits. The stratigraphic succession is summarized in Fig. 2.23 and is essentially based on Krainer (1993), as is the discussion below.

The succession starts with a polymict conglomerate at the base, which lies discordantly on top of the Devonian to Early Carboniferous sediments folded in the course of the Variscan orogeny. The Stangnock Formation that lies above this is interpreted as a fluviatile

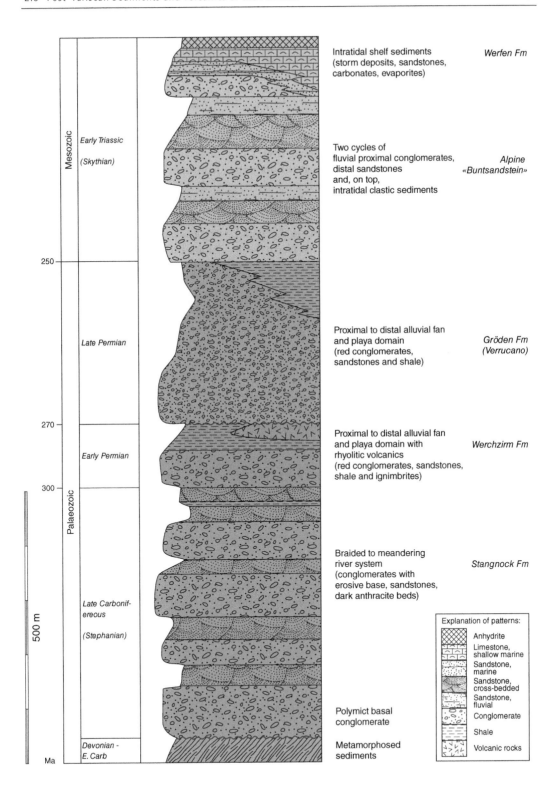

Intratidal shelf sediments
(storm deposits, sandstones,
carbonates, evaporites)

Werfen Fm

Two cycles of
fluvial proximal conglomerates,
distal sandstones
and, on top,
intratidal clastic sediments

*Alpine
«Buntsandstein»*

Proximal to distal alluvial fan
and playa domain
(red conglomerates,
sandstones and shale)

*Gröden Fm
(Verrucano)*

Proximal to distal alluvial fan
and playa domain with
rhyolitic volcanics
(red conglomerates, sandstones,
shale and ignimbrites)

Werchzirm Fm

Braided to meandering
river system
(conglomerates with
erosive base, sandstones,
dark anthracite beds)

Stangnock Fm

Polymict basal
conglomerate

Metamorphosed
sediments

Explanation of patterns:

Anhydrite

Limestone,
shallow marine

Sandstone,
marine

Sandstone,
cross-bedded

Sandstone,
fluvial

Conglomerate

Shale

Volcanic rocks

◀ **Figure 2.23** The
Carboniferous and Permian
sediments of the Gurktal
nappe (Austroalpine nappe
system), summarized after
Krainer (1993). Source: von
Raumer & Neubauer (Hrsg)
(1983). Reproduced with
permission of Springer
Science+Business Media.

rock suite deposited in an intramontane basin, built up in several megacycles. Each of these megacycles has conglomerates at the base (with an erosional contact at the base) deposited by a braided river system and, above this, cross-bedded conglomerate–sandstone associations of a more meandering river. This is capped by dark anthracitic shales with fossilized plant fragments, which can be interpreted as ox-bow lakes deposits accumulated on a floodplain. This rock suite is remarkably similar to that found in the older portion of the North Swiss Permo-Carboniferous Trough.

Coloration changes to red in the transition to Permian sediments, which is thought to be attributable to climate. The Early Permian Werchzirm Formation is composed of red conglomerates/breccias, immature sandstones and fine-grained red beds. This formation is interpreted as the deposits of a proximal to distal alluvial fan and a playa complex. Rhyolitic volcanics (ignimbrites and pyroclastics) are often found at the top of the formation.

The transition to the Gröden Formation of the Late Permian is characterized by a hiatus that was caused by faulting. This so-called Saalian Phase probably can be attributed to transpressive and transtensive movements in a pre-existing graben, which may be analogous with the interpretation of the deformation of the older trough fill in the North Swiss Permo-Carboniferous Trough (Fig. 2.18). Otherwise, the Gröden Formation is again a case of deposits from a proximal to distal alluvial fan giving way to a playa complex at the top. However, the coarse clastic sediments contain large quantities of reworked material from Early Permian volcanics. The Gröden Formation is also remarkable in its similarity

to the Permian of the North Swiss Permo-Carboniferous Trough and the sedimentary component of the Glarus Verrucano.

Sedimentation continues uninterrupted towards the top, but there is an abrupt change in depositional conditions and in the composition of the sediments in the Alpine Buntsandstein (Fig. 2.23). Krainer (1993) explains this change through a rapid climatic change to more humid conditions. Three megacycles can be distinguished in the Alpine Buntsandstein and in the Werfen Formation above it. Each cycle starts with proximal conglomerates deposited by a braided river system that then gives way to more distal sandstones. Each megacycle is completed by fine clastic, shallow-marine sediments. In the case of the Werfen Formation, storm deposits are present instead of the fluviatile sediments and the formation is completed by evaporites that gradually give way to the Alpine Muschelkalk.

The Permo-Carboniferous in the Southalpine Nappe System

In the Southalpine nappe system, the Permian strata are continuous from underlying formations, some of which were overprinted by metamorphism, as is the case for quartz phyllites. According to Krainer (1993), in many locations, these deposits are not associated with small grabens. Figure 2.24 shows a comparison between the rock suites in Lombardy and in the Dolomites (see Krainer 1993). There are parallels between most of the stratigraphic units, but these have been given different names due to language differences. Of note is the greater presence of volcanic components by comparison with the deposits discussed above.

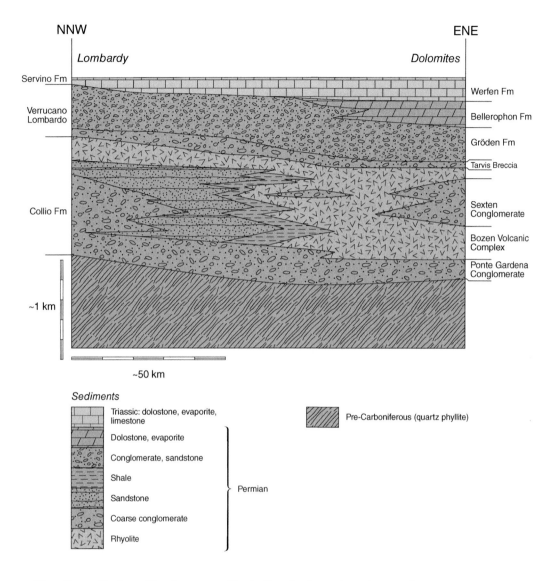

The Ponte-Gardena Conglomerate at the base of the formation exhibits substantial variation in thickness, which can be attributed to synsedimentary tectonics when the basin subsided. The Bozen Volcanic Complex comprises lavas, ignimbrites and tuffs that are intercalated with fluviatile and lacustrine sediments. The lacustrine sediments are interpreted as lake deposits within the active volcanic complex that were laid down during intervals between volcanic activity. The Tarvis Breccia above the volcanics also exhibits local variation in thickness, which probably can be attributed to block rotations associated with listric normal faults in the Saalian Phase. The Gröden Formation corresponds to the Verrucano Lombardo. With regard to age, both these formations can be allocated approximately to the Late Permian. In the east, in the Dolomites and the Carnic Alps, clastic sedimentation gives way to the shallow-marine deposits (including

▲ **Figure 2.24**
The Permian volcano-sedimentary sequence of Lombardy (Southalpine nappe system), summarized after Krainer (1993).

dolostones and evaporites) of the Bellerophon Formation. These marine sediments, as is the case for those in the Triassic Werfen Formation or Servino Formation, are evidence for a transgression from the southeast. The Permo-Carboniferous rocks are of continental-clastic and volcaniclastic nature and were deposited in intramontane troughs across the entire area of the (future) Alps. These troughs are graben structures that developed through thinning of the warm, thickened crust of the Variscan orogen and through dextral motion between Eurasia and Gondwana. The oldest troughs were formed previously in the Late Carboniferous and their fills were referred to as 'molasse' by some authors in the past. However, this name contradicts the definition that is now used, according to which molasse sedimentation refers to a foreland basin and is not *a priori* post-orogenic. The term 'Veruccano' also must be used with caution. As Krainer (1993) notes, it is almost impossible to distinguish between Early Permian and Late Permian graben formation in a geodynamic environment. The transition from dextral shearing between Eurasia (Baltica) and Gondwana (Africa), on the one hand, and the break-up of Pangaea and the associated opening of the Atlantic and (Neo)Thethys, on the other, would probably have had similar local effects. Furthermore, it is unclear whether the opening of the arm of the Tethys from a southeasterly direction started in the Permian or the Triassic.

References

Bonin, B., Brändlein, P., Bussy, F., Desmons, J., Eggenberger, U., Finger, F., Graf, K., Marro, Ch., Mercolli, I., Oberhänsli, R., Ploquin, A., von Quadt, A., von Raumer, J., Schaltegger, U., Steyrer, H. P., Visonà, D. & Vivier, G., 1993, Late Variscan magmatic evolution of the Alpine basement. In: von Raumer, J. & Neubauer, F. (eds), Pre-Mesozoic Geology in the Alps, Springer Verlag, 171–201.

Cortesogno, L., Dallagiovanna, G., Gaggero, L. & Vanossi, M., 1993, Elements of the Palaeozoic history of the Ligurian Alps. In: von Raumer, J. & Neubauer, F. (eds), Pre-Mesozoic Geology in the Alps, Springer Verlag, 257–277.

De Graciansky, P. C., 1993, Basement-cover relationship in the Western Alps: Constraints for pre-Triassic reconstructions. In: von Raumer, J. & Neubauer, F. (eds), Pre-Mesozoic Geology in the Alps, Springer Verlag, 7–28.

Diebold, P., Naef, H. & Ammann, M., 1991, Zur Tektonik der zentralen Nordschweiz: Interpretation aufgrund regionaler Seismik, Oberflächengeologie und Tiefbohrungen. NAGRA Technischer Bericht 90–04 (Text und Beilagenband).

Eisbacher, G. H., Lüschen, E. & Wickert, F., 1989, Crustal-scale thrusting and extension in the Hercynian Schwarzwald and Vosges, Central Europe. Tectonics, 8/1, 1–21.

Eisbacher, G. H., Linzer, H.-G., Meier, L. & Polinski, R., 1990, A depth-extrapolated structural transect across the northern Calcareous Alps of western Tirol. Eclogae geologicae Helvetiae, 83/3, 711–725.

Finger, F., Frasl, G., Haunschmid, B., Lettner, H., von Quadt, A., Schermaier, A., Schindlmayr, A. O. & Steyrer, H. P., 1993, The Zentralgneise of the Tauern Window (Eastern Alps): Insight into an intra-Alpine Variscan batholith. In: von Raumer, J. & Neubauer, F. (eds), Pre-Mesozoic Geology in the Alps, Springer Verlag, 375–391.

Franks, G. D., 1968, A study of Upper Paleozoic sediments and volcanics in the northern part of the eastern Aar massif. Eclogae geologicae Helvetiae, 61/1, 49–140.

Frisch, W., Vavra, G. & Winkler, M., 1993, Evolution of the Penninic basement of the Eastern Alps. In: von Raumer, J. & Neubauer, F. (eds), Pre-Mesozoic Geology in the Alps, Springer Verlag, 349–360.

Höck, V., 1993, The Habach-Formation and the Zentralgneis – A key in understanding the Palaeozoic evolution of the Tauern Window (Eastern Alps). In: von Raumer, J. & Neubauer, F. (eds), Pre-Mesozoic Geology in the Alps, Springer Verlag, 361–374.

Huber, M. & Huber, A., 1984, Das Kristallin des Südschwarzwaldes. NAGRA Technischer Bericht 84–30, 226 pp.

Krainer, K., 1993, Late- and Post-Variscan sediments of he Eastern and Southern Alps. In: von Raumer, J. & Neubauer, F. (eds), Pre-Mesozoic Geology in the Alps, Springer Verlag, 537–564.

Lammerer, B., 1986, Das Autochthon im westlichen Tauernfenster. Jahrbuch Geologische Bundes-Anstalt, 129/1, 51–67.

Lammerer, B. & Weger, M., 1998. Footwall uplift in an orogenic wedge: the Tauern Window in the Eastern Alps of Europe. Tectonophysics, 285, 213–230.

Maggetti, M. & Flisch, M., 1993, Evolution of the Silvretta nappe. In: von Raumer, J. & Neubauer, F. (eds), Pre-Mesozoic Geology in the Alps, Springer Verlag, 469–484.

Matter, A., 1987, Faciesanalyse und Ablagerungsmilieus des Permokarbons im Nordschweizer Trog. Eclogae geologicae Helvetiae, 80/2, 345–367.

Mercolli, I., Biino, G.G. & Abrecht, J., 1994, The lithostratigraphy of the pre-Mesozoic basement of the Gotthard massif: a review. Schweizerische Mineralogische und Petrographische Mitteilungen, 74, 29–40.

Naef, H., 2007, Geologische Profile durch das Gebiet von Atlasblatt Baden. In: Bitterli-Dreher, P., Graf, H. R., Naef, H., Diebold, P., Matousek, F., Burger, H. & Pauli-Gabi, Th., Erläuterungen zum Geologischen Atlas der Schweiz, Blatt 1070 Baden (Atlasblatt 120).

Neubauer, F. & Sassi F. P., 1993, The Austro-Alpine Quartzphyllites and related Palaeozoic Formations. In: von Raumer, J. &

Neubauer, F. (eds), Pre-Mesozoic Geology in the Alps, Springer Verlag, 423–439.

Niklaus, P.-A. & Wetzel, A., 1996, Faziesanalyse und Ablagerungsmilieu der fluviatilen Sedimentfüllung des Karbontroges von Salvan-Dorénaz. Eclogae geologicae Helvetiae, 89/1, 427–437.

Pfiffner, O. A., 1980, Displacements along thrust faults. Eclogae geologicae Helvetiae, 78/2, 313–333.

Pfiffner, O. A., 1993a, Palinspastic reconstruction of the pre-Triassic basement units in the Alps: the Central Alps. In: von Raumer, J. & Neubauer, F. (eds), Pre-Mesozoic Geology in the Alps, Springer Verlag, 29–39.

Pilloud, C., 1991, Structures de déformation alpines dans le synclinal de Permo-Carbonifère de Salvan-Dorénaz (masif des Aiguilles rouges, Valais). Mémoires de Géologie Lausanne, No. 9.

Ratschbacher, L. & Frisch, W., 1993, Palinspastic reconstruction of the pre-Triassic basement units in the Alps: the Eastern Alps. In: von Raumer, J. & Neubauer, F. (eds), Pre-Mesozoic Geology in the Alps, Springer Verlag, 41–51.

Schaltegger, U., 1994, Unravelling the pre-Mesozoic history of Aar and Gotthard massifs (Central Alps) by isotopic dating – a review. Schweizerische Mineralogische und Petrographische Mitteilungen, 74, 41–51.

Schaltegger, U. & Corfu, F., 1995, Late Variscan 'Basin and Range' magmatism and tectonics in the Central Alps: evidence from U–Pb geochronology. Geodinamica Acta, 8, 82–98.

Schmid, S. M., 1993, Ivrea Zone and adjacent Southern Alpine basement. In: von Raumer, J. & Neubauer, F. (eds), Pre-Mesozoic Geology in the Alps, Springer Verlag, 567–583.

Schönlaub, H. P. & Heinisch, H., 1993, The classic fossiliferous Palaeozoic units of the Eastern and Southern Alps. In: von Raumer, J. & Neubauer, F. (eds), Pre-Mesozoic Geology in the Alps, Springer Verlag, 395–422.

Thelin, P., Sartori, M., Burri, M., Gouffon, Y. & Chessex, R., 1993, The Pre-Alpine Basement of the Briançonnais (Wallis, Switzerland). In: von Raumer, J. & Neubauer, F. (eds), Pre-Mesozoic Geology in the Alps, Springer Verlag, 297–315.

Trümpy, R., 1980, Geology of Switzerland – a Guide Book. Schweizerische Geologische Kommission, Wepf & Co. Publishers, Basel.

Von Raumer, J., 1998, The Paleozoic evolution in the Alps: from Gondwana to Pangea. Geologische Rundschau, 87, 407–435.

Von Raumer, J. & Neubauer, F. (eds), 1993a, Pre-Mesozoic Geology in the Alps, Springer Verlag, 677 pp.

Von Raumer, J. & Neubauer, F., 1993b, Late Precambrian and Palaeozoic evolution of the Alpine basement – an overview. In: von Raumer, J. & Neubauer, F. (eds), Pre-Mesozoic Geology in the Alps, Springer Verlag, 625–639.

Von Raumer, J., Ménot, R. P., Abrecht, J. & Biino, G., 1993c, The pre-Alpine evolution of the external massifs. In: von Raumer, J. & Neubauer, F. (eds), Pre-Mesozoic Geology in the Alps, Springer Verlag.

Von Raumer, J. F., Bussy, F., Schaltegger, U., Schulz, B. & Stampfli, G. M., 2013, Pre-Mesozoic Alpine basements – their place in the European Paleozoic framework. Geological Society of America Bulletin, 125, 89–108, doi: 10.1130/B30654.1.

3   The Alpine Domain in the Mesozoic

The Mesozoic sediments in the Alps reflect the deposition conditions in the relatively small basins and microcontinents, from a plate tectonic perspective, that developed between the African and the Eurasian continents during the break-up of Pangaea. There is a corresponding diversity in the types of rock and rock suites that are present. A simple structure can be derived from the facies belts (or the deposition conditions) in the plate tectonic configuration.

Mesozoic sediments testify to the break-up of Pangaea

- The southern margin of Baltica (cf. Fig. 1.5), or Europe, evolved into a passive continental margin during the break-up of Pangaea. On this European continental margin, a distinction is made between the facies belts of the Jura Mountains, of the adjacent Dauphinois realm in the southeast (in the Western Alps or the Helvetic realm in the Central Alps) and of the Ultradauphinois or Ultrahelvetic realm that lies even farther to the southeast.
- The internally structured facies belt belonging to the Penninic nappe system extended between the European and the Adriatic continental margins. This corresponds to the oceanic arms that opened up between Baltica and Africa during the break-up of Pangaea.
- The northeasterly portion of the southern continent, Africa, was broken up into fragments. One of these fragments is called the Adriatic microcontinent (or Apulia). The facies belts of the Austroalpine and Southalpine realms can be distinguished in the northern margin of this microcontinent, the Adriatic continental margin.

First, the sedimentary successions in these different facies belts will be discussed below based on a simplified synthesis. This will be followed by a discussion of the plate tectonic evolution of the sedimentary environments and the lateral variation in the rock suites within the facies belts.

3.1 The Mesozoic Rock Suites

Mesozoic deposits were built up on a fairly heterogeneous substratum: the eroded Variscan mountain chain. In some locations, the Mesozoic strata were deposited directly on the metamorphic crystalline rock of the 'Altkristallin' or on granite, in others, on folded Palaeozoic sediments and, in some cases, the Mesozoic sediments were deposited apparently concordantly and directly on top of the Permian sediments and volcanics of the Permo-Carboniferous troughs. In most cases, there is a substantial hiatus (temporal gap) at the lower boundary of the Mesozoic, that is, the rocks immediately below the Mesozoic are significantly older than those at the base of the Mesozoic strata.

The upper boundary of the Mesozoic is also partially discordant with the overlying Cenozoic sediments, that is, a hiatus between rocks of the two eras. This boundary will be discussed in the next chapter within the context of the Cenozoic rock suites.

The European Continental Margin

Funk et al. (1987) provide a summary overview of the rock suites and their evolution in the European continental margin. For most of the Mesozoic, the European continent in the foreland of the Alps was submerged by a very shallow sea, in which limestones, among other sediments, were deposited. Towards the south, this continental platform gave way to a shelf sea with slightly deeper waters, in which mainly limestones were deposited. The Mesozoic rocks of the continental platform can now be studied in the Jura Mountains and its northern and western foreland. In contrast, the shelf deposits are now exposed in the Helvetic realm in the Central Alps and in its equivalent in the Western Alps, the Dauphinois realm.

Figure 3.1 shows the Mesozoic sedimentary rock suites in the Jura Mountains, which are representative of the European continental platform. The rock suite is illustrated in a highly simplified manner and is intended to convey only its gross characteristics. The Triassic sediments show the transition from continental to very shallow marine conditions. The sandstones, conglomerates, siltstones and shales of the Buntsandstein are continental formations. The dolostones and limestones that follow above were deposited in the intertidal zone, while the evaporites (anhydrite and halite) formed through evaporation of seawater in shallow basins. In addition to evaporites, coal and sandstone are also found in the Late Triassic Keuper Formation. The sandstones can be interpreted as the deposits of meandering rivers. This continental influence can also be seen in a similar fashion in Germany and France. These shallow-marine to continental formations are also summarized using the term 'Germanic Keuper'.

The main deposits during the Jurassic were limestone. The massive limestone deposits of the Late Jurassic (Malm) are much in evidence morphologically in the form of cuestas and escarpments. These make up large areas in the Jura Mountains. It is therefore no great surprise that the geological period covering its deposition is globally known as the Jurassic. On closer inspection, the Jurassic limestones exhibit two different conditions underlying their formation. Oolitic limestone (e.g. the 'Hauptrogenstein') and reef limestones indicate extremely shallow conditions in the northwest, while marl, oolitic ironstone and limestone of the same age in the southeast were deposited at depths below the storm wave base. The northwestern facies is also called Celtic or Rauracian, while the southwestern facies is called Swabian. Over time, the boundary between these facies zones migrated to the northwest, then to the southeast, indicating a dynamic history of subsidence for the continental platform.

Cretaceous sediments are found only in the western Jura Mountains (to the west of a line drawn between the towns of Bienne and Besançon) and are essentially limited to the Early Cretaceous. These are mainly shallow-water deposits with cross-bedding. Products of continental erosion were washed in and gave the limestones their yellow or ochre coloration. The youngest limestones at the transition to the Cenozoic are characterized by palaeokarstification. Products of continental erosion from the Cenozoic are to be found in in deep karst pockets. The region was clearly uplifted above sea level in the Late Cretaceous or in the Paleogene. The Cretaceous is absent in the eastern Jura Mountains but products of Cenozoic erosion are found in karst pockets in the limestones of the Late Jurassic (Malm). It is unclear whether no Cretaceous sediments were ever

▶ **Figure 3.1** The Mesozoic sedimentary sequence of the Jura Mountains – a simplified summary stratigraphic column.

Mesozoic of Jura Mountains

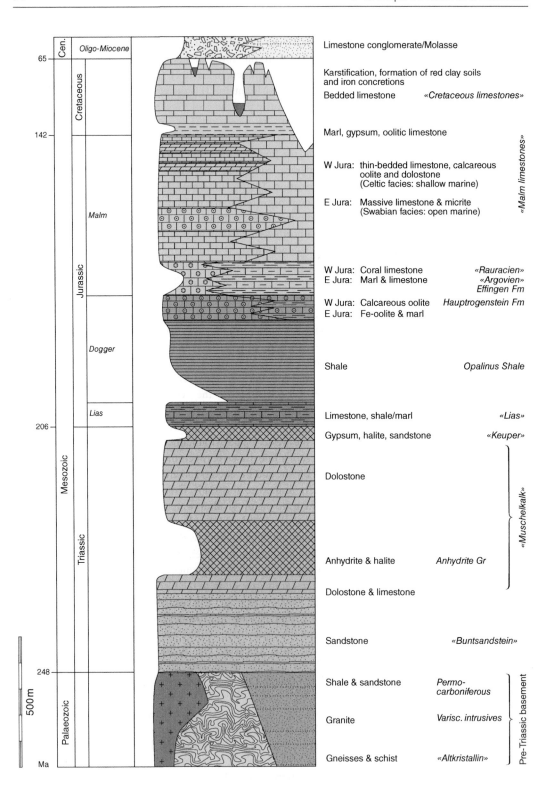

Limestone conglomerate/Molasse

Karstification, formation of red clay soils
and iron concretions

Bedded limestone   *«Cretaceous limestones»*

Marl, gypsum, oolitic limestone

W Jura: thin-bedded limestone, calcareous
         oolite and dolostone
         (Celtic facies: shallow marine)

E Jura: Massive limestone & micrite
         (Swabian facies: open marine)

W Jura: Coral limestone   *«Rauracien»*
E Jura: Marl & limestone   *«Argovien»*
                                   Effingen Fm
W Jura: Calcareous oolite   *Hauptrogenstein Fm*
E Jura: Fe-oolite & marl

Shale   *Opalinus Shale*

Limestone, shale/marl   *«Lias»*

Gypsum, halite, sandstone   *«Keuper»*

Dolostone

Anhydrite & halite   *Anhydrite Gr*

Dolostone & limestone

Sandstone   *«Buntsandstein»*

Shale & sandstone   *Permo-*
                             carboniferous

Granite   *Varisc. intrusives*

Gneisses & schist   *«Altkristallin»*

deposited in the eastern Jura Mountains or whether these were eroded during a subaerial phase.

The overall thickness of the Mesozoic sediments in the Jura Mountains is between 1000 and 2000 metres. The thickness of the Triassic sediments, alone, fluctuates between 200 and 1000 metres. The maximum Triassic thickness is found in the Neuchâtel Jura Mountains.

The Mesozoic rock suites in the Helvetic realm reveal the transition from a continental platform in the Triassic to a shelf platform in the Jurassic and Cretaceous. The stratigraphic summary overview in Fig. 3.2 is another simplified, synoptic illustration. The Triassic starts with quartzitic sandstone, the Mels Sandstone. A basal conglomerate is present in some locations. The crystalline basement below it exhibits a marked weathered horizon. However, a concordant transition from continental red sediments to Mels Sandstone is recognizable in locations where the Mels Sandstone overlies Permian Verrucano. The Mels Sandstone is the product of a marine transgression, and its age is likely to be Middle Triassic. The Middle Triassic also comprises a dolomitic rock suite (Röti Dolostone; Fig. 3.3A) that contains evaporites at its base in the palaeogeographical southern part. The intertidal deposition conditions for the Röti Dolostone are revealed, for example, by the presence of algal mats and monomictic primary breccias (Fig. 3.3B). With reference to its composition and the conditions underlying its formation, this rock suite is by all means comparable to that of the Jura Mountains but it is not nearly as thick. The Late Triassic includes the 'Quartenschiefer' Formation, a succession of continental red beds and sandstones that is typical of the Central Alps.

Towards the Western Alps, the clastic sediments become less prevalent and are replaced by evaporites. Overall, as is the case in the Jura Mountains, the 'Germanic Keuper' facies is present.

As a rule, marine sediments are found in the Jurassic strata. The lower parts of the Early Jurassic (Liassic) sediments contain marl, limestone and sandstone. Shallow marine conditions prevailed, as is demonstrated by the cross-stratification and wave ripples in Fig. 3.3C. Limestones and breccias are present in the upper portion of the Liassic. The breccia contains components of Röti Dolostone (Trümpy 1949), indicating a considerable submarine relief. However, the Liassic is entirely absent in the northern part of the Helvetic realm and the Dogger follows directly after the Trias, which Trümpy (1949) explained by the presence of a temporary high zone, the Alemannic Land, which was delimited by synsedimentary faults. Although deposits of shale followed by ferruginous sandstone are a consistent feature in the Middle Jurassic (Dogger), the regional differences in thickness indicate a heterogeneous subsidence history. The thick Late Jurassic (Malm) limestones, the Quinten Limestone, exhibit far greater homogeneity. These micritic limestones were deposited in a subeuxinic basin. The sparse fauna in the lower portion is more pelagic in nature and there are occasional corals in the upper portion, once again indicating relatively little depth of water. The Quinten Limestone can be up to 500 metres thick (cf. Fig. 3.3D).

In the Cretaceous, limestones predominate that are intercalated with variable proportions of marl. The Early Cretaceous limestones can be interpreted as neritic sediments that were deposited in a coastal environment. The

▶ **Figure 3.2** The Mesozoic sedimentary sequence of the Helvetic realm of the Central Alps – a simplified synoptic stratigraphic column.

Mesozoic of Helvetic realm

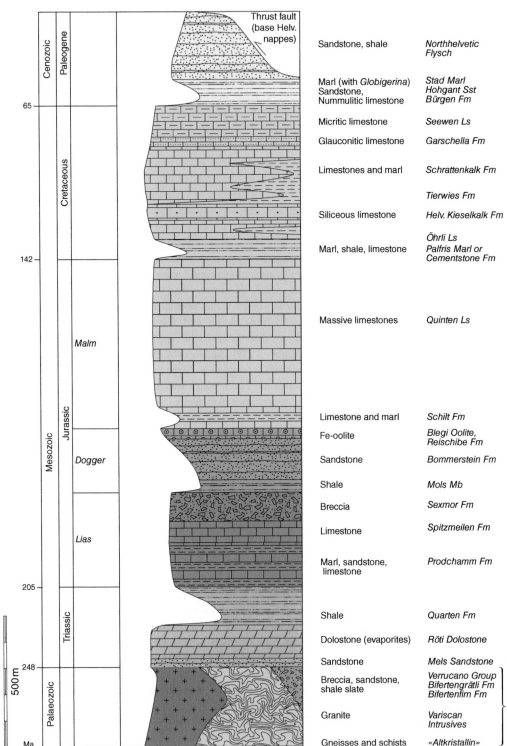

(A)

(B)

Figure 3.3 Mesozoic sediments of the Helvetic realm at outcrop.

(A) Triassic and Jurassic in the Wenden valley (northern margin of the Aar massif, canton of Bern, Switzerland). Above the crystalline basement (lower right) follows a thin yellowish layer (Röti Dolostone) and a thin dark band (Middle Jurassic/Dogger) and finally, forming a high cliff, Quinten Limestone (Late Jurassic/Malm).

(B) Monomict primary breccia within the Röti Dolostone (eastern Aar massif, Punteglias Glacier, canton of Graubünden, Switzerland).

(C) Cross-bedding and ripples in the Early Jurassic/Lias in the Flumerberge (Prodchamm, canton of St. Gall, Switzerland).

(C)

(D) Jurassic and Cretaceous in the Tamina and Calfeisen valleys near Vättis (eastern Aar massif, canton of St. Gall, Switzerland). The Quinten Limestone forms the major part of the high cliffs. The Cretaceous above is made of limestones mainly and may be distinguished from the Quinten Limestone by its slightly brownish color. The gentle slope higher up is formed by Cenozoic sediments. A klippe of Permian Verrucano thrust onto the Mesozoic-Cenozoic sediments can be recognized at the right margin of the photograph. The dam of Gigerwald is visible on the left margin.

(D)

considerable proportion of oolitic limestones in the Öhrli Limestone and the Schrattenkalk Formation indicates shallow water depth. Fluctuating thickness indicates a subsiding shelf, while the multiple shifts between shallow-water limestones and clays and marl containing deep-water fauna indicate a rhythmic subsidence. A low sedimentation rate is a characteristic of the Middle Cretaceous Garschella Formation. In the Late Cretaceous, pelagic limestone (Seewen Limestone) was deposited first, followed by pelagic marl and clay later on. The latest Cretaceous clays and marls (Wang strata) are restricted to the most southern part palaeogeographically and lie unconformably on the underlying strata.

There is a long hiatus indicated at the boundary to the Cenozoic, as Eocene marine sediments follow directly above the Cretaceous sediments. The local palaeokarstification of the Cretaceous strata indicates an uplift of the marine sediments above sea level. In the palaeogeographical northern part, the autochthonous cover of the Aar massif, the Eocene lies on top of Early Cretaceous (Bernese Oberland) or even on top of Late Jurassic (eastern end of Aar massif) strata.

The facies belt of the shelf region in the Western Alps is known as the Dauphinois realm. The Dauphinois realm is essentially a direct lateral equivalent of the Helvetic realm. As the summary stratigraphic overview in Fig. 3.4 shows, the sedimentary rock suites in the Triassic are very similar and only the names of the individual formations differ. The sandstones in the 'Werfen Formation' correspond either to the Buntsandstein in the Jura Mountains or to the Mels Sandstone in the Helvetic realm. It is only the name that is perhaps slightly misleading, in that it has been borrowed from the Southalpine facies zone (Dolomites). Parallels can easily be drawn between the Muschelkalk Dolostone and the Röti Dolostone of the Helvetic realm and the evaporites and marl in the Late Triassic indicate affiliation with the 'Germanic Keuper' facies zone. Neritic limestone is specific to the Dauphinois realm in this context, which occurs in the latest Triassic, in the Rhaetian.

The Jurassic sediments exhibit manifold changes between limestones, marls and marly limestones. Thick limestone dominates in the Early Jurassic, with thick marl dominating at the transition

from the Middle to Late Jurassic. Individual strata are subjected to considerable lateral changes (cf. Debelmas 1974), and overall the Jurassic sediments can have a thickness of up to 2000 metres, with variations the result of deposition on a subsiding shelf that was divided into small basins. The massive, coarsely bedded limestones of the 'Tithonique' form a morphologically distinctive Late Jurassic formation, and together with the Kimmeridgian limestone that lies below can be equated to the Helvetic Quinten Limestone. Subsidence was far less in the southern Chaînes subalpines of Provence, where the Liassic is absent and the Tithonian limestones are coralliferous (i.e. deposited in shallow water), following which, in the earliest Cretaceous, part of the shelf was uplifted to above sea level.

The Cretaceous sediments are also mainly present in the form of marl and limestone. Marl dominates in the Early Cretaceous, while within the limestones, at least one distinctive and almost continuous layer is discernible: the Urgonian Limestone, which is made of neritic shallow-water limestone and can be more or less correlated with the Helvetic Schrattenkalk Formation. In the Late Cretaceous, pelagic marly limestones followed by massive limestones dominate. Interestingly, however, the younger portion of the Late Cretaceous (Senonian) in the middle part of the Chaînes subalpines, in the vicinity of Devoluy-Diois, lies unconformably on top of Cenomanian strata. It appears that initial folding had already taken place here in the Turonian–Conacian (around 90 million years ago).

The Cenozoic follows above the Cretaceous sediments, but after a long hiatus between. Analogous to the neighbouring Helvetic realm, the entire

▶ **Figure 3.4** The Mesozoic sedimentary sequence of the Dauphinois realm of the Western Alps. Source: Debelmas (1974).

Mesozoic of Dauphinois realm

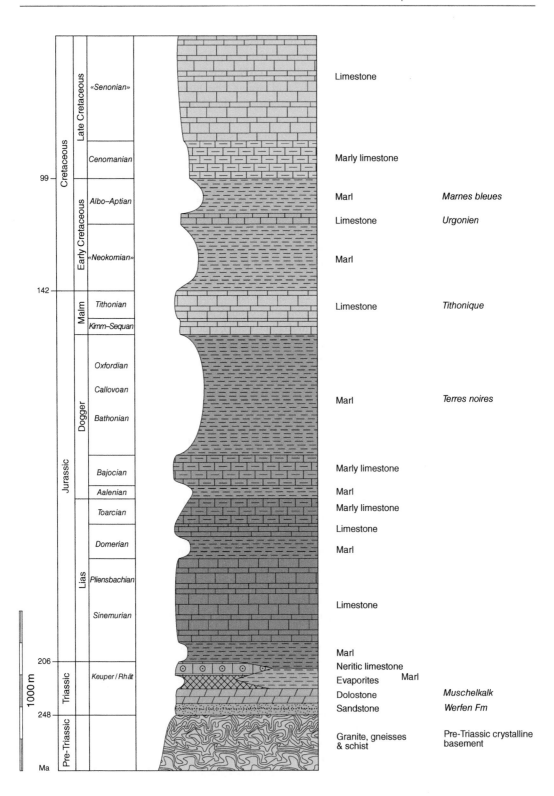

shelf region of the Dauphinois realm appears to have been uplifted to above sea level.

Water depth increased towards the continental slope in the southernmost portion of the shelf region in the Helvetic and Dauphinois realms. Accordingly, marls and deep-water clays were deposited here instead of carbonates, forming a facies beltknown as the Ultrahelvetic and Ultradauphinois realm. However, the boundary to the Helvetic and Dauphinois realms is not sharp but is characterized by a transition zone. Accordingly, a Southhelvetic realm is distinguished between the Helvetic and the Ultrahelvetic realms in the eastern Central Alps. In many cases, the South- and Ultrahelvetic realms contain only parts of the Mesozoic rock suites: the Blattengrat and Sardona nappes in eastern Switzerland contain only marly Cretaceous and the Cenozoic deposits, but in the case of the Sex-Mort nappe in western Switzerland, Cenozoic deposits lie immediately above the Triassic and Jurassic; a complete, but disrupted series of rock formations can be identified in the Bex (Triassic), Arveyes (Jurassic), Anzeinde (Jurassic–Cretaceous) and Plaine Morte (Cretaceous and Cenozoic) nappes. The sediments in all of these nappes indicate deposition in deeper water, with the exception of the Triassic. There are also oceanic sediments in the Ultradauphinois realm, but the rock suites do not reach beyond the Early Cretaceous.

In the extreme east, in the Tauern Window in the Eastern Alps, the Mesozoic rock suite in the autochthonous cover of the Tauern crystalline rock is attributed to the Helvetic realm. Both the Triassic sediments and the Late Jurassic Hochstegen Limestone can easily be correlated with the Triassic

sediments and the Quinten Limestone of the Helvetic realm (Lammerer 1986, Vesela et al. 2008). In spite of their schuppen structure, formed with the Rhenodanubian Flysch, the sediments in the Ultrahelvetic nappes of the northern margin of the Eastern Alps can be followed almost seamlessly from the Central Alps through to the east.

Oceanic Arms between the Baltic and Africa

The facies zone in the marine basins to the south of the European and to the north of the Adriatic continental margins is subdivided by great differences in water depth. The associated Mesozoic sedimentary rock suites are allocated to the Penninic Zone, in which three quite different sedimentary sequences can be distinguished: the Northpenninic sediments of the Valais Trough, the mid-Penninic sediments of the Briançon Rise and the Southpenninic sediments of the Piemont Ocean. Figure 3.5 shows a simplified illustration of the corresponding rock suites in juxtaposition.

The sediments of the Valais Trough are generally sheared off their pre-Triassic substratum. However, at their base, a mélange is usually found that is made up of gneisses, Permian clastics, Triassic dolostones and evaporites, Liassic limestones and bands of basaltic volcanics with pillow lava (Steinmann 1994). The actual base of the Valais Trough was formed by continental crust from the thinned continental margin and true oceanic crust in the central portion of the basin. The basalts in the mélange are effusive in nature (cf. Fig. 3.6) and can be compared geochemically to mid-oceanic-ridge basalt: Steinmann & Stille (1999) demonstrated that the basalts did not originate in a subcontinental mantle, but are true oceanic crust. The layers of melange themselves

▶ **Figure 3.5**
Schematic overview of the Mesozoic sedimentary sequences of the Penninic realm. The sedimentary sequence of the Valais Trough corresponds to the Tomül nappe of eastern Switzerland, and the ophiolitic sequence of the Piemont Ocean to the Platta nappe of eastern Switzerland.

Mesozoic of Penninic realm: Valais Trough, Briançon Rise, Piemont Ocean

Valais Trough: thick sequence of deep marine sediments

Stratigraphic successions of the Penninic realm

North-Penninic: Deep-marine facies
Valais Trough

Middle Penninic: Shallow-marine facies
Briançon Rise

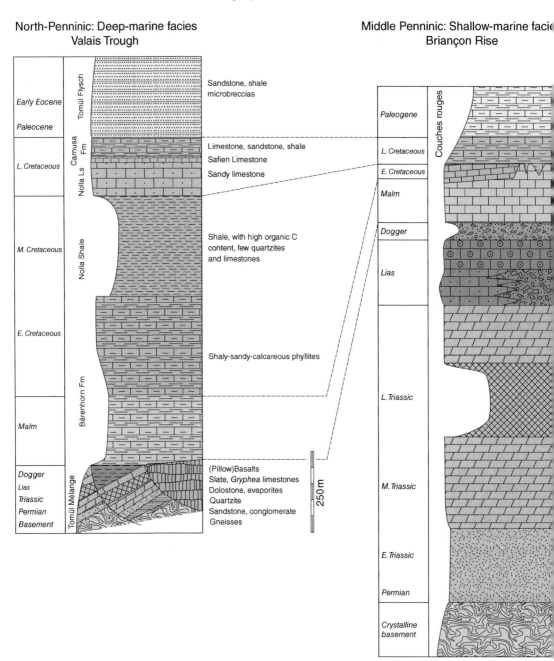

South Pennic: Oceanic facies
Piemont Ocean

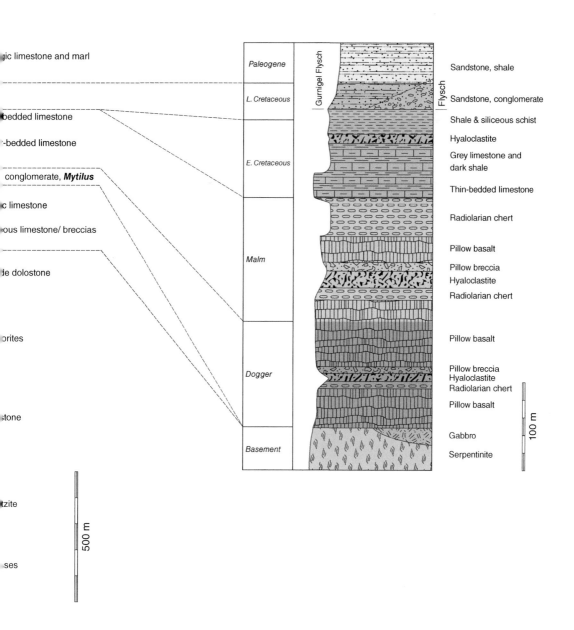

ic limestone and marl

bedded limestone

-bedded limestone

conglomerate, **Mytilus**

c limestone

ous limestone/ breccias

le dolostone

orites

stone

Paleogene — Sandstone, shale

L. Cretaceous — Sandstone, conglomerate

Shale & siliceous schist

Hyaloclastite

E. Cretaceous — Grey limestone and dark shale

Thin-bedded limestone

Radiolarian chert

Pillow basalt

Malm — Pillow breccia
Hyaloclastite
Radiolarian chert

Pillow basalt

Dogger — Pillow breccia
Hyaloclastite
Radiolarian chert

Pillow basalt

Basement — Gabbro
Serpentinite

Gurnigel Flysch

Flysch

100 m

500 m

zite

ses

Figure 3.6 Basalts in the 'Bündnerschiefer' of the Valais Trough.

(A)

(A) Fractured basalts. The gaps formed by fracturing are filled with secondary calcite. Photograph from the Tomül nappe (Vals, canton of Graubünden, Switzerland).

(B) Pillow lavas in basalts. Photograph from the Tomül nappe (Vals, canton of Graubünden, Switzerland).

(B)

are the result of both Mesozoic extensional history, which split the continental crust into fragments and produced oceanic crust in certain locations, and of Alpine thrust tectonics, which sheared and imbricated adjacent layers.

The actual fill of the Valais Trough is made up of a Jurassic–Cretaceous argillaceous–sandy–calcareous sequence produced in a hemipelagic–turbiditic environment, referred to as **Bündnerschiefer** (also called 'schistes lustrés' or 'calcescisti'). These are now present mainly in the form of metamorphic, argillaceous–sandy calcareous phyllites.

The Bündnerschiefer Group is over 1000 metres thick, almost impossible to subdivide and stratigraphically poorly dated. Steinmann (1994) showed that the majority of this rock suite dates from the Cretaceous.

The rock suite in the Tomül nappe in Central Graubünden that was investigated by Steinmann (1994) will be used as an example and briefly described below. The Bärenhorn Formation, which overlies the basal Tomül Mélange, is composed mainly of sandy-calcareous beds that are 10–20 centimetres thick, intercalated with bands of shale that are 5–10 centimetres thick. Turbiditic sandstone beds can be observed in various locations. The overlying Nollaton Formation (Nolla Shale) is substantially more argillaceous: thick layers of shale are typical in the lower portion, while the upper portion also contains turbiditic deposits. Based on the high proportion of organic carbon and its variation within the formation, Steinmann (1994) suggested that the Nollaton Formation should be placed in the Middle Cretaceous. The overlying Nollakalk Formation (Nolla Limestone) is composed mainly of sandy limestone and is assigned to the Cenomanian. The uppermost unit of the Bündnerschiefer Group, the Carnusa Formation, contains turbiditic limestone in its lower portion ('Safierkalk'), and turbiditic limestone intercalated with sandstone containing shale in its upper portion.

The Tomül Flysch then overlies the Bündnerschiefer Group. The lower boundary of the flysch is marked by the 'main conglomerate', a succession of matrix-supported conglomerates that is 40 metres thick, and can possibly be dated to the Coniacian. Clasts in the conglomerates can be up to 50 centimetres in size and the majority are reworked from the immediately underlying Mesozoic. However, isolated dolostone pebbles and reworked breccia

pebbles containing dolostone pebbles are also observed. The Tomül Flysch itself is composed of slightly pelitic calcschists that contain fine-grained breccias followed by a transition into sandstones rich in quartz at the top. The age of the Tomül Flysch is debatable, being possibly Coniacian and younger (Steinmann 1994).

Input of detritus into the Bärenhorn Formation occurred from a southerly direction (Steinmann 1994). Crystalline rocks are assumed to have been present in the source area in order to explain the high quartz content. The prevalence of turbidite deposits reduces gradually towards the top, indicating waning tectonic activity and progressive subsidence of the Valais Trough. The younger rock suites, the Nollakalk and Carnusa formations and the Tomül Flysch, can be interpreted as a gradually prograding submarine fan system. Steinmann (1994) suggests that areas uplifted above sea level were present in the south at this time and acted as the sediment-source area.

A completely different story can be discerned from the sedimentary sequences of the Briançon Rise. A detailed analysis of the individual sedimentary strata can be found in the publications by Baud & Septfontaine (1980) for the Central Alps and by Lemoine (1988) for the Western Alps. Sartori et al. (2006) established a systematic classification for the formations in the internal part of the Central Alps (Bernhard nappe complex) and compared this to the externally located Prealps (Klippen nappe).

Sedimentation in the Triassic essentially starts with basal quartzites, followed by dolostones and, in certain locations, evaporites. A thick layer of anhydrite was then deposited in the Middle Triassic, which is overlain by

Briançon Rise:
Shallow marine
carbonate
sequence

yellowish dolostones from the Late Triassic – in contrast to the 'Germanic Keuper' described above. The gradual subsidence and deposition of these platform sediments, which kept pace with the subsidence, probably can be attributed to thermal subsidence linked to the opening of the Meliata Ocean in the east.

The Jurassic and Cretaceous sedimentary sequences also exhibit a different pattern of subsidence when compared with the European continental margin, with subsidence being interrupted on two occasions: Mosar et al. (1996) provide a detailed discussion of these processes. In the Early Jurassic, some portions of the Briançon Rise were uplifted, raised above sea level and karstified. The presence of breccias (Fig. 3.7) indicates local erosion (Mettraux & Mosar 1989). These processes were caused by the opening of the Piemont Ocean, a oceanic arm that opened in the Early Jurassic through oblique rifting characterized by transform faults, immediately to the southeast of the Briançon Rise. In transpressive segments, individual blocks were forced upwards along steeply dipping thrust faults. Due to the subsequent thermal subsidence, over 1500 metres of sediments (mostly oolitic limestone) accumulated in the northwestern part of the Briançon Rise, in the so-called Subbriançonnais. Continental conditions prevailed in the southeastern part, the actual Briançon Rise, which are indicated, for example, by the presence of conglomerates and coal seams. Figure 3.5 displays the sedimentary sequence as encountered on the actual rise. A second cessation in subsidence occurred at the transition of the Middle to Late Jurassic, which is explained by the opening up of the Valais Trough immediately to the

Piemont Ocean: Ophiolites and oceanic sediments

northwest of the Briançon microcontinent. This led to substantial emersion and erosion in the Western Alps, which can be observed in the surroundings of the little town of Briançon (Tricart et al. 1988), as a result of the oblique opening up of the Valais Trough and involved transform faults with transpressive segments. The thermal subsidence associated with the opening up of the Valais Trough led to deposition of 30–300 metres of massive, partially oolitic, limestones. In the Early Cretaceous, pelagic conditions became established, eventually replaced towards the end of the Cretaceous by sediment starvation conditions that even led to interruptions in deposition. The so-called 'Calcaire plaqueté' was deposited in the Early Cretaceous pelagic conditions, but the red beds of the overlying 'Couches rouges' were deposited during Late Cretaceous and Palaeocene times when sediment starvation conditions and interruptions in deposition prevailed. Figure 3.7D shows the transgressive contact of Cretaceous red beds on karstified Late Jurassic limestone.

Deposition generally remained interrupted during the Palaeocene, but there was a rapid increase in the rate of subsidence in the Eocene that led to accumulation of flysch deposits. The basin was incorporated into an accretionary prism that also included the youngest sediments of the Piemont Ocean, which was then closing. This rapid subsidence resulted from downflexing of the plate under the weight of the Penninic nappes that were forming in the southeast.

The Southpenninic rock suite associated with the Piemont Ocean reflects the opening and closing of the oceanic arm. The oldest rocks are ophiolites, such as serpentites, that are regarded as hydrated mantle peridotites, alongside gabbros and

(A)

Figure 3.7 Breccias from the margins of the Briançon Rise.

(A) Polymict breccia of the Middle Jurassic (Dogger) of the Breccia nappe (Seeberg, Simmental, canton of Bern, Switzerland). The components are mostly rounded and are derived from Triassic (light) and Early Jurassic (dark) carbonates.

(B)

(B) Polymict breccia of the Jurassic of the Suretta nappe (Cröt, canton of Graubünden, Switzerland). The pebbles are strongly deformed by Alpine overprint and are probably derived from Triassic carbonates mainly.

(C) Polymict breccia of the Jurassic of the Suretta nappe (Val Ferrera, canton of Graubünden, Switzerland). The pebbles are strongly deformed by Alpine overprint and are derived from Permian intrusives (light green) and Triassic carbonates.

(C)

basalts, which formed during spreading of the mid-oceanic ridge. At many locations there are outcrops of Triassic quartzites and dolostones at the base of the rock suite above the pre-Triassic crystalline basement. These rocks can be interpreted as fragments of the continental margin that was subjected to extension on both sides of the Piemont Ocean. In the Western Alps, the Late Triassic dolostones exhibit a notable similarity to those of the Adriatic continental margin

(D)

(D) Palaeokarst filling in the Klippen nappe (Bunschlere, Simmental, canton of Bern, Switzerland). The karst filling consists of red marly shales of the Paleocene Couches Rouges and angular fragments of Late Jurassic limestones.

(Debelmas 1974). The same applies to the Triassic–Jurassic rock suites in the Arosa Zone of the Central Alps (Eberli 1988).

In the Platta nappe of the Central Alps, Dietrich (1970) describes ophiolites with gabbro, serpentinite, dykes of former peridotite–pyroxenite gabbro and mafic volcanics with massive lava, pillow lavas, meta-hyaloclastites and breccias comprising the rock suite. The volcanics lie parallel to the stratification in the enclosing sediments and are chemically comparable to mid-oceanic-ridge basalts (MORB). Whereas the sediments above the ophiolites of the Early to the Middle Jurassic contain beds of phyllite, calcschists and breccia, the Late Jurassic strata are dominated by radiolarian chert interlayered with ophiolites. The radiolarian chert was overprinted by Alpine metamorphism and is now present in the form of sericite–chlorite quartzite. This was also deposited under true oceanic conditions below the calcite compensation depth (CCD). Layers of basaltic material are present in the Cretaceous rocks, but the radiolarian chert is replaced by calcareous phyllite

and phyllites (see Dietrich 1970, who dated the youngest volcanics to 113 million years old).

Marthaler (1984) divided the Combin Zone into two nappes (Lower and Upper Combin zones) and dated the lithological successions in both nappes. There is a restricted Triassic layer at the bottom of both nappes that contains quartzite and cargneule, or dolostone (Evolène, or Frilihorn Formation). Above this, Jurassic marble and phyllitic marble of possible Cretaceous age, can be distinguished within the Frilihorn Formation, and the younger Late Cretaceous strata are more like 'Bündnerschiefer'. The 'Série Rousse' in the Lower Combin Zone takes its name from the rusty brown colour of sandy marble, and is phyllitic in some locations and similar to the 'Couches rouges': conglomeratic beds contain dolostone, marble and quartzite pebbles; sandy beds are intercalated with dark argillaceous beds and are turbiditic in nature. The rock suite can be placed in the Turonian–Senonian. Marthaler (1984) dated the 'Série Grise' in the Upper

▶ **Figure 3.8** The
Mesozoic sedimentary
sequence of the Dolomites
(Southalpine realm). Source:
Adapted from Doglioni &
Fiores (1997) and Bossellini
et al. (2003).

Combin Zone as Cenomanian to Senonian: it is similar in composition to the Bündnerschiefer Group in the Valais Trough and is made up of sandy shale and marble, calcschist and calcareous phyllite. The sandstone and microbreccia indicate turbiditic conditions, which is why the term 'calcareous flysch' has sometimes been used. The top of the formation is composed of a prasinite horizon with questionable basalt pillow lavas. Thus the 'Série Grise' can by all means be compared with the Cretaceous rocks that Dietrich (1970) describes for the Platta nappe. However, in contrast to the Platta nappe, no ophiolites are present at the base of the rock suites in the Combin Zone and therefore, although not for this reason alone, a palaeogeographical proximity to the Briançon Rise is assumed. Escher et al. (1997) view the Tsaté nappe, a complex folded unit containing basaltic pillow lavas at its base that are probably of Jurassic age, as containing the oceanic portions of the Combin Zone.

Actual oceanic rock suites, metaperiodites, gabbros and basalts with a tholeiitic chemical composition, are to be found in the Zermatt–Saas Fee and Antrona zones in the Central Alps and in the Lanzo Zone in the Western Alps (Escher et al. 1997). The metaperiodites are lherzolitic and the basalts are comparable to MORBs. In some locations oceanic sediments are present in the form of calcschists and metamorphosed radiolarian cherts. In the case of the Lanzo Zone, a large lherzolitic peridotite body is linked to serpentinite, metagabbro and metabasalt, and local occurrences of radiolarian chert are also observed.

The Adriatic Continental Margin

Sedimentation on the Adriatic margin is characterized by a very thick sequence of Triassic dolostones. The Jurassic sediments reflect intensive extension of the continental margin and reveal a correspondingly diverse structure composed of locally of zones of basins and highs. Based on the present-day tectonic position, a distinction is made between the Southalpine realm and the Austroalpine realm on the Adriatic continental margin.

The stratigraphic succession in the Dolomites (South Tyrol) is illustrated in Fig. 3.8 as an example of the Southalpine realm. This is based on Doglioni & Flores (1997) and Bossellini et al. (2003). The Werfen Formation, a calcareous and clastic sequence that was deposited in a shallow shelf region, overlies the Late Permian Bellerophon Formation that contains evaporites and dolostones. Subsequent fluctuation in sea level in the Anisian may have been caused by tectonic processes and led to localized erosion. During the transgression that followed, a carbonate platform became established with the deposition of the shallow-water limestones of the Contrìn Formation. Subsidence in the Ladinian was subject to substantial local variation. Carbonate sedimentation kept pace with the subsidence in some locations and produced thick successions of dolostones (e.g. the Sciliar Dolomite). In other locations, steep submarine escarpments, with a 40–45° slope angle developed due to differential subsidence, on which slope breccias were deposited. A magmatic–tectonic event in the Ladinian is characterized by the intrusion of basaltic dykes and collapse structures. During this process, heterogeneous megabreccias formed ('Caotico eterogeneo' in Fig. 3.8). Calmer conditions prevailed in the Carnian and allowed larger carbonate platforms to develop again (Cassian Dolomite) that prograded over the regions with greater water depths. The result of this was a generally shallower basin in which coral reefs became

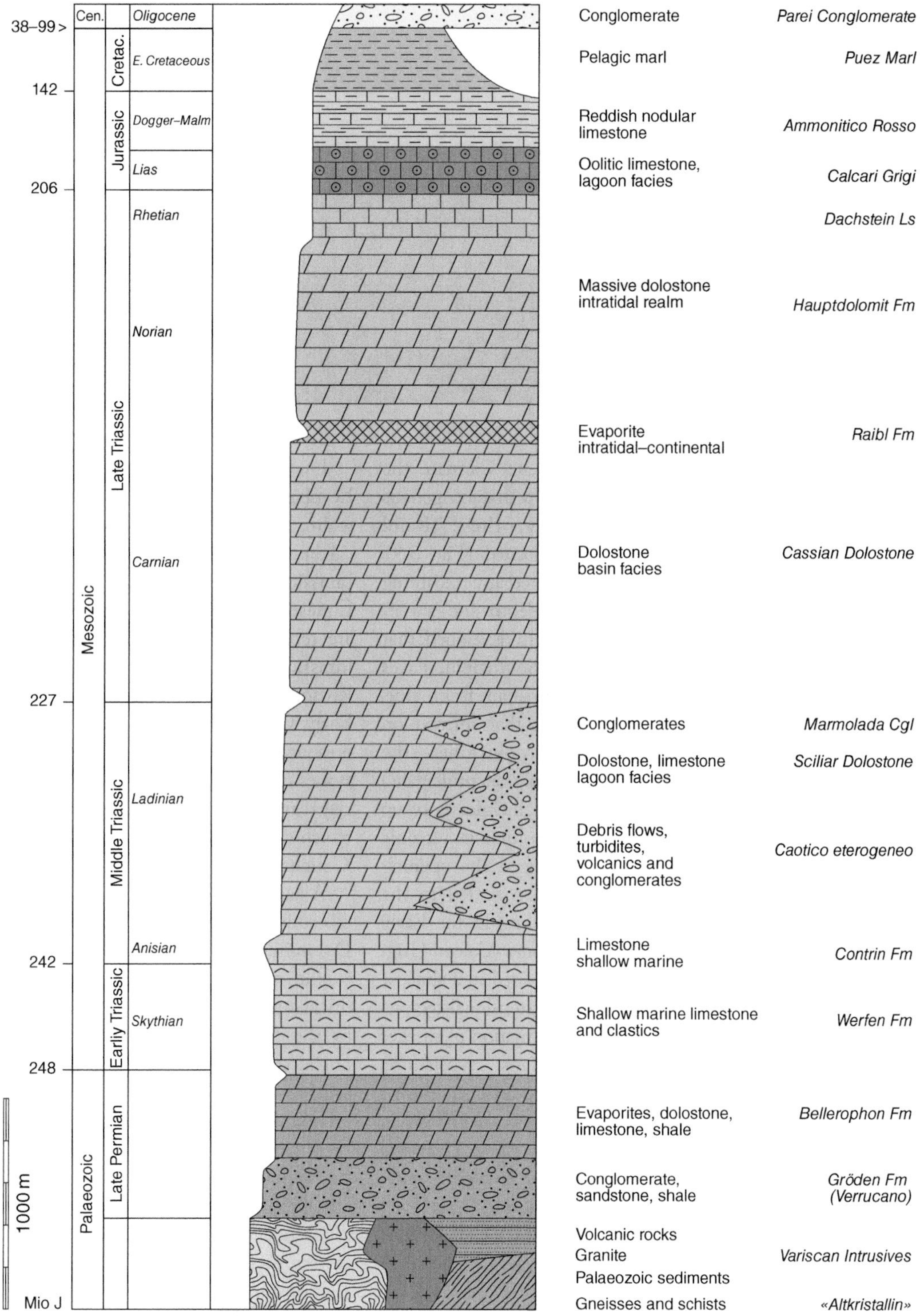

38–99 >	Cen.		Oligocene	Conglomerate	*Parei Conglomerate*
142		Cretac.	E. Cretaceous	Pelagic marl	*Puez Marl*
	Jurassic		Dogger–Malm	Reddish nodular limestone	*Ammonitico Rosso*
206			Lias	Oolitic limestone, lagoon facies	*Calcari Grigi*
			Rhetian		*Dachstein Ls*
	Late Triassic		Norian	Massive dolostone intratidal realm	*Hauptdolomit Fm*
				Evaporite intratidal–continental	*Raibl Fm*
227			Carnian	Dolostone basin facies	*Cassian Dolostone*
	Middle Triassic		Ladinian	Conglomerates	*Marmolada Cgl*
				Dolostone, limestone lagoon facies	*Sciliar Dolostone*
				Debris flows, turbidites, volcanics and conglomerates	*Caotico eterogeneo*
242	Early Triassic		Anisian	Limestone shallow marine	*Contrin Fm*
			Skythian	Shallow marine limestone and clastics	*Werfen Fm*
248	Late Permian			Evaporites, dolostone, limestone, shale	*Bellerophon Fm*
				Conglomerate, sandstone, shale	*Gröden Fm (Verrucano)*
Mio J				Volcanic rocks Granite Palaeozoic sediments	*Variscan Intrusives*
				Gneisses and schists	*«Altkristallin»*

Mesozoic

Palaeozoic

1000 m

Figure 3.9 The Southalpine Triassic in the Sella Mountains (South Tyrol, Italy). The thin Raibl Formation forms a clearly visible ledge in the upper part of the high cliff.

Mesozoic of Southalpine realm

▶ **Figure 3.10** The Mesozoic sedimentary sequence of the Austroalpine realm of the Central Alps. LAA, UAA: Lower Austroalpine, Upper Austroalpine.

established. In the Late Carnian, the water became so shallow that terrigenous sediments and evaporites (Raibl Formation) were deposited (Fig. 3.9). This completely removed any remaining differences in water depth and a very large carbonate platform developed, that not only covered the Southalpine facies zone, but also the Austroalpine facies zone. The typical sediment that formed on this platform is dolostone, the 'Hauptdolomit' (or 'Dolomia principale') Formation. It reaches thicknesses of up to 1 kilometre. In the Jurassic, tectonic movements that presaged the opening up of the Piemont Ocean led to the break-up of the platform. In the Early Jurassic, oolitic limestones (Calcari Grigi in Fig. 3.8) were deposited in a shallow sea. Synsedimentary extensional tectonics led to the rapid formation of smaller basins through tectonic subsidence, in which reddish nodular limestone (Ammonitico Rosso) was deposited in the Middle and Late Jurassic. From the Cretaceous onwards, the passive continental margin had subsided to such an extent through thermal subsidence that pelagic conditions prevailed, in which, for example, the Puez Marl was deposited.

Further to the west, in the Lombardian part of the Austroalpine realm, turbidite deposits of varying thickness were deposited in the Early Jurassic (the Lombardian Siliceous Limestone), and true radiolarian cherts are recorded in the Middle and Late Jurassic, in addition to the Ammonitico Rosso. The pelagic conditions of the Early Cretaceous are revealed by the presence of the micritic beige limestones of the 'Maiolica' and the pelagic clays and limestones of the 'Scaglia', which can be compared to the Penninic 'Couches rouges'.

A large hiatus in the sedimentary record indicates uplift of this part of the Dolomites above sea level, while further to the west, in the Late Cretaceous, thick submarine fans were deposited on to the flat, deep-sea floor from the north: these flysch sediments probably provide evidence of the first orogenic processes in the Eastern Alps.

A similar development is noted in the facies of the Austroalpine realm, but there are also substantial regional differences between the western and eastern parts. Figure 3.10 illustrates the relationships in the western part, the present-day Central Alps. Continental

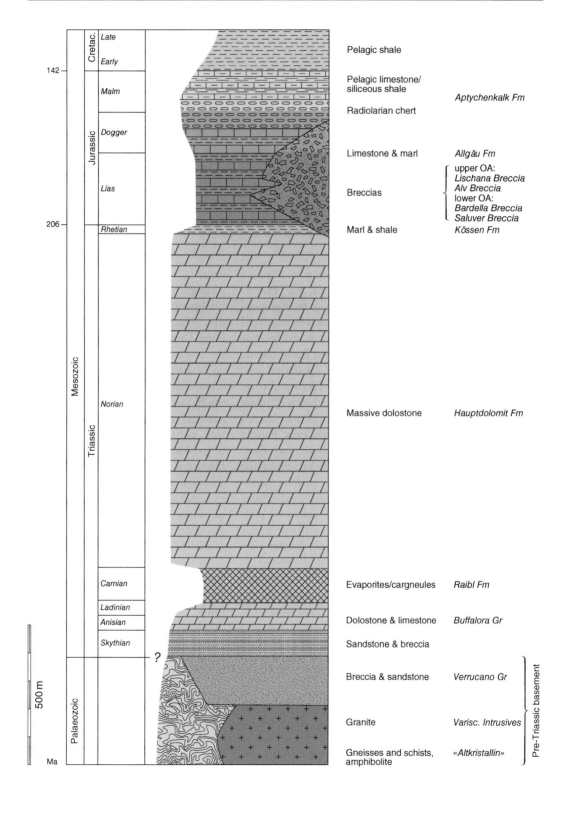

▲ Figure 3.11 Landscape
dominated by Austroalpine
carbonates in the Nordkette
north of Innsbruck (Austria).
Source: Photograph courtesy
of Michaela Ustaszewski.
Reproduced with permission.

*Mesozoic of
Austroalpine realm*

clastic sediments represent Permian graben fill, similar to the Permo-Carboniferous troughs on the European continental margin that were described in Chapter 2. Sandstones and breccias form the transition into Early Triassic clastics, which are also considered continental formations (Buntsandstein): the exact boundary between Permian and Triassic in these formations has not been defined stratigraphically. Farther to the east, the Permian and Early Triassic formations are present in the form of continental to shallow-marine clastics and evaporites (Mandl 2000). Later, in the Anisian, carbonate platforms developed across the entire Austroalpine facies zone: the Buffalora Group limestones and dolostones in Fig. 3.10. A substantial drop in sea level led to the formation of evaporites and dark marine clays, and also resulted in subaerial exposure of some areas and consequent deposition of sediments containing siliciclastics from the European continent (Mandl 2000). The sediments are grouped together in the Raibl Beds and their equivalent is also found in the Southalpine realm (Fig. 3.8). Above the Raibl Beds, there is a thick suite of dolostones ('Hauptdolomit' Formation) in the western facies zone that interlocks

laterally towards the east with the prograding carbonate platform of the Dachstein Formation. The dolostone of the 'Hauptdolomit' Formation was formed in the intertidal zone. Similar to the Southalpine realm, this formation is up to 1000 metres thick, its structure is homogeneous and it indicates a stable equilibrium between subsidence and sedimentation, at a constant subsidence rate of 0.1 millimetres per year (Fig. 3.11). The overlying Kössen Formation contains marls and clays and points once again to a continental influence. Overall, the thick carbonate successions of the Triassic can be explained by slow thermal subsidence that kept pace with sedimentation.

In comparison with the Triassic deposits, the Jurassic sediments contrast enormously. In the Early Jurassic, thick breccias emanating from debris flows occur in the western facies of the Austroalpine realm (Bardella, Saluver, Lischana and Alv breccias in Fig. 3.10). In addition to 'Hauptdolomit', the components in these breccias also include crystalline rocks. These debris flows occurred on steep submarine escarpments that were formed by synsedimentary normal faults and were a result of the extension of the Adriatic continental margin. Limestones intercalated with marls were deposited (Allgäu Formation) away from these steep escarpments. These relationships indicate rapid sinking of the sea floor, that is, tectonic subsidence. In the Middle and Late Jurassic, radiolarian cherts were deposited that indicate true oceanic conditions below the CCD, but the exact lower and upper boundaries of the radiolarian cherts have not been defined stratigraphically. According to Mandl (2000), the radiolarian cherts in the eastern facies of the Austroalpine realm span the Callovian and Oxfordian and reflect the maximum water depth in the basin. These great water depths persisted throughout the latest Jurassic and the Early Cretaceous when pelagic limestones ('*Aptychus* limestone', comparable to the 'Maiolica' of the Southalpine realm) and siliceous shales were deposited. Pelagic clays then followed in the Late Cretaceous (to Cenomanian) that exhibit pronounced similarity to the 'Scaglia' of the Southalpine realm or the 'Couches rouges' of the Briançon Rise.

The Cretaceous Gosau sediments that formed in intramontane basins in the Eastern Alps are of particular importance. These sediments are evidence for the earliest (the so-called Eo-Alpine) orogenesis in the Alps and will be discussed in more detail later in the context of the tectonic evolution (in Chapter 6).

3.2 Plate Tectonic Evolution

As has been indicated several times, the chronological successions in the rock suites of the different facies realms reflect both the formation of sediments of the same ages under different conditions and also the changes in these conditions. These differences and changes are closely linked to the plate tectonic relationships in the Alpine region, which is characterized by the break-up of Pangaea and the associated opening of oceanic arms. The closing of one of the oceanic arms, the Piemont Ocean, plays an important role even during the first phase in its evolution in the Cretaceous. In the following, this Mesozoic plate tectonic evolution will be discussed based on several snapshots in time. This evolution makes a seamless transition to that of the Cenozoic, which will be discussed in the next chapter.

Triassic: Epicontinental Platforms

During the Triassic, the future Alpine region was dominated by large carbonate platforms. These became established between the continental masses of Pangaea in the north, west and southwest (Baltica, Greenland, North America, Africa), on the one hand, and oceanic basins in the east and south (Meliata, Palaeo-Tethys, Neo-Tethys), on the other (cf. Fig. 3.12, based on Stampfli & Borel 2002). Although these continental masses were tied to one plate, they contained individual areas that were above sea level, such as the Vindelician High, which passed through the middle of the future Alpine region. In the Late Triassic (Norian in Fig. 3.12), a rift was already present between Baltica and Greenland due to the initiation of the break-up of Pangaea, but this did not yet have any effect on the Alpine region. In contrast, another rift in the region of the future North Sea influenced the ingression of the sea in a southerly direction into the 'Germanic Keuper' facies zone.

Figure 3.12 Plate configuration in the Norian (220 Ma). The Vindelician High separates the facies realm of the 'Germanic Keuper' that is linked to the North Sea rift and the 'Alpine Trias' that is linked to the Meliata Ocean. Gr, Greece; It, Italy. Source: Stampfli & Borel (2002). Reproduced with permission of Elsevier.

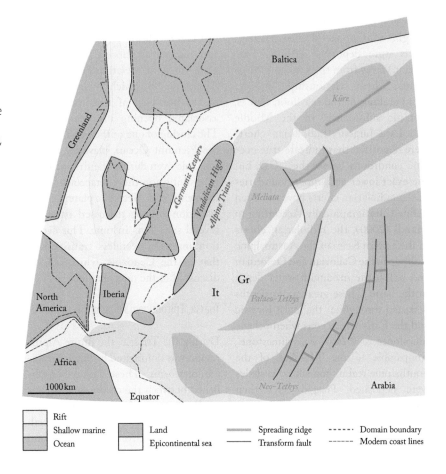

	Rift				
	Shallow marine		Land	Spreading ridge	Domain boundary
	Ocean		Epicontinental sea	Transform fault	Modern coast lines

Vindelician High separates 'Germanic Keuper' from 'Alpine Trias'

The Vindelician High separated a facies characterized by continental conditions in the west, the so-called 'Germanic Keuper', from a facies dominated by marine conditions in the east, the so-called 'Alpine Trias'. The marine 'Alpine Trias' was linked to a transgression from the east, that originated in the Meliata Ocean, a branch of the Palaeo-Tethys.

Figure 3.13 shows a more detailed palaeogeographical map of the Alpine region in the Norian (ca. 215 million years ago): outlines of the pre-Triassic basement blocks shown in Fig. 2.17 have been transferred to this map. In addition, the boundaries between the future Alpine nappe systems are also

given (Helvetic–Dauphinois, Penninic, Austroalpine, Southalpine). The division into the facies zones of the 'Germanic Keuper' in the northwest and the 'Alpine Trias' in the southeast is clearly visible. In the northeast, this separation is marked by the land mass of the Vindelician high and, in the southwest, by a high zone in the region of Corsica and adjacent crystalline blocks of the future Penninic nappes. A transitional form with Late Triassic yellowish dolostones is found between these highs in those parts of the Penninic realm that are later differentiated into the Briançon microcontinent: the Bernhard nappe complex, composed of the Monte Rosa, Maggia, Tambo and Suretta nappes. It is unclear to what extent the Adula nappe is part of this, as the Triassic cover of this nappe was later completely sheared off.

An additional distinction can be made within the Austroalpine realm between the shallow-marine carbonate platform containing the 'Hauptdolomit' and Dachstein formations and the Hallstatt facies, a deeper shelf region containing Hallstatt Limestone (Mandl 2000). This subdivision of the Austroalpine Triassic is illustrated in Fig. 3.14, where the prograding reef belts of the Wetterstein and Dachstein formations are shown, as well as the interdigitation between the shallow platform carbonates and the basin facies, represented by the Pötschen and Hallstatt limestones: the different evolution in the Early and Middle Triassic is also evident. Whereas the Gutenstein and Steinalm limestones, which were deposited in a shallow-marine environment, can be followed laterally across the entire zone, the basin facies of the Reifling Formation occur separately: in the east, in the Hallstatt facies, and in the middle of the western facies zone, within the carbonate platform of the Wetterstein Formation. The differential basin evolution continued through to the fall in sea level at the Carnian–Norian transition, when more uniform carbonate platform development prevailed, as found in the 'Hauptdolomit' and Dachstein formations. In contrast, the Hallstatt facies in the east prevailed across the Ladinian–Norian period, but in this facies there was local development of high zones over Permian evaporites as a result of diapirism.

Jurassic: Opening up of Oceanic Arms

In the Jurassic, plate movements became more pronounced due to the break-up of Pangaea. The Central Atlantic opened

up and a connection to the Neo-Tethys was established, which passed between Iberia and Africa (present-day Gibraltar). Figure 3.15 shows the situation in the Oxfordian (154 million years ago), based on the reconstruction by Stampfli & Borel (2002). A dead-end oceanic arm, the Piemont Ocean, opened in the region of the future Penninic realm, between the Dauphinois–Helvetic and the Austroalpine–Southalpine realms (Stampfli & Borel 2002 call this oceanic arm the 'Alpine Tethys'). The definitive establishment of two separate passive continental margins, one in the northwest (Baltica, or Europe) and one in the southeast (Adriatic) was thus achieved. The Vardar Ocean succeeded the Meliata Ocean in the east of the Austroalpine realm and the Pindos Basin succeeded the Palaeo-Tethys. Relative to Baltica (or Europe), the African plate migrated to the east while the Central Atlantic was opening up. The oceanic arms that were opening up thus grew obliquely to their shorelines in a transtensive regime. Consequently, the mid-oceanic ridges of the oceanic arms must have been associated with numerous transform faults.

The palaeogeographical map in Fig. 3.16 once again shows the locations of the future crystalline blocks in the Alpine region. The Austroalpine and Southalpine realms were shifted several hundred kilometres to the southeast, in order to make way for the opening up of the Piemont Ocean. Extensional faults that are more or less north–south oriented are visible on the Adriatic continental margin in the Southalpine and Austroalpine realms and these compensated the east–west oriented extension linked to the oblique opening up of the Piemont Ocean. Breccias were deposited at these faults from the Early Jurassic onwards, which will be discussed in more

▶ **Figure 3.13**
Palaeogeographical map of the future Alpine realm in the Norian (220 Ma). The present-day outcrop shapes of the various basement units are given as reference frame. Also indicated are the boundaries of the future Foreland, Dauphinois–Helvetic, Penninic and Southalpine-Austroalpine realms. The Triassic of the Simano and Antigorio nappes is nowhere exposed. Thus its attribution to any of the Triassic facies types remains uncertain.

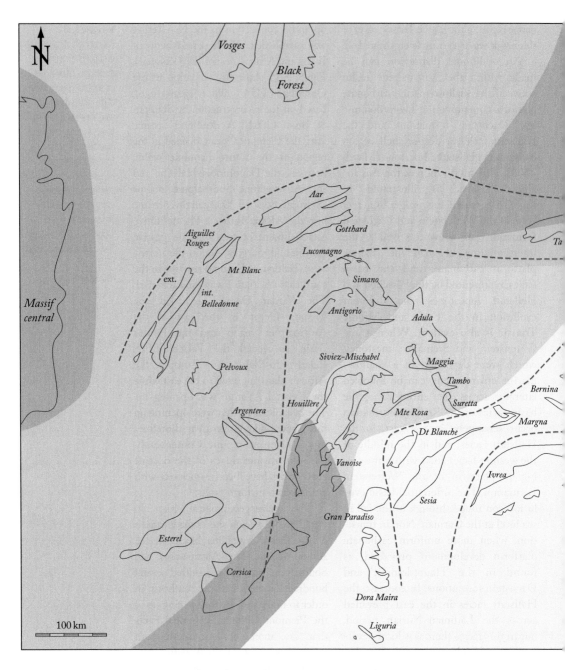

detail later. There were also active extensional faults on the northwestern side of
the Piemont Ocean. However, the extensional processes followed a more complex
pattern here and breccia deposits indicate
only some of the faults. It is usually only
the differences in thickness between the
individual Jurassic formations that
indicate extension-related tectonic
subsidence: a more detailed discussion
follows later. The opening of the Piemont
Ocean itself was accompanied by a

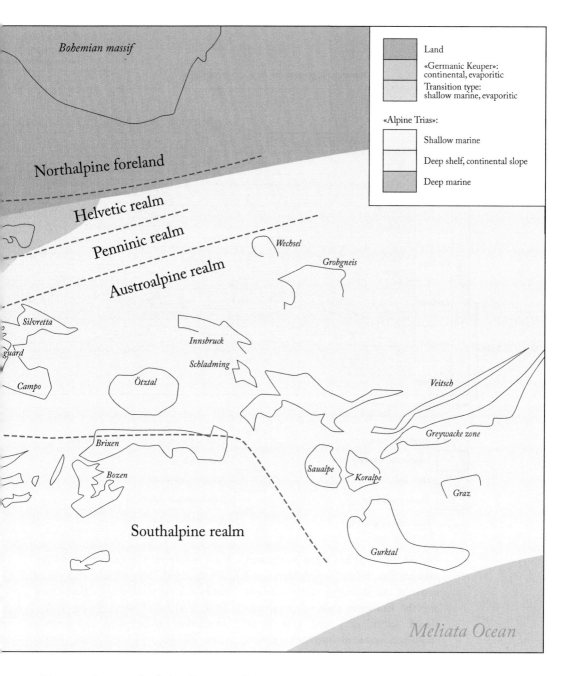

Land

«Germanic Keuper»:
continental, evaporitic

Transition type:
shallow marine, evaporitic

«Alpine Trias»:

Shallow marine

Deep shelf, continental slope

Deep marine

Bohemian massif

Northalpine foreland

Helvetic realm

Penninic realm

Austroalpine realm

Wechsel

Grobgneis

Silvretta

guard

Campo

Innsbruck

Schladming

Ötztal

Veitsch

Greywacke zone

Brixen

Bozen

Saualpe

Koralpe

Graz

Southalpine realm

Gurktal

Meliata Ocean

subhorizontal normal fault (master décollement) that exhumed parts of the subcontinental mantle (Manatschal et al. 2003). A mid-oceanic ridge then formed during the subsequent opening through spreading, and magmas associated with this process were asthenospheric in origin, including, for example, the gabbros that intruded in the Dogger (166–161 million years ago).

The Briançon microcontinent with its associated submarine rise effectively

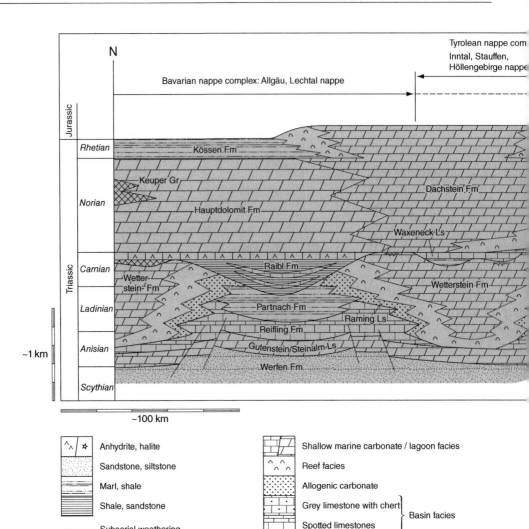

formed the distal European continental margin. The Valais Trough developed between the Briançon microcontinent and the European shelf due to extensional subsidence, and this basin continued westwards into the Vocontian Trough in the Dauphinois realm. Synsedimentary faults were activated at the southern margin of the Valais Trough, or at the northern margin of the Briançon microcontinent, whereas usually only differences in thickness of the individual layers are observed at the northern margin of the trough. Breccias

with dolostone pebbles and polymict breccias with crystalline components indicate local rises within the Valais Trough (Adula Rise of Probst 1980). These rises can be interpreted as tilted fault blocks associated with listric normal faults. Furthermore, new oceanic crust was formed in the Middle to Late Jurassic (Steinmann 1994), but this made up only a small portion of the overall area of the Valais Trough (see below).

In the following, some of the more important observations that provide

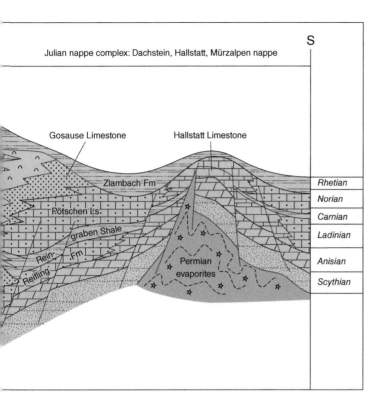

Julian nappe complex: Dachstein, Hallstatt, Mürzalpen nappe

S

Gosause Limestone   Hallstatt Limestone

Zlambach Fm

Pötschen Ls.

graben Shale

Rein- Fm

Reifling

Permian
evaporites

Rhetian

Norian

Carnian

Ladinian

Anisian

Scythian

Figure 3.14
Correlation scheme
of the Triassic facies types in
the Northern Calcareous Alps
(Austroalpine nappe
complex). Source: Adapted
from Mandl (2000).
Reproduced with permission
of the Austrian Geological
Society ÖGG.

evidence for the history of the opening up of the Piemont Ocean and the associated extension of the two continental margins are idiscussed in greater detail.

Trümpy (1949) and Dollfuss (1965) demonstrated that synsedimentary tectonics affected the shelf sequences in the Early and Middle Jurassic (Lias and Dogger) in the **Helvetic nappe system** in eastern Switzerland. In addition to abrupt fluctuations in thickness, breccia deposits are also noted in the latest Liassic. Components that are present also include Triassic dolostones, indicating a notable palaeorelief. Figure 3.17 has been redrawn and amended according to Trümpy (1980) and shows the situation in a north–south cross-section that approximately covers the present-day outline of the Aar massif and the eastern part of the Mont Blanc massif: the high zone in the north, where the Lias is absent, is called the 'Alemannic Land'. The fact that fault tectonics continued in the Dogger can be discerned only from the abrupt fluctuations in thicknesses. The formations in the extreme south, Stgir, Inferno and Coroi,

Figure 3.15 Plate
configuration in the
Oxfordian (154 Ma). The
boundaries between the
Dauphinois–Helvetic,
Penninic and Austroalpine–
Southalpine facies realms
are shown as dashed lines in
red. A narrow sea branching
off the Central Atlantic
extends into the Penninic
realm. Gr, Greece; Grld,
Greenland; It, Italy. Source:
Stampfli et al. (2002).
Reproduced with permission
of Elsevier.

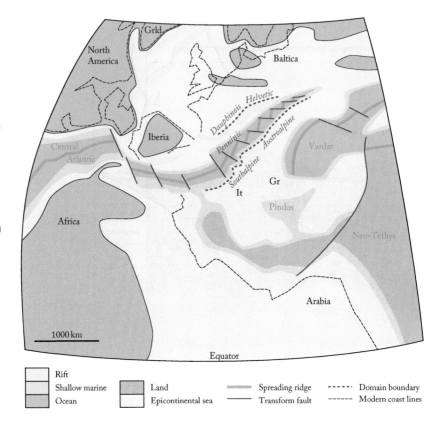

are attributed to the Ultrahelvetic realm: these now lie in an inverted position on the southeastern Gotthard massif. Some of the synsedimentary normal faults were also still active as 'persistent faults' later on in the Early Cretaceous (cf. Fig. 3.25). Wildi et al. (1989) investigated the subsidence rates on the European continental margin in the Mesozoic and noted that these subsidence rates exhibit neither a temporally nor a spatially uniform pattern: in the Early and Middle Jurassic they were between 10 and 40 metres per million years (0.01–0.04 millimetres per year), that is, substantially slower than the rates on the Adriatic continental margin in the Late Triassic (0.1 millimetres per year; cf. above); they then accelerated to 0.02 millimetres per year in the

Late Jurassic in the proximal (Jura Mountains), and 0.05 millimetres per year in the distal, southern shelf of the Helvetic realm.

The **Vocontian Trough** in the French Subalpine chains of the Dauphinois realm is characterized by fairly large fluctuations in thickness. Figure 3.18 shows these conditions in a schematic illustration in a north–south cross-section that is based on Debelmas (1974). The great thicknesses of the black shales ('Terres noires') in the Middle Jurassic and the marl successions in the Early and Middle Cretaceous ('Marnes bleues') are of note. According to Debelmas et al. (1983), the Vocontian Trough was already formed in the Early Jurassic, as evidenced by the great thickness of marly limestones containing fossilized cephalopods in the central portion of the trough (not indicated in Fig. 3.18).

The sedimentary sequences of the **Briançon microcontinent**, or the **Briançon Rise**, have been a much-discussed subject concerning the geology of the Alps. As early as 1877, Kaufmann recognized the fact that different rock types in central Switzerland crop out as laterally discontinuous blocks and that the individual blocks are in contact with rock types of a completely different nature. Many people found it difficult to accept the correlation of the deposits of Briançon-derived sediments between the Engadiner Window in the Eastern Alps, through the klippen of the Central Alps and into the Penninic nappes of the Western Alps, as being representative of a coherent palaeogeographical zone. The elongate rise can be divided transversely to its longitudinal axis into an actual Briançon microcontinent or rise and a Subbriançonnais realm, as shown in Fig. 3.19A. The cross-section in Fig. 3.19A is based on a retro-deformation of a balanced cross-section of the Klippen nappe in a transect of the Simmental based on Wissing & Pfiffner (2002) and shows a microcontinent that is about 40 kilometres wide. In the Triassic, evaporites were deposited that later functioned as detachment horizons during Alpine nappe formation. The evaporites are stratigraphically at the base of the Triassic dolostones in the actual Briançon Rise and on top of these dolostones in the Subbriançonnais realm. The Triassic dolostones of the Subbriançonnais have therefore largely remained attached to their pre-Triassic basement. A reduced sedimentary sequence was deposited on the actual Briançon Rise, in which the pelagic platform limestones of the Late Jurassic

almost directly overlie the Triassic. Middle and Late Jurassic sediments progressively appear towards the northwest, and their thickness increases abruptly at synsedimentary faults. The Dogger sediments are at their thickest in the Subbriançonnais, and the pelagic limestones of the Early Cretaceous and the 'Couches rouges' of the Late Cretaceous and the Paleocene also appear here. There was a steep submarine escarpment at the southeastern edge of the Briançon Rise that can be explained by a synsedimentary normal fault, steep at the top and levelling off downward where breccias were deposited. In the cross-section in Fig. 3.19, the main components of the breccias were Triassic dolostones (Fig. 3.7B). A local horst is found at the northwestern edge of the microcontinent, which can be followed from the Haute Savoie, through western Switzerland into central Switzerland, and is accordingly as the Môle–Moléson–Mythen horst (MMM in Fig. 3.19A). This horst rapidly gives way to the Valais Trough: in Fig. 3.19A it is unclear whether breccias are also associated with this transition. However, further to the east, in eastern Switzerland, breccias are found in the Falknis nappe and in the Schamser nappes at the northern edge of the Briançon Rise: these breccias also contain crystalline components and form quite large submarine fans (Gruner 1981, Rück 1995).

According to Septfontaine (1995), compressive fault structures also contribute to the structure of the Briançon Rise at the transition to the Western Alps, as illustrated at the bottom of Fig. 3.19B. An asymmetric high with an anticline in the north-northwest and a normal fault in the south-southeast are visible in the northwestern part of the actual Briançon

▶ **Figure 3.16**
Palaeogeographical map of the future Alpine realm in the Oxfordian (154 Ma). The present-day outcrop shapes of the various basement units are given as a frame of reference. Also indicated are the boundaries of the future Foreland, Dauphinois–Helvetic, Penninic and Southalpine–Austroalpine realms. The Briançon Rise separates the Valais and Vocontian troughs from the Penninic Ocean.

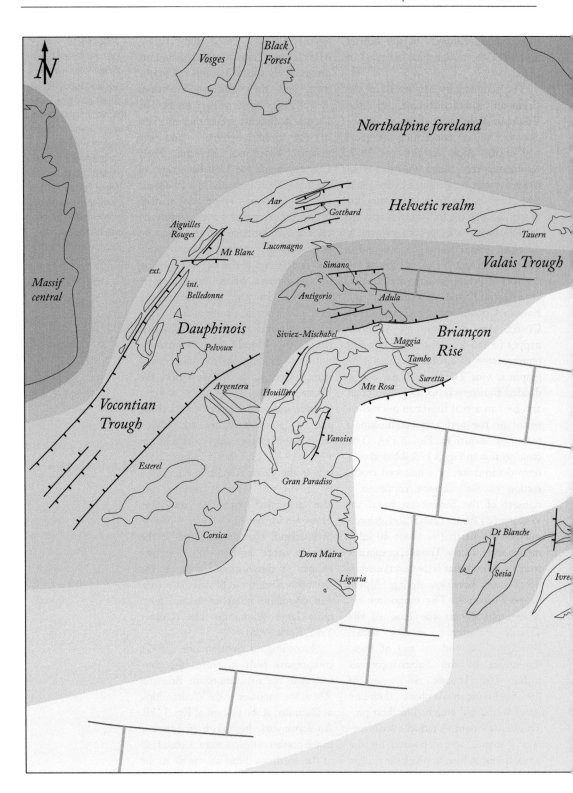

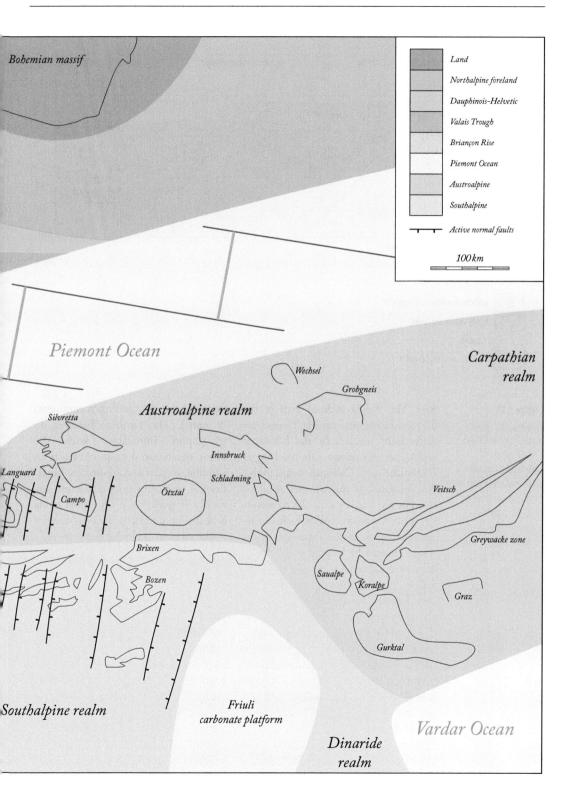

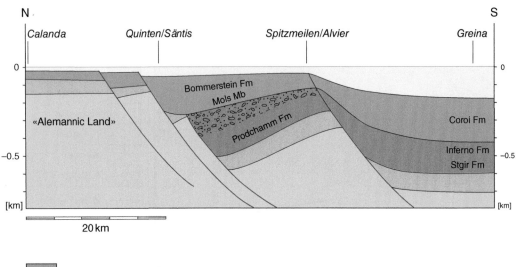

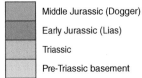

▲ **Figure 3.17**
Palinspastic cross-section
through the Helvetic nappe
system of eastern
Switzerland for the Middle
Jurassic. Synsedimentary
normal faults are associated
with thickness and facies
changes of the Early (Lias)
and Middle (Dogger) Jurassic
sediments. Source: Based on
Trümpy (1980).

Rise. The steeply inclined beds of the Triassic, Liassic and earliest Dogger are discordantly overlain by the Sommant and *Mytilus* formations of the late Dogger (Bathonian and Callovian respectively). The formation of the anticlinal structure could be explained by a specific, transpressive orientation of a palaeofault during the opening up of the Valais Trough, as was postulated by Schmid et al. (1990). However, according to Septfontaine (1995), parts of the rise were already above sea level in the Bajocian and were thus subject to karstification and erosion. The uplift of the rise and the associated emersion reached a further high point in the Callovian–Oxfordian. The subaerial conditions in some locations on the rise continued into the Late Jurassic and even into the Cretaceous in the southernmost part of the actual Briançon Rise.

The shallow water depth on the rise is also evident in the Western Alps, in the vicinity of the little town of Briançon

that gave the rise (and microcontinent) its name. The famous 'Brêches du télégraphe' intercalate with the above-mentioned deposits of breccia. In addition, angular unconformities, karstification and erosional channels are visible (Figure 3.20). The tilted strata beneath the unconformity were formed by the rotation of blocks of pre-rift sediments during movement on listric normal faults, and were then covered by syn-rift sediments. Channel breccias indicate erosion, and karstification points to intermittent periods of emersion of the rise.

On the **Adriatic continental margin**, extension is revealed by a pronounced subdivision into rises and basins. Figure 3.21, based on Bernoulli (2007), shows how the geometry of the synsedimentary normal faults produces half-grabens with wedge-shaped syn-rift sedimentary fills. This applies, in particular, to the Lombardian Siliceous

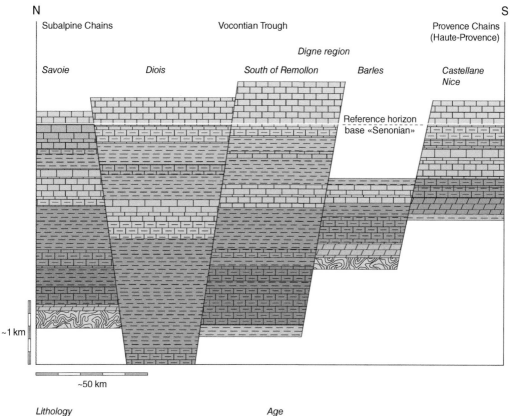

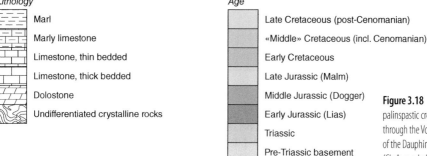

Lithology

Marl

Marly limestone

Limestone, thin bedded

Limestone, thick bedded

Dolostone

Undifferentiated crystalline rocks

Age

Late Cretaceous (post-Cenomanian)

«Middle» Cretaceous (incl. Cenomanian)

Early Cretaceous

Late Jurassic (Malm)

Middle Jurassic (Dogger)

Early Jurassic (Lias)

Triassic

Pre-Triassic basement

Figure 3.18 Schematic palinspastic cross-section through the Vocontian Trough of the Dauphinois realm (Chaînes subalpines, France). The base of the Late Cretaceous limestones was chosen as the reference horizon. The thicknesses of the Early and Middle Jurassic as well as the Early and Middle Cretaceous sedimentary sequences are greater in the centre of the trough as compared with the northern and southern rims. Source: Based on Debelmas (1974).

Limestone of the Early Jurassic. Breccias are found in the form of fissure fills in many locations immediately above and in the broken up Triassic platform carbonates. These include, for example, the 'Macchiavecchia' breccias that are quarried and used for decorative purposes (Fig. 3.22). After the Early Jurassic phase of tectonic subsidence, pelagic marly limestones ('Ammonitico Rosso') and radiolarian cherts were deposited as post-rift sediments during the Middle Jurassic.

Numerous synsedimentary faults were also active in the Austroalpine portion. These produced tilted fault blocks with fissures and steep submarine escarpments, in and on which breccias were deposited. Mader (1987) determined the simultaneous juxtaposition of breccia and pelagic deep-sea deposits in the Upper

(A)

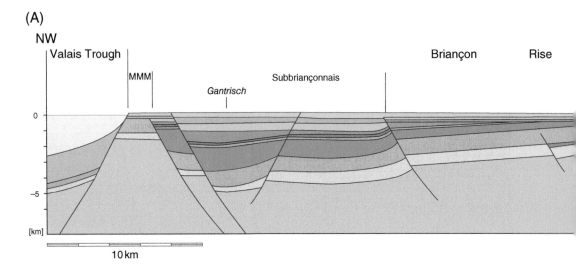

(B)

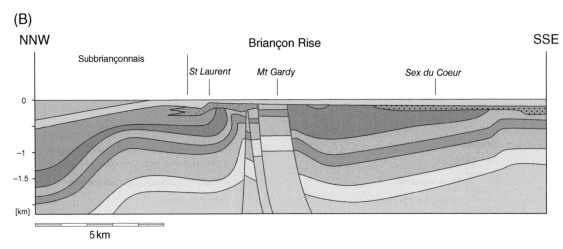

▲ **Figure 3.19** Palinspastic cross-section through the Briançon Rise (Penninic realm) for the end of the Cretaceous (65 Ma).

Austroalpine S-charl nappe. Furthermore, he managed to classify and date the breccias. As shown in Fig. 3.23, some of the breccias are fissure fills and overlie the Triassic dolostones of the 'Hauptdolomit' Formation with an angular unconformity. The unconformity indicates tilting and the fissures indicate extension and brittle break-up, both of which are consistent with the formation of tilted fault blocks in the hanging wall of listric normal faults. The basin geometry in Fig. 3.23, produced through retro-deformation of Mader's (1987) geological cross-section, shows that in the southeast, Jurassic (Rims Formation, Blais Radiolarian Chert,

Aptychus Limestone) and Cretaceous (Triazza Formation) pelagic sediments follow above a thin layer of Early Jurassic Lischana Breccia, whereas in the northwest this breccia is far thicker and extends into the Late Jurassic. Furthermore, this classification of the breccia by Mader (1987) revealed that fracturing also continued after the initial deposition of breccias (breccia types A–C in Fig. 3.23 have, themselves, been broken up again). This indicates episodicity of sedimentation and extension along with the development of submarine relief on the Adriatic continental margin. The breccias are chaotic in structure and contain components up to

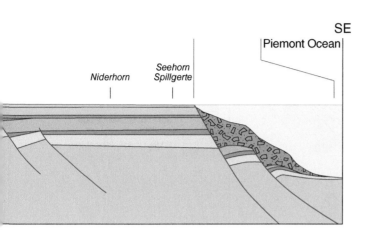

(A) The Briançon Rise in western Switzerland is bounded on either side by synsedimentary normal faults that caused enormous lateral facies changes. This same phenomenon is observed even within the rise. Source: Based on Wissing & Pfiffner (2002).

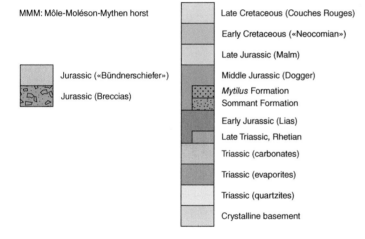

MMM: Môle-Moléson-Mythen horst

Jurassic («Bündnerschiefer»)

Jurassic (Breccias)

Late Cretaceous (Couches Rouges)

Early Cretaceous («Neocomian»)

Late Jurassic (Malm)

Middle Jurassic (Dogger)

Mytilus Formation

Sommant Formation

Early Jurassic (Lias)

Late Triassic, Rhetian

Triassic (carbonates)

Triassic (evaporites)

Triassic (quartzites)

Crystalline basement

(B) In Savoie (France) synsedimentary normal faults also initiated abrupt lateral facies changes. Some of the faults are considered to be thrust faults and suggest local compression or transpression. Source: Based on Septfontaine (1995).

the size of a house. Components from the Triassic (mainly dolostones of the 'Hauptdolomit' Formation) predominate in the lowest part (types A and B), but from type C onwards, Jurassic limestones also appear in the spectrum of components and only at the top do Triassic components once again predominate, in types E and F. Overall, the Lischana Breccia was mainly formed by submarine landslides (Mader 1987). These landslides descended down steep escarpments that were the result of the steep normal faults in the upper part of the escarpment. The dip of the beds of the 'Hauptdolomit' Formation in the

footwall of the Lischana Breccia leads us to suspect that the normal faults were dipping towards the southeast, that is, towards the continent.

Eberli (1988) recognized a system of Jurassic half-grabens in the Lower Austroalpine nappes. As shown in Fig. 3.24, the listric normal faults that delimit these half-grabens with a steep escarpment dip towards the east in the direction of the continent in the eastern portion, and towards the west in the direction of the Piemont Ocean in the western portion – therefore a submarine rise existed in between. As demonstrated by Eberli (1988), the normal

▶ **Figure 3.21** Palinspastic cross-section through the Southalpine domain in Lombardy (Italy). The continental margin was fragmented into basins and rises corresponding to (half) grabens and horsts in an east–west direction. The resulting thickness variations are particularly striking in the Early Jurassic syn-rift deposits. Source: Bernoulli (2007). Reproduced with permission of Swiss Bulletin of Applied Geology.

Figure 3.20 Detailed outcrop sketches from the Briançon Rise in the vicinity of Briançon (Vallée de la Durance, France) – redrawn from Debelmas et al. (1979, 1983). Angular unconformities, karst features, breccias and channel breccias point to Jurassic-Cretaceous synsedimentary tectonic movements. Source: Based on Debelmas (1974).

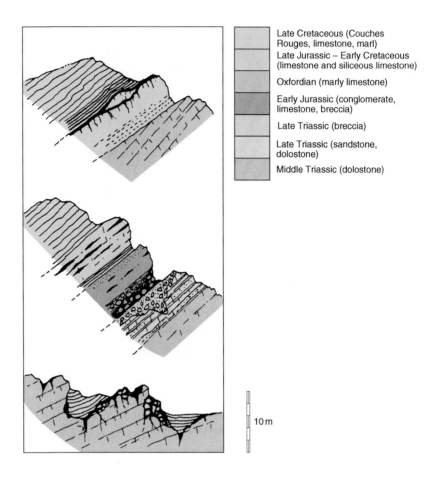

Late Cretaceous (Couches Rouges, limestone, marl)

Late Jurassic – Early Cretaceous (limestone and siliceous limestone)

Oxfordian (marly limestone)

Early Jurassic (conglomerate, limestone, breccia)

Late Triassic (breccia)

Late Triassic (sandstone, dolostone)

Middle Triassic (dolostone)

10 m

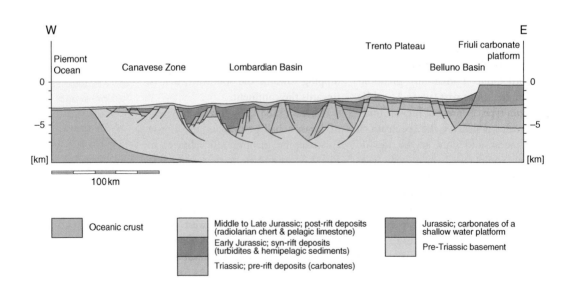

W   E

Trento Plateau   Friuli carbonate platform

Piemont Ocean   Canavese Zone   Lombardian Basin   Belluno Basin

0   0

−5   −5

[km]   [km]

100 km

Oceanic crust

Middle to Late Jurassic; post-rift deposits (radiolarian chert & pelagic limestone)

Early Jurassic; syn-rift deposits (turbidites & hemipelagic sediments)

Triassic; pre-rift deposits (carbonates)

Jurassic; carbonates of a shallow water platform

Pre-Triassic basement

Figure 3.22 Breccias of the 'Brocatello' or 'Macchiavecchia', which were formed by synsedimentary fracturing of the carbonate platform.

(A)

(A) Detailed view of the breccia. Fragments of the carbonate platform were deposited in open fractures. A reddish fine-grained matrix filled the open spaces between the coarse components of the breccia.

(B) Large-scale view showing varying sizes of coarse components, which are cut by veins in part, indicating that they had been fractured previous to their deposition in the larger open fracture.

(B)

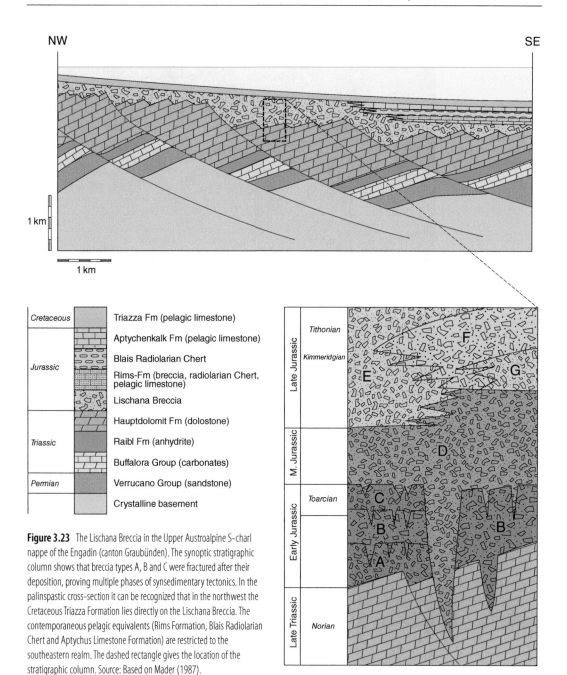

Figure 3.23 The Lischana Breccia in the Upper Austroalpine S-charl nappe of the Engadin (canton Graubünden). The synoptic stratigraphic column shows that breccia types A, B and C were fractured after their deposition, proving multiple phases of synsedimentary tectonics. In the palinspastic cross-section it can be recognized that in the northwest the Cretaceous Triazza Formation lies directly on the Lischana Breccia. The contemporaneous pelagic equivalents (Rims Formation, Blais Radiolarian Chert and Aptychus Limestone Formation) are restricted to the southeastern realm. The dashed rectangle gives the location of the stratigraphic column. Source: Based on Mader (1987).

faults in the east (Ortler nappe) were active in the Early Jurassic, based on the fact that the half-grabens are filled with Early Jurassic sediments. This is in agreement with the relationships in the S-charl nappe that were discussed above (Mader 1987). The normal faults to the west of the rise were not active until the Middle Jurassic and the half-grabens are filled with Middle Jurassic sediments. The sedimentary fill in the half-grabens exhibits a sequence containing reworked

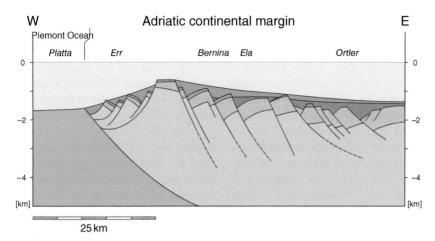

W   Adriatic continental margin   **E**

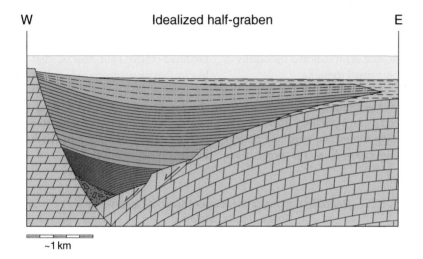

W   Idealized half-graben   **E**

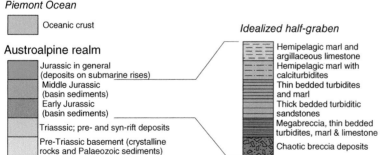

~1 km

Figure 3.24 Basin model of the passive Adriatic margin. The synsedimentary faults in the eastern part were active earlier, in the Early Jurassic, while in the western part faulting started in the Middle Jurassic only. The sedimentary fill of the half-grabens covers succesively more and more of the subsiding tilted block. Source: Based on Eberli (1988).

Penninic realm

Piemont Ocean

Oceanic crust

Austroalpine realm

Jurassic in general
(deposits on submarine rises)

Middle Jurassic
(basin sediments)

Early Jurassic
(basin sediments)

Triasssic; pre- and syn-rift deposits

Pre-Triassic basement (crystalline
rocks and Palaeozoic sediments)

Idealized half-graben

Hemipelagic marl and
argillaceous limestone

Hemipelagic marl with
calciturbidites

Thin bedded turbidites
and marl

Thick bedded turbiditic
sandstones

Megabreccia, thin bedded
turbidites, marl & limestone

Chaotic breccia deposits

sediments that become finer and are more thinly bedded towards the top. Chaotic beds of breccias were first deposited at the foot of the escarpment. Younger strata then successively prograde away from the active normal fault towards the subsiding tilted fault block. Hemipelagic marls and argillaceous limestones that are devoid of reworked sediments indicate the end of rift-related sedimentation. The components in the different breccias and turbidites stem exclusively from the Triassic and Jurassic sediments. The pre-Triassic basement in this region was apparently not exposed by the extensional tectonics. Eberli (1988) observed two megacycles in the Ela nappe, each one showing fining upward sequences, which once again indicates episodicity in the extensional tectonics.

When the **Piemont Ocean** opened up, the sublithospheric mantle was subjected to upwelling beneath the thinning continental crust and thereby became exposed on the ocean floor.

During this process, the peridotite was hydrated and transformed into serpentinite. At the Adriatic continental margin, upwelling was accompanied by a large-scale normal fault that spanned a large distance, subparallel to the contact between the pre-Triassic crystalline rocks and the Triassic sediments. Due to this very shallow dipping normal fault, the thinned crust was basically pulled away from the mantle as it welled upwards, thus facilitating the exhumation of the mantle. The situation is outlined in Fig. 3.25, based on the illustration by Manatschal et al. (2003). Numerous listric normal faults merge at depth with the subhorizontal main normal fault (the 'master décollement') and produced tilted fault blocks, that is, blocks that were successively rotated as they moved along the listric surface of the normal faults. The rocks along the master décollement were subjected to brittle deformation under the relatively low prevailing temperatures and are

Figure 3.25 Model for the evolution of the transition from the thinned passive margin of the Austroalpine realm to the Piemont/Penninic Ocean. The palinspastic cross-section is vertically exaggerated. Owing to the listric shape of the synsedimentary normal faults, some of the tilted blocks came to lie directly on the exhumed, hydrated, serpentinized mantle. Source: Manatschal et al. (2003). Reproduced with permission of Swiss Geological Society SGG.

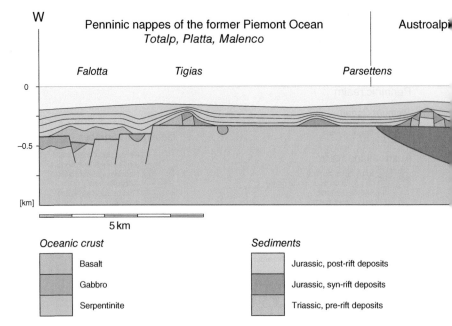

now present in the form of cataclasite. Further extension led to the formation of a mid-oceanic ridge with gabbros intruding at depth and pillow basalts on the ocean floor, but some of these rocks were also affected by the ongoing extension and hence were broken up and turned into breccia.

During the upwelling of the mantle, high-density rocks briefly ended up close to the surface. However, they rapidly sank back down due to isostatic equilibrium adjustments (a process known as tectonic subsidence, which is a direct result of tectonic thinning and extension of the continental crust), which resulted in subsidence of the sea floor to below the CCD, such that radiolarian cherts were deposited on top of the exhumed, sertpentinized mantle. Figure 3.26A (redrawn from Manatschal et al. 2003) illustrates the ideal succession that can be expected on the ocean floor: above the gabbros and serpentinites, containing zones of brittle deformation and

fissures, are found breccias containing components of serpentinite, gabbro and pillow lava, following which are found deep-sea clays and radiolarian cherts – the breccias indicate a submarine relief that developed on the ocean floor through active fault tectonics.

Figure 3.26B shows the rock suite that Peters & Dietrich (2008) observed in the Platta nappe of eastern Switzerland, which probably corresponds to the situation in the abyssal plain away from the active faults: the mantle rocks are at the base (serpentinized peridotite (lherzolite) cross-cut by dykes of dolerite and gabbro); above this there are localized gabbro sills, followed by thick pillow basalts that contain beds of radiolarian cherts and red shales (deep-sea clays) as well as debris from pillow lavas (hyaloclastites and pillow lava breccias); finally, a succession of deep-sea sediments made up of black shales and siliceous shales, as well as phyllites, completes the rock suite.

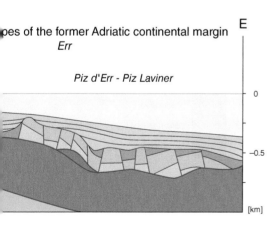

pes of the former Adriatic continental margin
Err

Piz d'Err - Piz Laviner

E

0

−0.5

[km]

Pre-Triassic crystalline basement

Granites

Gneisses and schist

Figure 3.26B demonstrates that eruptions of pillow basalts were episodic and that this was interrupted twice for lengthy periods during the Jurassic, as indicated by the two beds containing radiolarian chert and shale. The oldest pillow basalts erupted in the Callovian, at about the same time as the younger half-grabens in the sub-Austroalpine nappes were opening up (Ela nappe, see above, Fig. 3.24). Interestingly, gabbros in the Gets nappe in western Switzerland, which contains ophiolite components in mélanges, were also dated to the Callovian (166 million years ago) by Bill et al. (1997). The two younger, Late Jurassic basalt beds in the Platta nappe (Fig. 3.26B) are the same age as the youngest Lischana Breccias in the Upper Austroalpine nappes (see Fig. 3.23), which leads to the conclusion that both the extensional tectonics and spreading were episodic in nature, separated by periods of tectonic quiescence.

Cretaceous: Opening and Closing of Oceanic Arms

The oceanic arms that had opened in the Jurassic continued to evolve in the Early Cretaceous, but from the Middle Cretaceous onwards the direction of movement between Baltica (Europe) and Africa changed, such that the oceanic arms that had just formed were slowly closed again by convergence and subduction. In the Early Cretaceous, Africa continued to move to the east relative to Baltica, but Iberia and its appendage in the east, the Briançon microcontinent, moved westwards with Africa, which created a transtensive rift basin between Iberia and Baltica that

(A)

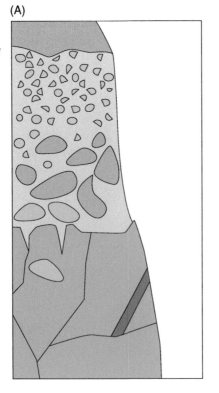

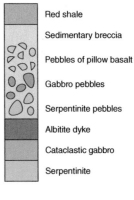

Red shale

Sedimentary breccia

Pebbles of pillow basalt

Gabbro pebbles

Serpentinite pebbles

Albitite dyke

Cataclastic gabbro

Serpentinite

10 m

Figure 3.26 Formation of oceanic crust in the Piemont/Penninic Ocean.

(A) Sequence of ophiolites and oceanic sediments in a model case of mantle exhumation followed by tectonic subsidence. Cataclastic gabbros are overlain by sedimentary breccias containing pebbles of serpentinite (stemming from the exhumed mantle), followed by gabbro (new oceanic crust) and basalt (uppermost new oceanic crust). Red deep sea clays were finally deposited onto the subsided ocean floor. Source: Manatschal et al. (2003). Reproduced with permission of Swiss Geological Society SGG.

(B)

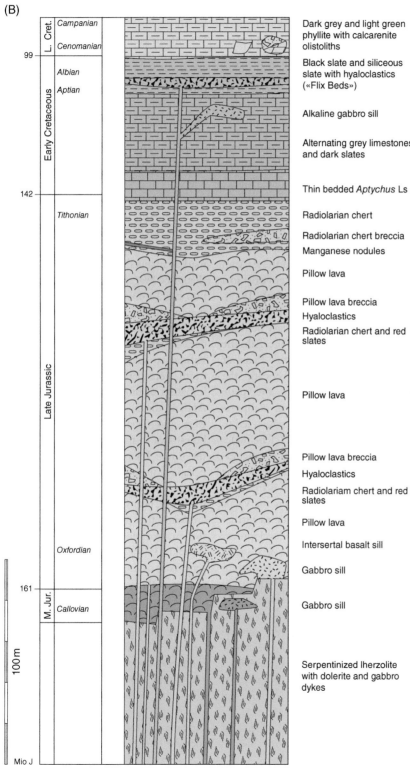

Dark grey and light green phyllite with calcarenite olistoliths

Black slate and siliceous slate with hyaloclastics («Flix Beds»)

Alkaline gabbro sill

Alternating grey limestones and dark slates

Thin bedded *Aptychus* Ls

Radiolarian chert
Radiolarian chert breccia
Manganese nodules

Pillow lava

Pillow lava breccia
Hyaloclastics
Radiolarian chert and red slates

Pillow lava

Pillow lava breccia
Hyaloclastics
Radiolariam chert and red slates

Pillow lava

Intersertal basalt sill

Gabbro sill

Gabbro sill

Serpentinized lherzolite with dolerite and gabbro dykes

(B) Sequence of ophiolites and oceanic sediments in the Platta nappe (Penninic nappe system of the Central Alps). Source: Adapted from Peters & Dietrich (2008).

opened the Bay of Biscay and the Valais Trough. Figure 3.27, redrawn from Stampfli & Borel (2002), shows the plate configuration in the Aptian (125 million years ago): the Atlantic Ocean had opened up further and now separated Iberia from North America; the Valais Trough had continued to broaden and joined up with the Piemont Ocean in the east to form a Penninic Ocean; the Dauphinois–Helvetic and Southalpine–Austroalpine realms formed the flanks of this ocean on either side; the connection between the Atlantic and the Neo-Tethys had closed.

In the Middle Cretaceous, 92 million years ago, the movement of the African plate changed from an eastern drift to a northern drift (Dewey et al. 1989). This led to the development of a subduction zone at the boundary between the Piemont Ocean and the Adriatic continental margin and the Piemont Ocean was subducted beneath the Adriatic plate. However, the plate movements also resulted in compression of the Adriatic plate, which led to a first Cretaceous orogeny (cf. below).

The palaeogeographical configuration of the crystalline blocks in the Alpine region in the Aptian is shown in Fig. 3.28. In this illustration, the Valais Trough appears broader due to further opening through oceanic development than at the time of the Oxfordian (Fig. 3.16) and still exhibits its western continuation into the Vocontian Trough. The Nollaton Formation was deposited in the Valais Trough in the Aptian, and Steinmann (1994) demonstrated that these clays could be attributed to a global anoxic event. Anoxic events are characterized by increased concentrations of organic carbon and are the result of globally raised temperatures and the associated collapse of oceanic currents. The 'Marnes bleues',

which reach their maximum thickness in the centre of the neighbouring Vocontian Trough in the Dauphinois realm, can be correlated to the Nollaton Formation. Although the Helvetic shelf subsided at laterally varying speeds (Funk 1985), which will be discussed in more detail below, a clear slow-down in subsidence occurred during the Aptian.

The southeast edge of the Piemont Ocean is shown as a subduction plate boundary in Fig. 3.28, but subduction started a little later, in the Turonian, once the African plate movement had changed direction. In the central region of the Austroalpine realm, there is a marked hiatus that includes the Aptian (Faupl & Wagreich 2000) and extends into the Turonian, but at the northern margin of the Austroalpine realm, in the future Northern Calcareous Alps, deep-water clastics and course clastic sediments sourced mainly from the south were deposited, which indicate initiation of orogenesis. Finally, in the Aptian, distal, terrigenous turbidites were also deposited over deep-water limestones in the south, in the area of the future Lienzer Dolomites. Mainly pelagic sediments were deposited in the Southalpine region at this time, but in the north, near Cortina, flysch deposits are found that indicate a source area in the Austroalpine region (Doglioni & Bosellini 1987): Figure 3.28 provides a schematic overview of the depositional centres of these clastic sediments.

A few aspects relating to tectonic evolution during the Cretaceous are considered in greater detail in the following. The **shelf of the Helvetic realm** is once again used as an example, as for the Jurassic in Fig. 3.17. The same cross-section in Fig. 3.29 shows the situation in the Early Cretaceous, based on Funk (1985). An increase in the thickness of the earliest

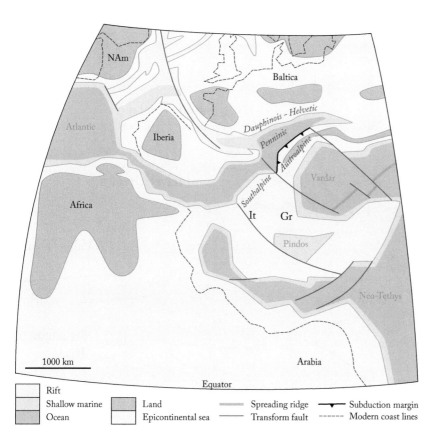

Figure 3.27 Plate configuration in the Aptian (125 Ma). The Atlantic had opened further and the narrow seas in the Penninic realm had also become wider. Gr, Greece; It, Italy; Nam, North America. Source: Stampfli & Borel (2002). Reproduced with permission of Elsevier.

Rift
Shallow marine
Ocean
Land
Epicontinental sea
Spreading ridge
Transform fault
Subduction margin
Modern coast lines

▶ **Figure 3.28** Palaeogeographical map of the future Alpine realm in the Aptian (125 Ma). The present-day outcrop shapes of the various basement units are given as a frame of reference. Also indicated are the boundaries of the future Foreland, Dauphinois–Helvetic, Penninic and Southalpine–Austroalpine realms. The Southeastern margin of the Penninic Ocean was the site of subduction, where this ocean plunged successively into the mantle beneath the Adriatic Plate. Clastic and coarse clastic sediments were shed from highs within the Austroalpine realm into local basins. These highs bear witness to an early Alpine mountain chain.

Cretaceous Palfris Shale and of the siliceous limestone of the Kieselkalk Formation is clearly visible: between these, oolitic limestones (Öhrli and Betlis limestones) are replaced by marly shale towards the south, and a micritic limestone (Diphyoides Limestone) occurs in their place in the extreme south. The uppermost oolitic limestone, the Schrattenkalk Formation, is the equivalent of the 'Urgonian' in the Western Alps and manifests the enormous expansion of this carbonate platform. Immediately below the upper oolitic limestone, the Drusberg Marl also increases in thickness towards the south. Large changes indicating synsedimentary tectonics are observed at two locations, which include the first occurrence of marls

instead of oolitic limestones and the sudden occurrence of micritic limestones. According to Wildi et al. (1989), subsidence rates varied between locations, as was the case in the Jurassic, but were higher in the distal part of the shelf, in the Helvetic realm, than in the proximal part (0.02–0.04 compared with 0.005–0.01 millimetres per year). One of the locations exhibiting large local differences in subsidence rates is shown in the cross-section in Fig. 3.29. Later, in the Albian (108–96 million years ago), the subsidence rates in the Helvetic realm reduced to values that were usually below 0.005 millimetres per year.

Figure 3.29 also shows the Late Jurassic limestones (mainly Quinten Limestone) with their associated

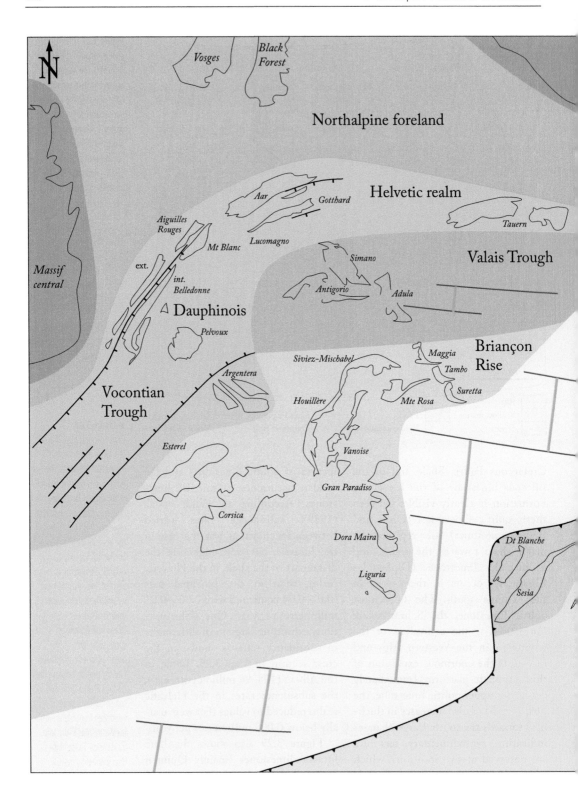

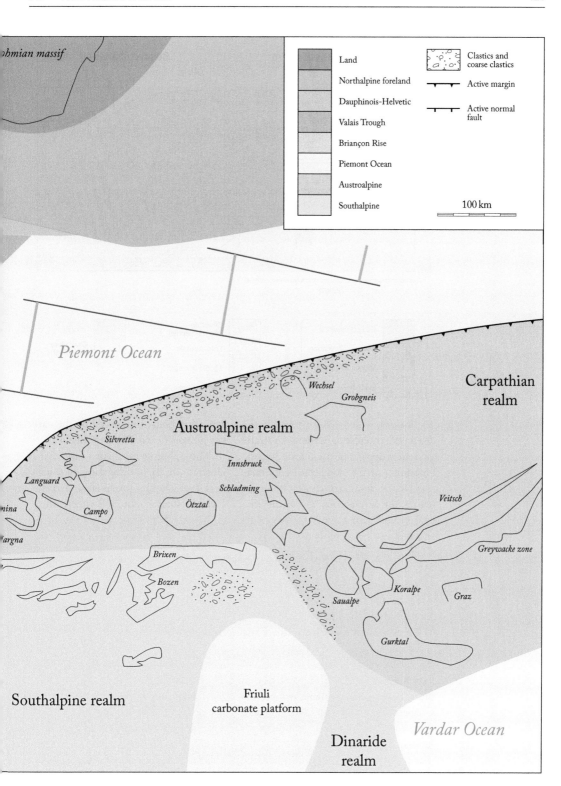

Land

Northalpine foreland

Dauphinois-Helvetic

Valais Trough

Briançon Rise

Piemont Ocean

Austroalpine

Southalpine

Clastics and
coarse clastics

Active margin

Active normal
fault

100 km

Bohmian massif

Piemont Ocean

Wechsel

Grobgneis

Carpathian
realm

Austroalpine realm

Silvretta

Innsbruck

Languard

Schladming

Veitsch

Bernina

Campo

Ötztal

Margna

Greywacke zone

Brixen

Bozen

Koralpe

Graz

Saualpe

Southalpine realm

Friuli
carbonate platform

Gurktal

Dinaride
realm

Vardar Ocean

Figure 3.29 Palinspastic cross-section through the Helvetic realm of eastern Switzerland for the Early Cretaceous. Synsedimentary normal faults are associated with thickness and facies changes. Limestones are replaced by marl and shale towards the south. Source: Adapted from Funk (1985).

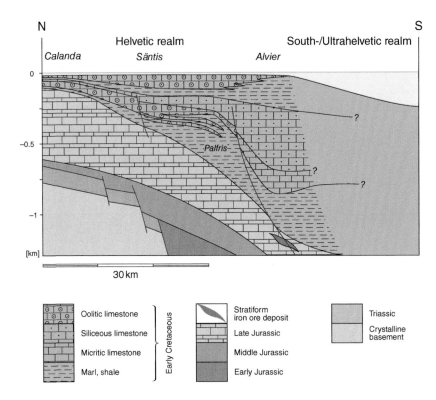

thicknesses, which reduce towards the distal end. Interestingly, there is a stratiform iron deposit in these Late Jurassic limestones at the downward continuation of the southernmost fault shown within the Cretaceous strata, the formation of which was substantially affected by synsedimentary tectonics. Fault zones facilitated the passage of hot fluids, which according to Pfeifer et al. (1988) originated from the pre-Triassic crystalline basement, but local depressions on the shelf prevented mixing of these fluids with seawater. Early and Middle Jurassic sediments are also shown schematically at the base of Fig. 3.29, where it is demonstrated that the northernmost fault shown in the Cretaceous rocks almost exactly forms a continuation of the southernmost Jurassic fault. It is therefore possible that persistent faults accompanied the tectonic evolution, with the Cretaceous

activation being influenced by the earlier Jurassic (? and Permian) history.

The photographs in Fig. 3.30 illustrate the large differences in thickness between the proximal and distal Helvetic shelf in eastern Switzerland.

A second example shows the basin geometry and basin dynamics of the **Valais Trough** with the thick deposits of Bündnerschiefer Group over thinned continental crust and new oceanic crust. Figure 3.31 is based on the work by Steinmann (1994). The model of the basin shows two areas of oceanic crust where the rocks were formed from mantle melts in two separate pull-apart basins. The northern basin had already opened in the Jurassic, but the southern basin did not open till the Early Cretaceous. The adjacent continental crust was thinned by extension and broken up into smaller tilted fault blocks. These surface irregularities covering small areas played a role

(A)

(B)

Figure 3.30 Two photographs of the Helvetic Cretaceous in eastern Switzerland, taken near Flims (canton Graubünden) and in the Churfirsten (canton St Gall) highlight the contrasting thicknesses: the Schrattenkalk Formation is present as a very thin bed within the cliff in the Flimserstein, whereas in the Churfirsten this limestone makes up the entire peaks. Ver, Verrucano; QL, Quinten Limestone; C, Cementstone Formation; Pa, Palfris Shale; ÖL, Öhrli Limestone; KK, Helvetic Kieselkalk Formation; Db, Drusberg Marl; Sr, Schrattenkalk Formation.

▶ **Figure 3.31** The 'Bündnerschiefer' of the Penninic nappe complex in Graubünden. In the stratigraphic columns, the Nolla Shale Formation, as an example, can be traced from Prättigau (Valais Trough) all the way into the Falknis and Schams nappes (Briançon Rise). The basin model displays two small oceanic domains. The northern one opened in the Jurassic, the southern one in the Early Cretaceous. Source: Adapted from Steinmann (1994).

later on during nappe formation. In the basin itself, the Nollaton Formation was deposited in the region of the future Grava and Tomül nappes. According to Steinmann (1994), these former clays now transformed into slates can be equated to the Valzeina Formation of Prättigau in the north and the clay phyllites of the Briançon Rise (Schams nappes) in the south (Fig. 3.31, top).

Sedimentation continued in the Valais Trough, first with the limestones of the Nollakalk Formation and the sandy limestones of the Carnusa Formation, and then peaked with the deposition of flysch sediments. The Tomül Flysch was shed from a source in the south, the Prättigau Flysch from the north. Sedimentation continued into the Eocene in the case of the Prättigau Flysch.

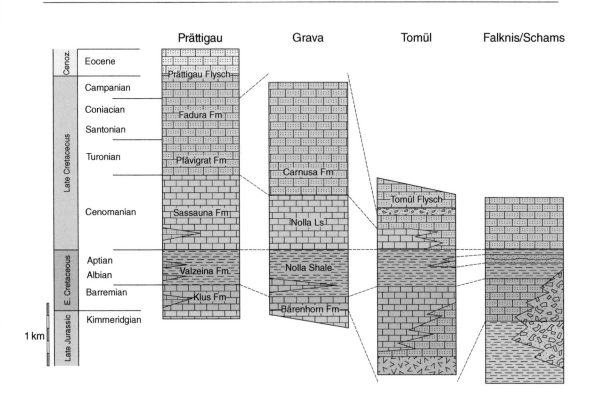

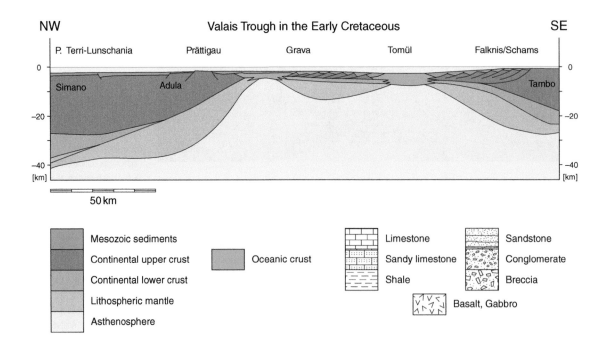

Based on the above descriptions, it emerges that limestones and marls were deposited on the European continental margin on the shelf (Jura Mountains, Helvetic and Dauphinois realms), in the adjacent Valais Trough and on the Briançon Rise in the Late Cretaceous. In contrast, the clastic sediments that were deposited on the Adriatic continental margin reflect the presence of areas elevated above sea level and even a topography with some relief. Furthermore, the flysch sediments from the region of the Piemont Ocean indicate that this ocean was in the process of being closed by east-southeast-directed subduction. This juxtaposition of shelf sedimentation, subduction and orogeny represents the transition between the break-up of Pangaea and the onset of convergent plate movement, which peaked in the Cenozoic with the collision between the European and Adriatic continental margins.

References

Baud, A. & Septfontaine, M., 1980, Présentation d'un profil palinspastique de la nappe des Préalpes médianes en Suisse occidentale. Eclogae geologicae Helvetiae, 73, 651–660.

Bernoulli, D., 2007, The pre-Alpine geodynamic evolution of the Southern Alps: a short summary. Bulletin für angewandte Geologie, 12/2, 3–10.

Bill, M., Bussy, F., Cosca, M., Masson, H. & Hunziker, J. C., 1997, High-precision U–Pb and 40Ar/39Ar dating of an Alpine ophiolite (Gets nappe, French Alps). Eclogae geologicae Helvetiae, 90/1, 43–554.

Bossellini, A., Gianolla, P. & Stefani, M., 2003, Geology of the Dolomites. Episodes, 26, 181–185.

Debelmas, J., 1974, Géologie de la France, doin Editeurs, Paris, 544 pp.

Debelmas, J., Arnaud, H., Caron, C., Gidon, M., Kerckhove, C., Lemoine, M. &

Vialon, P., 1979, Guides géologiques régionaux: Savoie et Dauphiné. Masson, Paris, 213 pp.

Debelmas, J., Arnaud, H., Gidon, M. & Kerckhove, C., 1983, Alpes du Dauphiné. Guides géologiques régionaux, Masson, Paris, 198 pp.

Dietrich, V., 1970, Die Stratigraphie der Platta Decke. Fazielle Zusammenhänge zwischen Oberpenninikum und Unterostalpin. Eclogae geologicae Helvetiae, 63/2, 631–671.

Doglioni, C. & Flores, G., 1997, An introduction to the Italian geology. Verlag Lamisco, 93 pp.

Dollfuss, S., 1965, Über den helvetischen Dogger zwischen Linth und Rhein. Eclogae geologicae Helvetiae, 58/1, 453–554.

Eberli, G., 1988, The evolution of the southern continental margin of the Jurassic Tethys Ocean as recorded in the Allgäu Formation of the Austroalpine Nappes of Graubünden. Eclogae geologicae Helvetiae, 81/1, 175–214.

Escher, A., Hunziker, J. C., Marthaler, M., Masson, H., Sartori, M. & Steck A., 1997, Geologic framework and structural evolution of the Western Swiss–Italian Alps. In: Pfiffner, O. A., Lehner, P., Heitzmann, P., Mueller, S. & Steck, A. (eds), Deep Structure of the Swiss Alps: Results of NRP 20. Birkäuser Verlag, Basel, 205–222.

Faupl, P. & Wagreich, M., 2000, Late Jurassic to Eocene palaeogeography and geodynamic evolution of the Eastern Alps. In: Neubauer, F. & Höck, V. (eds), Aspects of Geology in Austria, 79–94. Also published in den Mitteilungen der Österreichischen Geologischen Gesellschaft, 92 (1999).

Funk, Hp., 1985, Mesozoische Subsidenzgeschichte im Helvetischen Schelf der Ostschweiz. Eclogae geologicae Helvetiae, 78/2, 249–272.

Funk, Hp., Oberhänsli, R., Pfiffner, A., Schmid, S. & Wildi, W., 1987, The evolution of the northern margin of Tethys in eastern Switzerland. Episodes, 19/2, 102–106.

Gruner, U., 1981, Die jurassischen Breccien der Falknis-Decke und altersäquivalente Einheiten in Graubünden. Beiträge geologische Karte Schweiz, Neue Folge 154, 136 pp.

Kaufmann, F. J., 1877, Geologische Beschreibung der Kalkstein- und Schiefergebiete der Kantone Schwyz, Zug und des Bürgenstocks bei Stanz. Beiträge geologische Karte Schweiz, 14/2. Abteilung, 180 pp.

Lammerer, B., 1986, Das Autochthon im westlichen Tauernfenster. Jahrbuch Geologische Bundes-Anstalt, 129/1, 51–67.

Lemoine, M., 1988, Des nappes embryonnaires aux blocs basculés: évolution des idées et modèles sur l'histoire mésozoïque des Alpes occidentales Bulletin Société géologique de France, 8ème serie 4, 787–797.

Mader, P., 1987, Die Jura- und Kreideablagerungen im Lischana-Gebiet (Oberostalpine S-charl-Decke, Unterengadin). Eclogae geologicae Helvetiae, 80/3, 633–653.

Manatschal, G., Müntener, O., Desmurs, L. & Bernoulli, D., 2003, An ancient ocean–continent transition in the Alps: The Totalp, Err-Platta, and Malenco unites in the eastern Central Alps (Graubünden and northern Italy). Eclogae geologicae Helvetiae, 96, 131–146.

Mandl, G. W., 2000, The Alpine sector of the Tethyan shelf Examples of Triassic to Jurassic sedimentation and deformation from the Northern Calcareous Alps. In: Neubauer, F. & Höck, V. (eds), Aspects of Geology in Austria, 61–77. Also published in den Mitteilungen der Österreichischen Geologischen Gesellschaft, 92 (1999).

Marthaler, M., 1984, Géologie des unités penniques entre le Val d'Anniviers et le Val de Tourtemagne. Eclogae geologicae Helvetiae, 77/2, 295–448.

Mettraux, M. & Mosar, J., 1989, Tectonique alpine et paléotectonique liasique dans les Préalpes Médianes en rive droite du Rhône. Eclogae geologicae Helvetiae, 82/2, 517–540.

Mosar, J., Stampfli, G. M. & Girod, F., 1996, Western Préalpes Médianes romandes: timing and structure. A review. Eclogae geologicae Helvetiae, 89/1, 389–425.

Peters, Tj. & Dietrich, V., 2008, Erläuterungen zum Geologischen Atlas der Schweiz 1:25 000, Blatt 1256 Bivio. Bundesamt für Landestopografie.

Pfeifer, H.-R., Oberhänsli, H. & Epprecht, W., 1988, Geochemical evidence for a synsedimentary origin of Jurassic iron-manganese deposits at Gonzen (Sargans, Helvetic Alps, Switzerland). Marine Geology, 84, 257–272.

Probst, Ph., 1980, Die Bündnerschiefer des nördlichen Penninikums zwischen Valser Tal und Passo die San Giacomo. Beiträge geologische Karte Schweiz, Neue Folge 153, 63 pp.

Rück, Ph., 1995, Die Schamser Decken: Teil I: Stratigraphisch-sedimentologische Untersuchungen der Schamser Decken. Beiträge geologische Karte Schweiz, Neue Folge 167, 78 pp.

Sartori, M., Gouffon, Y. & Marthaler, M., 2006, Harmonisation et définition des unités lithostratigraphiques briançonnaises dans les nappes penniques du Valais. Eclogae geologicae Helvetiae, 99, 363–407.

Schmid, S. M., Rück, Ph. & Schreurs, G., 1990, The significance of the Schams nappes for the reconstruction of the paleo-tectonic and orogenic evolution of the Penninic zone along the NFP-20 East traverse (Grisons, eastern Switzerland). In: Roure, F., Heitzmann, P. & Polino, R. (eds), Deep Structure of the Alps. Mémoire Société géologique France, no. 156, 263–287.

Septfontaine, M., 1995, Large scale progressive unconformities in Jurassic strata of the Prealps S of Lake Geneva: Interpretation as synsedimentrary inversion structures; Paleotectonic implications. Eclogae geologicae Helvetiae, 88(3), 553–576.

Stampfli, G. & Borel, G., 2002, A plate tectonic model for the Paleozoic and Mesozoic constrained by dynamic plate boundaries and restored synthetic oceanic isochrons. Earth and Planetary Science Letters, 191, 17–33.

Steinmann, M., 1994, Ein Beckenmodell für das Nordpenninikum der Ostschweiz. Jahrbuch der Geologischen Bundesanstalt, Bd 137/4, 675–721.

Steinmann, M. & Stille, P., 1999, Geochemical evidence for the nature of the crust beneath the eastern North Penninic basin of the Mesozoic Tethys Ocean. Geologische Rundschau, 87, 633–643.

Tricart, P., Bourbon, M., Chenet, P. Y., Cros, P., Delorme, M., Dumont, T., de Graciansky, P.-C., Lemoine, M., Mégrard-Galli, J. & Richez, M., 1988, Tectonique synsédimentaire triasico-jurassique et rifting téthysien dans la nappe briançonnaise de Peyre-haute (Alpes occidentales). Bulletin Société géologique de France, (8e s.) IV/4, 669–680.

Trümpy, R., 1949, Der Lias der Glarner Alpen Denkschriften der Schweizerischen Naturforschenden Gesellschaft, 79.

Trümpy, R., 1980, Geology of Switzerland a guide book. Schweizerische Geologische Kommission, Wepf & Co. Publishers, Basel.

Vesela, P., Lammerer, B., Wetzel, A., Söllner, F. & Gerdes, A., 2008, Post-Variscan to Early Alpine sedimentary basins in the Tauern Window (eastern Alps). In: Siegesmund, S. Fügenschuh, B. & Froitzheim, N. (eds), Tectonic Aspects of the Alpine–Dinaride–Carpathian System. Geological Society London, Special Publications, 298, 83–100.

Wildi, W., Funk, Hp., Loup, B., Amato, E. & Huggenberger, P., 1989, Mesozoic subsidence history of the European marginal shelves of the Alpine Tethys (Helvetic realm, Swiss Plateau and Jura). Eclogae geologicae Helvetiae, 82/3, 817–840.

Wissing, S. B. & Pfiffner, O. A., 2002, Structure of the eastern Klippen nappe (BE, FR): Implications for Alpine tectonic evolution. Eclogae geologicae Helvetiae, 95, 381–398.

4 The Alpine Domain in the Cenozoic

Sedimentation of mainly clastic sediments prevailed in the Alpine region during the Cenozoic, and there were also localized phases of erosion. In addition, magmatic activity led to the placement of plutonic and volcanic rocks locally.

- Many of the clastic sediments were produced more or less as a direct result of Alpine orogeny, in particular its uplift into a mountain chain. One older sedimentary sequence was deposited under marine conditions, with a younger one developing under fluctuating marine and terrestrial conditions. The sedimentary basins were relatively narrow troughs that subsided along the plate boundary at the time.
- However, there were also lengthy phases of erosion in the Cenozoic. As a result of this, clastic sediments accumulated in basins within the developing mountain chain.
- In addition to the Cenozoic sediments, igneous rocks also formed in the Alps, including granitic intrusives and volcanics that are closely linked to Alpine orogenesis.

Figure 4.1 gives a schematic overview of the different rock suites, illustrated in sequence according to age and origin. This shows that the deposition of certain sedimentary sequences apparently occurred only during one period within the Cenozoic. In addition, the sediments were deposited in locally confined basins. For example, on the left in Fig. 4.1, which represents the north side of the Alps, we see an almost seamless chronological transition from sedimentation in the flysch basins of the Penninic realm (Prättigau Flysch), through the Helvetic realm (Sardona and Northhelvetic flyschs), to sedimentation in the Molasse Basin. Although sedimentation is continuous on the south side of the Alps, flysch deposits are limited to the Late Cretaceous (Lombardian Flysch), and give way to pelagic marl. The subsequent Oligocene and Miocene coarse clastic sediments (the Gonfolite Lombarda) were deposited in deep water and therefore cannot be compared with the Molasse on the north side of the Alps. The Vienna and Styrian basins are intramontane basins in which parts of an already existing mountain chain subsided. Deposits of limestone, conglomerate and gravel are highly localized in the Jura Mountains. Igneous rocks make up a tiny proportion of the volume of the Alps: granodioritic intrusions intruded along the peri-Adriatic fault system in the south of the Alps; andesitic dykes intruded at the same time as the younger intrusions, and volcanic bombs and fine detritus in the Northhelvetic Flysch are evidence for this volcanic activity. In spite of their sometimes limited occurrence with reference to volume, the Cenozoic rocks provide us with important information on the formation of the Alps.

In the sections below, the different types of sediments, their sequences and depositional conditions will be discussed first, followed by an interpretation of the sedimentary sequences within the plate tectonic framework of the Alpine orogeny, including interpretation of the igneous rocks.

Geology of the Alps: Revised and updated translation of Geologie der Alpen, Second Edition. O. Adrian Pfiffner.
© 2014 John Wiley & Sons, Ltd. Published 2014 by John Wiley & Sons, Ltd.

4.1 The Cenozoic Sedimentary Sequences

The Cenozoic sediments are invariably associated with the terms 'flysch' and 'molasse', but the use of both these terms has changed in the past.

'Flysch' originated in the Simmental (canton Bern, Switzerland), where farmers used this name for fields in areas with slaty, 'bad' rock. Early on, Studer (1827) used the term flysch for a suite of clays and grey-blackish sandstones, but he used the term to refer to a sequence of layers or a formation. Flysch then became an Eocene rock suite after Studer had visited the Entlebuch (canton Lucerne, Switzerland) with A. Escher, where he became persuaded of this by the local stratigraphic observation that the clays and sandstones overlie Eocene limestones (see Studer 1853). Later on, other flysch units were dated in the Alps, leading to the realization that even older flysch formations existed. When the nappe theory became generally accepted, the different flysch units were allocated to the individual nappes and the term flysch was almost used as a name for nappes. Later, once the origin of turbidity currents was understood, flysch was effectively used as a synonym for suites of turbidite deposits. Hsü & Briegel (1991) provide a detailed discussion of the term flysch.

The present-day use of the term flysch must comply with the following criteria:

- a thick, well-bedded sedimentary sequence is present that is composed of marine sandstones, calcareous arenites or conglomerates, intercalated with marine claystones or shales;
- the deposits were laid down by turbidity currents or mass flows (with a slower flow rate);
- the deep marine basin is a foreland basin to an active orogen;
- the basin is underfilled, that is, sediment accumulation did not keep pace with subsidence.

The photograph in Fig. 4.2A shows a succession of turbidites in the Northhelvetic Flysch near Elm (canton Glarus, Central Alps) where bedding is highlighted by the preferred erosion and removal of the fine-grained upper parts of the turbidites beneath the massive coarse-grained base of the succeeding turbidite. With regard to age, Late Cretaceous to Paleogene flyschs in the realms of the Valais Trough and the Piemont Ocean are allocated to one group and Paleogene flyschs on the European continental margin are allocated to a second group. On the south side of the Alps, Late Cretaceous flyschs were deposited in the Lombardian Basin and Paleogene flyschs from the Dinarides.

In this context, the term **'wildflysch'** requires a special mention. This term refers to rock formations with a matrix of dark shale, in which blocks of sandstone, limestone, conglomerate and granite float in chaotic orientation. The matrix is often honeycombed with shiny sliding surfaces and contains fragments of calcite and quartz veins. The term was used by Kaufmann (1877, 1886), but at the time very little was understood regarding its genesis, as is illustrated by this citation from Kaufmann (1886, p. 553):

Flysch:
deep water clastics

'The cause of these extreme changes in orientation cannot simply be searched for in the general dislocation pressure; this is because simpler relationships usually occur again immediately, as soon as we leave the wildflysch (black shale) cross-sections. It appears that

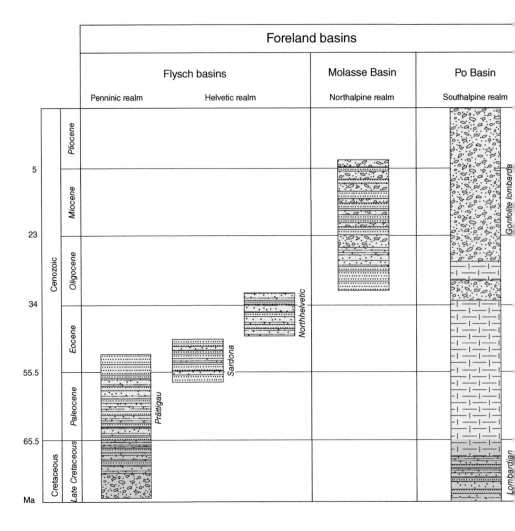

this wildflysch achieved an irresistible power to enlarge its volume through some chemical quality and thus make space through worm-like gyri.'

The present-day explanation is based on observations of the exposed sections described above and takes into consideration the fact that the different 'Wildflysch' formations in the Alps are always found along the sides of important nappe contacts or thrust faults, typically at the base of Penninic nappes comprising sediments. 'Wildflysch' is regarded as a tectonic melange that

formed when the nappe was emplaced onto the 'Wildflysch', namely, where the basal thrust emerges at the surface of the Earth (see below). Figures 4.2B and c display the features typical for 'Wildflysch' as observed in an Ultra-helvetic unit that crops out near Mellau (Vorarlberg, Austria) in the Eastern Alps: in a zone about 30 metres wide, fragments of limestones and sandstones are dispersed in a dark shaley matrix.

'Molasse' is derived from the French word 'meule', the name used for soft sandstones that could be used for making

Molasse: marine and fluvial clastics

Intramontane basins		Alpine magmatic provinces
Vienna Basin Styrian Basin	Jura Mountains	

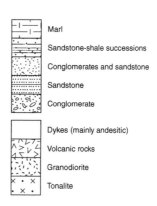

Marl

Sandstone-shale successions

Conglomerates and sandstone

Sandstone

Conglomerate

Dykes (mainly andesitic)

Volcanic rocks

Granodiorite

Tonalite

Hegau volcancis

Dyke swarms

Taveyannaz volcanics

Peri-Adriatic intrusives

millstones. Studer (1853) had already distinguished between four large-scale units in the Molasse Basin of the Central Alps: distinctions that remain in use today. These are two marine suites containing sandstones and shales, each followed by a continental suite containing sandstones, conglomerates and marl, and Studer (1853) interpreted the (coarse) clastic sediments as the erosional debris from a nearby mountain range. The term molasse was subsequently used outside the Alps, also in some cases it generally referred to sequences of sandstones and conglomerates, but in others it was extended to clastics in the core of a mountain chain or to graben fills in the foreland of a mountain chain. In time, however, more detailed knowledge about the depositional conditions became available and the term molasse was thereafter restricted to a facies association of continental, mainly fluviatile and shallow-water marine, deposits that are encountered in the foreland of a mountain chain.

Modern usage of the term is associated with several conditions and molasse has thus become, as is the case

▲ **Figure 4.1** Schematic overview of the Cenozoic rock suites in the Alps. Sediments accumulated in the foreland of the Alps (Flysch basins, Molasse Basin and Po Basin) as well as in intramontane basins. Some of the sedimentary sequences developed continuously from the underlying Cretaceous sequences. Magmatic rocks include the Peri-Adriatic plutons and associated dyke swarms, as well as volcanic rocks in the northern foreland.

Figure 4.2 Flysch and 'Wildflysch' at outcrop.

(A) Regularly bedded sequence of turbidites within the Taveyannaz Formation of the Northhelvetic Flysch of the Central Alps. Photograph taken near Elm (Canton Glarus, Switzerland).

(A)

(B) Broken up pieces of limestone and sandstone in a dark shaly matrix in a 30-metre-thick layer of ultrahelvetic 'Wildflysch'. Photograph taken near Mellau (Vorarlberg, Austria).

(B)

(C) Detail of (B) showing chaotic internal structure.

(C)

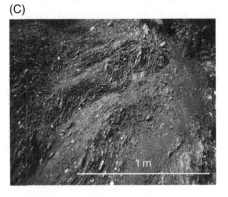

for flysch, a genetic term. These conditions can be summarized as follows:

- a continental–fluviatile and marine sequence of clastic sediments is present;
- the marine suites contain shallow-water deposits showing the effects of tides or waves;
- the continental–fluviatile suites contain 'fining-upward' cycles (Sohlbank

cycles), 'coarsening-upward' megacycles (prograding delta deposits) and fine clastics (floodplain deposits);
- the basin is in the foreland of an uplifting orogen;
- the basin is at times completely filled with sediments.

The photograph in Fig. 4.3A shows a succession of conglomerates that accumulated in the proximal part of

(A)

Figure 4.3 Molasse at outcrop.

(A) Regularly but poorly bedded sequence of conglomerates from the proximal part of the Rigi fan (Lower Freshwater Molasse). Photograph taken on the southwestern flank of Rigi (central Switzerland).

(B)

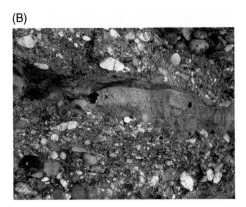

(B) Detail of (a) showing a lens of sandstone–mudstone. The large spectrum of pebbles in the conglomerate provides the strikingly distinct colours at outcrop.

the Rigi fan (Lower Freshwater Molasse, central Switzerland). The detail shown in Fig. 4.3B highlights the polymict character of these conglomerates along with a lens of sandstone and mudstone preserved within the coarse-grained fraction. With reference to age, the sediments in the Molasse Basin comprise the Paleogene and Neogene (Oligocene to Miocene).

4.2 Late Cretaceous and Paleogene Flyschs

An older series of flyschs is found scattered across the Penninic nappe system. In some cases, the flysch units are nappes in their own right with a basal shear zone. In other cases, there is a gradual stratigraphic transition from the underlying Mesozoic sediments. With very few exceptions, only approximate ages are known for these flysch sediments. The flysch basins evolved both from the Valais Trough and the Piemont Ocean. The palaeogeographical allocation of individual flysch units to these sea basins has been highly controversial for a long time and still has not been fully clarified. A variety of different aspects were used as criteria for this correlation: age, lithological composition, composition of the components in the coarse clastic rocks, the

direction of turbidity currents and the tectonic position in the nappe pile. Wildi (1985) provides an overview of the flysch basins that takes into consideration the palaeogeography at different points in time, the origin and composition of the clastic components and the flow directions within the basins.

The formation of the Cretaceous flysch basins is linked to the onset of subduction and closing of the Piemont Ocean and to the Eo-Alpine orogeny and/or the subsequent closure of the Valais Trough in the Paleogene. The basins were, at least in some cases, long, narrow and deep troughs. Turbidity currents, triggered at the basin margins, usually ran parallel to the basin axis in the centre of the basin. In other cases, accumulation of the clastic sediments occurred in a system of submarine fans. In all cases, these flysch sediments were deposited in a deep marine environment. Conclusions on the origin of the detrital material can be made based on the presence of heavy minerals (zircon, chrome spinel, garnet, etc.) in the arenites. Wildi (1985) demonstrated that zircon, tourmaline and TiO_2 groups predominate in the flysch basins of the Valais Trough and the Piemont Ocean in the Western Alps (Ligurian Basin), indicating an origin from the granitic rocks of the Western Alps. In contrast, garnet predominates in the flysch basins of the Piemont Ocean of the Central Alps, which probably stems from the Southalpine realm and the Bohemian massif. Finally, chrome spinel is typical for Late Cretaceous flyschs in the region of the Piemont Ocean, with the oceanic ophiolites being viewed as the source of the chrome spinel.

The flyschs that are allocated to the Piemont Ocean (or the Southpenninic nappes) include the Voirons, Gurnigel,

Schlieren and Wägital flyschs, the fly-schs of the 'Nappe supérieure' (Sarine, Dranses, Simme and Gets flyschs), which was thrust onto the Klippen nappe, as well as the 'Série grise' and the Arblatsch Flysch in the inner part of the Alps. In the case of the Arblatsch Flysch, Ziegler (1956) found a certain similarity with the Ruchberg Sandstone of the (Northpenninic) Prättigau Flysch. However, grain size decreases from south to north in the Arblatsch Flysch, which points to a source in the south, while the scource area of the Prättigau Flysch was located in the north, according to Nänny (1948). In addition, Thum & Nabholz (1972) noted that although the mean total carbonate contents are similar for the Ruchberg Sandstone and the Arblatsch Flysch, the Arblatsch Flysch differs from the Ruchberg Sandstone due to its considerable dolostone content. The allocation of the Arblatsch Flysch to the Piemont Ocean therefore appears more plausible, as is also indicated by the maps produced by Ziegler (1956). Both a North and a Southpenninic origin are considered for the Rhenod-anubian Flysch, but as the Briançon Rise, which separated the two, most likely disappeared towards the east to create a unified Penninic Ocean, the question is rather irrelevant. The Niesen and Prättigau flyschs are allocated to the Valais Trough as both share a Mesozoic substratum by a stratigraphic transition.

In order to gain spatial and temporal insight of flysch development and dis-tribution, Fig. 4.4 provides a schematic overview of the sedimentary sequences in the Southpenninic flyschs. The indi-vidual units are placed in their respec-tive palaeogeographical positions. It should be noted that the Voirons, Gurnigel, Schlieren and Wägital flyschs are very similar (Winkler et al. 1985) and grade into each other laterally from west to east and for this reason, among others, the Gurnigel Flysch has been selected as a representative example in Fig. 4.4: the age and formation condi-tions of which are particularly well known thanks to the work of van Stuijvenberg (1979). The four flyschs in the 'Nappe supérieure' (Gets, Simme, Dranses and Sarine flyschs) are sum-marized based on the publication by Caron (1972). It is unclear where exactly within the Southpenninic nap-pes the 'Série grise' in the Tsaté nappe (described by Marthaler 1984) and the Arblatsch Flysch (mapped by Ziegler 1956) should be allocated: in particular, the 'Série grise' has been subject to considerable metamorphosism and it is therefore impossible to meaningfully either determine its age or subject it to sedimentological analysis.

At the base of the **Gurnigel Flysch**, with a total thickness of about 800 metres, van Stuijvenberg (1979) iden-tified a Late Cretaceous formation (the Maastrichtian Hellstätt Formation) of turbidites with conglomerates, sand-stones and marls, but little limestone and shale content. Above this, there are Cenozoic units containing varying proportions of sandstone and shale. Palaeocurrents initially flowed in a westerly direction in the Cretaceous, and then mainly in a southeasterly direction in the Cenozoic. According to van Stuijvenberg (1979), the sedi-ments were deposited in deep sea fan, and a hinterland containing exposed granites and Mesozoic sediments must have existed as the source area. The fly-sch basin was about 40 kilometres wide, and the source area is considered to have been a belt about 75–100 kilo-metres wide, with a shelf area about 20 kilometres wide. Based on the

Flysch basins in the terminal stage of closure of the Piemont Ocean

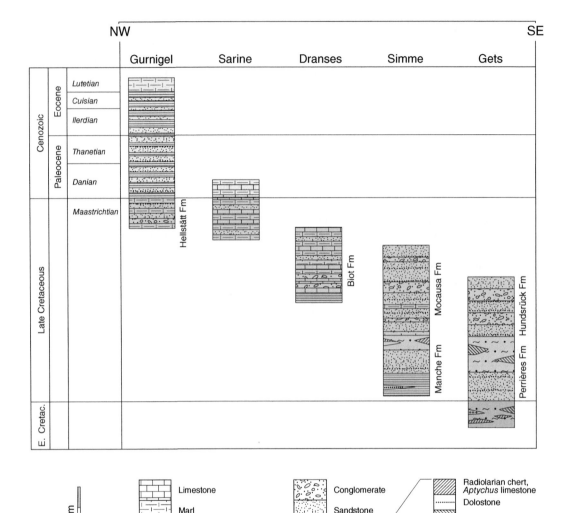

▲ **Figure 4.4**
Sedimentary sequence of the Southpenninic flyschs compiled after van Stuijvenberg (1979) and Caron (1972). Flysch sedimentation started in Cretaceous times. Cenozoic flysch deposits are present only in the Gurnigel and Sarine flyschs. Source: Adapted from Van Stuijvenberg (1979) and Caron (1972). Reproduced with permission of Swiss Topo.

information provided by van Stuijvenberg (1979), it appears reasonable to assume that the Briançon Rise was the supply area.

The **Sarine** and **Dranses flyschs** are calcareous flyschs containing sandstone–limestone–marl successions, but with occasional deposits of conglomerates. Fucoids (branched impressions of seaweed) are known from the Sarine Flysch, and feeding and grazing trace fossils in the Dranses Flysch were

responsible for the term 'helminthoid flysch' used in the past.

The **Simme Flysch** is characterized by: a basal sequence of shale that contains isolated conglomerate lenses and served as the detachment horizon for the Simme nappes; followed by a sandstone–shale succession and, above, by wildflysch, containing lenses of radiolarian chert, *Aptychus* limestone, dolostone and granite; above which is the Mocausa Formation, comprised of conglomerates

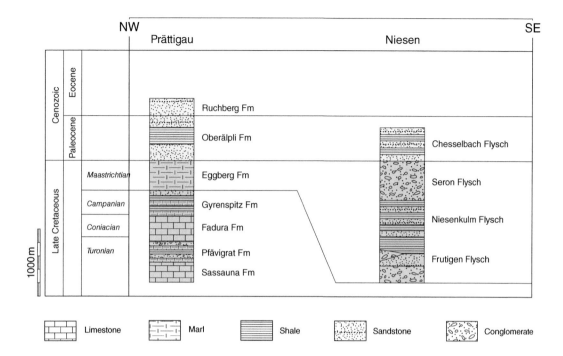

▲ **Figure 4.5**

Sedimentary sequence of the Northpenninic flyschs, compiled after Nänny (1948) and Ackermann (1986). The section Pfävigrat Formation to Eggberg Formation in the Prättigau Flysch is also known as 'Pre-flysch'. Source: Adapted from Nänny (1948) and Ackermann (1986).

and sandstones 300–400 metres thick, with a rise on the Adriatic continental margin in the southeast being indicated as the source area – as is also indicated by the lenses in the underlying wildflysch.

In the Early Cretaceous **Gets Flysch**, the oldest, portion is made up of shales and thinly bedded limestones, and within the flysch there are scattered layers of ophiolitic components and granites. A second horizon containing lenses of ophhiolitic components occurs above an overlying sandstone–shale succession with isolated conglomerate beds. The Late Cretaceous part comprises polymict conglomerates and quartzitic sandstones (the Hundsrück Formation).

The Northpenninic Prättigau and Niesen flyschs are Late Cretaceous–Paleogene in age and both succeeded a Mesozoic sedimentary sequence in the Valais Trough. In the **Prättigau Flysch** there is a gradual transition from the Bündnerschiefer Group to the flysch, and accordingly Trümpy et al. (1970)

proposed calling the Late Cretaceous formations (Pfävigrat, Fadura, Gyrenspitz and Eggberg in Fig. 4.5) 'pre-flysch', and limiting the use of the name Prättigau Flysch to the Paleocene–Eocene formations. Thick successions of limestones, marly and sandy limestones and marls predominate in the Cretaceous formations (Nänny 1948, Trümpy et al. 1970), with subsidiary polymict conglomerates also incorporated. Thum & Nabholz (1972) observed a very homogeneous heavy mineral spectrum across the entire suite, dominated by tourmaline and zircon, with apatite and rutile less abundant. Garnet is rare and chrome spinel is even rarer. The Paleocene to Early Eocene formatiosn (Oberälpli and Ruchberg, Fig. 4.5) exhibit greater flysch-like character and are composed of sandstones and shales. According to Nänny (1948), the palaeocurrents flowed north to south, and Thum & Nabholz (1972) found the heavy mineral spectrum unchanged.

The **Niesen Flysch** is composed of terrigenous sediments: thick layers of conglomerate are widespread and accompanied by alternating layers of turbiditic sandstone and shale. As for the Prättigau Flysch, the youngest unit, the Paleocene Chesselbach Flysch, also evolved from the Late Cretaceous flysch below it. Ackermann (1986) demonstrated that the hemipelagic pelites were deposited at depths below the CCD. The turbiditic successions in the older portion, the Frutigen Flysch, formed a debris fan fed from a submarine canyon with a southerly to southeasterly source area. Sediments from the continental slope predominate in the middle portion, the Niesenkulm and Seron flyschs, and palaeocurrents flowed mainly in a northerly direction. In contrast, the Paleogene Chesselbach Flysch represents a submarine fan with a source area situated in the north. According to Ackermann (1986), the predominant heavy mineral is tourmaline, with zircon and rutile present to a lesser extent, although apatite is more common in the Paleocene Chesselbach flysch. The pebble spectrum is rather reduced for the conglomerates in the older Frutigen Flysch, but contains sediments (in particular, dolostones and limestones), metamorphic rocks (namely, gneisses and mica schist) and igneous rocks (granite and rhyodacite), however, it becomes successively broader upwards, as is revealed, in particular, by the diversity of volcanics in the younger Seron Flysch. Ackermann (1986) interprets this decrease in the degree of maturity (or the increase in the prevalence of the more unstable components) as being due to more intensive tectonic activity in the hinterland, which successively exposed new rock types to erosion. The total change in the direction of the palaeocurrents in the Paleocene Chesselbach Flysch, and thereby also the supply area, is consistent with the change in the degree of maturity of the sediments, but in turn, the pebble spectrum again became more restricted.

In addition to the Southpenninic flyschs with a Cretaceous base, a different type of Cretaceous clastics must also be mentioned here, which were deposited by palaeocurrents from the uplifting Alps in a southerly direction in the Southalpine realm and are known as Lombardian Flysch of the Eastern Alps, which is composed of terrigenous sediments that built up a system of deep marine fans (Bichsel & Häring 1981). Strata underlying the Lombardian Flysch, as well as the younger sediments above the fans, are hemipelagic calcareous and argillaceous marls of different colours (Scaglia rossa, Scaglia variegata, Scaglia bianca and Scaglia cinerea). Deposition of the Scaglia occurred at water depths that were still above the CCD. Monotonic successions of turbidites dominated in the Lombardian fan system itself. At times, these turbidites were fed from deep marine canyons. In the Santonian, conglomerates with almost identical pebble spectra (crystalline rocks, dolostones, radiolarian cherts, pelagic and shallow-water limestones) were deposited at different locations. The directions of the palaeocurrents and the pebble types indicate that there was already a high zone in the north of the basin that was probably composed of Austroalpine and early formed Southalpine nappes.

4.3 Eocene–Oligocene Flyschs

A younger series of flyschs was deposited in the Helvetic and Dauphinois realms. Similar to the case in the Penninic nappe system, sedimentation occurred in long, narrow troughs that extended along the northern margin of the uplifting orogen. However, a phase of emersion and karstification preceded flysch sedimentation, in both, the Helvetic and Dauphinois realms, The subsequent Eocene neritic sediments, referred to as 'Neritic Eocene', therefore overlie the Mesozoic formations with angular unconformity (Herb 1988), or on a palaeorelief (Sinclair, 1997): at some locations, the entire Cretaceous was eroded during this phase of emersion. A typical sequence is a trilogy comprising neritic limestones at the base, followed by marl and then a thick suite of flysch made up of sandstones and shales, and it extends in age from the Eocene into the earliest Oligocene, but is extremely heterochronous. The diachronous evolution of the Eocene sediments was investigated in depth by Herb (1988) and Menkveld-Gfeller (1995). Marl containing *Globigerina* (Stad Marl) was deposited in slightly deeper water than the underlying neritic limestones and sandstones (Einsiedeln to Sanetsch formations in Fig. 4.6), which often contain *Nummulites* or *Assilinas*. There are local lenses of conglomerates (the so-called Wängen Limestone) in the Stad Marl that have been shed from synsedimentrary fault scarps, and these precede a turbiditic sequence of microbreccia, sandstone and shale, known as Northhelvetic Flysch, (Fig. 4.6). Three larger geographical complexes of **Northhelvetic Flysch** are exposed: Champsaur (Western Alps), Savoie-western Switzerland (boundary between Western and Central Alps) and the Glarus Alps (Central Alps). Figure 4.6 shows the vertical sequence that is typical for the Central Alps.

A property that stands out is the clear prevalence of volcanic detritus in the microbreccia of the conglomerates.

Detailed sedimentary and petrographic investigations conducted by Vuagnat (1952) showed that the presence of volcanic detritus was limited to the older strata within the flysch and that these give way to more sedimentary and granitic material towards the top. Andesitic detritus predominates in the Taveyannaz Formation and disappears gradually within the Elm Formation (called the Val-d'Illiez Sandstone in Vuagnat's 1952 publication). In addition, Vuagnat (1952) observed unaltered augite and hornblende in the lower portion of the Taveyannaz Formation, which leads to the conclusion that the volcanic debris was deposited shortly after the eruptions and transported over short distances only. Although the origin of this volcanic material is quite difficult to explain, the existence of a volcanic belt in the hinterland of the flysch trough could actually be expected at this time. However, none of the immediately adjacent tectonic units (Helvetic and Penninic nappes) exhibit any traces of volcanic vents. Late Eocene magmatites are found only further to the south, along the Peri-Adriatic fault system, in the form of granodioritic and tonalitic intrusives or andesite dykes. Vuagnat (1952) also detected diabase, chloritite and albitite as components in the Elm Formation, in addition to the andesites. These probably stem from the ophiolites in the Penninic nappes. Radiolarian cherts, which start to appear from the younger Taveyannaz sandstones onwards, and unmetamorphosed dolomitic limestone and gneisses probably stem from the border region between the Penninic and Austroalpine nappe systems. Granitic components are present in the oldest portion of the Taveyannaz Formation, then increase in prevalence in the younger portion and are widespread in

Flysch in foreland basins

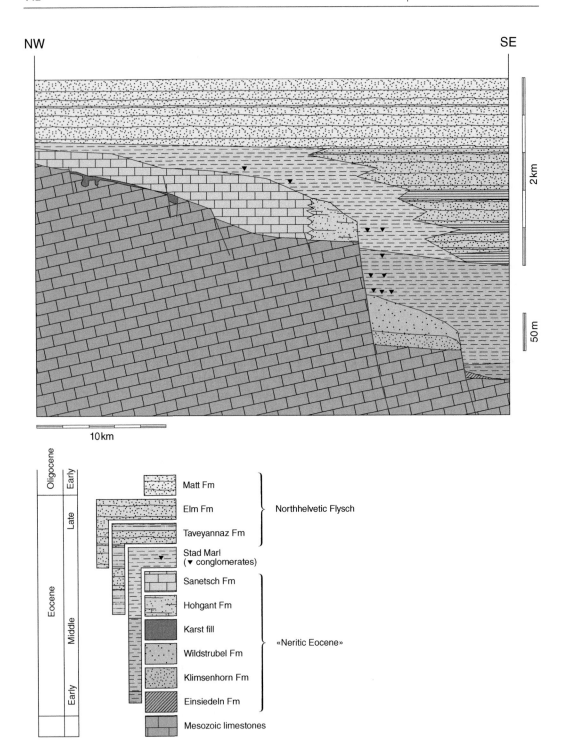

the Elm Formation. Their origin probably lies in the Austroalpine, possibly Penninic, crystalline basement. Siegenthaler (1974) investigated the directions of palaeocurrents in the Northhelvetic Flysch of the Glarus Alps and demonstrated that the palaeocurrents in the Taveyannaz Formation were first towards the north and then towards the northeast. A fairly uniform east-northeast-directed flow pattern, that is, parallel to the trough, was then present in the Elm Formation. Siegenthaler (1974) suspects that the larger volcanic components were deposited directly into the turbidity currents or into their catchment area, as these large components float chaotically in the otherwise graded turbidites.

A change occurred at the transition to the youngest formation, the Matt Formation. The components that are present stem from granites, gneisses, radiolarites and limestones, but components from volcanics are completely absent (Vuagnat 1952). Furthermore, the palaeocurrents that had been parallel to the trough have a radial pattern, indicating development of fluviatile deltas and a transition from flysch to molasse sedimentation. This change occurred at the Eocene–Oligocene boundary in the eastern Central Alps, but did not take place until the Middle Oligocene in the western Central Alps.

An absence of three groups of rocks in the spectrum of components is of note in the Northhelvetic Flysch: (i) Alpine metamorphosed sediments containing garnet, etc. (such rocks were present deep beneath the palaeosurface); (ii) sediments from the Northhelvetic facies zone (no palaeocurrents from the north); (iii) older Northhelvetic Flysch sediments (the incoming nappes had already overthrust the internal, older portions of the flysch trough).

A very similar succession is found in the Western Alps (see Sinclair 1997), but the neritic limestones were deposited on a palaeorelief containing gorges up to 200 metres deep, which are filled with conglomerates (the so-called Poudingues d'Argens), sandstones and clays (Sinclair 1997, Lickorish & Ford 1998 and references cited therein). Water depths of between 500 and 100 metres were estimated for the sedimentation of the Stad Marl that overlies the neritic limestones (Sinclair 1997, Lickorish & Ford 1998 and references cited therein), and the Champsaur and Annot sandstones are included in the flysch sandstones that are found at the top. The Champsaur Sandstone of the southern Subalpine chains is less andesitic, but rather is more ophiolitic in character. Even further to the south, in the Maritime Alps and the Haute Provence, the Annot Sandstone contains a broad spectrum of components that stem from the surrounding crystalline rocks and their sedimentary cover. The directions of palaeocurrents in the Western Alps indicate supply areas in the southeast, that is, the uplifting Alps, and in the south, in Corsica–Sardinia (Sinclair 1997, Lickorish & Ford 1998 and references cited therein) or the Provence platform.

4.4 Oligocene–Miocene Molasse in the Northalpine Foreland Basin

The Northalpine foreland basin extended along the edges of the Central and Eastern Alps during the Oligocene and the Miocene. It became narrower in the west, at the transition to the Western Alps and was wider towards the East. The sedimentary fill in the basin exhibits large natural lateral fluctuations. In spite of this, the strata can be categorized into

◀ **Figure 4.6** The Cenozoic sedimentary sequence of the Helvetic–Dauphinois realm may be subdivided into a 'trilogy': Neritic Eocene beds overly karstified Mesozoic limestones with an angular unconformity; the Neritic Eocene beds are in turn overlain by hemipelagic marls interspersed with conglomerates; flysch sediments overlie the marls. The boundaries between the three units of the 'trilogy' are highly diachronous. Source: Adapted from Herb (1988), Menkveld-Gfeller (1995) and Kempf & Pfiffner (2004).

four groups: Lower Marine Molasse, Lower Freshwater Molasse, Upper Marine Molasse and Upper Freshwater Molasse. The names themselves indicate the changes between marine and continental conditions.

The molasse sediments mainly overlie a marine Mesozoic suite, the top edge of which was affected by Early Cenozoic karstification. This karstification occurred when this portion of the European continent was uplifted above sea level, probably as a result of the foreland bulging upwards during the growth of the Alpine orogen. A basin formed between this bulge and the uplifting Alps, the Molasse Basin, which was successively filled and shifted towards the north over time. The oldest group, the **Lower Marine Molasse** (LMM), was deposited in the internal, southernmost region of the basin and originated from the precursor basin, the Northhelvetic Flysch. Accordingly, turbidite deposits shed from the slope of the uplifting Alps in the south are prevalent in the lower portion. The thickest LMM unit comprises a clay–marl suite that served as a detachment horizon during later formation of the Subalpine Molasse. The upper portion of the LMM comprises coastally dominated sandstone suites (Diem 1986), which indicates that the basin was completely filled. Subsidence of the LMM basin started in the east, from the Vienna Basin (Kempf & Pross 2005).

Fluviatile–continental sediments characterize the subsequent **Lower Freshwater Molasse** (LFM)and the change from a marine to a continental environment coincides with a global fall in sea level. Multiple fans built up next to each other in the LFM (Berger 1996, Schlunegger et al. 1996, Kempf et al. 1999), in which proximal coarse clastic conglomerates, called 'Nagelfluh' in

Switzerland, were shed from the south, that is, close to the edge of the Alpine mountain chain at that time. The pebble spectrum in these conglomerates corresponds to the uppermost nappes in the Alpine hinterland, the Austroalpine nappes, which are composed of Mesozoic sediments and a crystalline basement: accordingly, the older 'Nagelfluh' contains a spectrum of components including dolostones, limestones, sandstones and radiolarian cherts. Crystalline rocks appear only as pebbles in the younger 'Nagelfluh', in which the pebble spectra reflects the geological composition of the source area in the hinterland fairly directly. Sandstones and marls predominate in the distal region of the fans, where the transport power of the ancient rivers was weaker, with marls also deposited in the floodplains next to the fans. The thickness of the fans also decreases with decreasing grain size, from 3 to 4 kilometres in the proximal region to about 300 metres in the distal region.

Over time, the fans prograded into the foreland basin, leading to the deposition of coarse clastic proximal sediments, in a vertical profile, over the older distal fine clastic sediments, forming what is known as 'coarsening-upward megacycles'. Overall, the LFM displays a fairly complex pattern of differing, juxtaposed sediments at all points in time.

A narrow strait, the so-called Burdigalian Strait, characterized the **Upper Marine Molasse** (UMM). This strait was about 100 kilometres wide and connected the Pannonian Basin (Para-Tethys) with the Mediterranean (Tethys). Sandstones deposited in this strait exhibit a coastally dominated facies (Allen et al. 1985), with typical indications of currents (cross-stratification), tidal currents (flood channels and megaripples), shell deposits (containing marine seashells) and fossilized dessication fissures and

Molasse basin: Foreland basin with alternating marine and continental conditions

bird tracks (tidal flat environment). The most prominent UMM layer in the Swiss part of the Molasse Basin includes the St Gallen and Lucerne formations and the Bern Sandstone, which were deposited mainly in a belt that was more external than the LMM. In the south of the Burdigalian Strait, smaller fans developed that were shed from the Alpine hinterland. These fans built up thick suites of conglomerate that, in turn, give way to sandstones and marls distally and laterally. The basin was thereby overfilled and became dry land. The spectrum of components in the 'Nagelfluh' now also included rocks from deeper units in the partially eroded Alpine nappe pile (Penninic nappes with crystalline rocks).

As for the LMM, the fans of the UMM also prograded out into the foreland basin to form the **Upper Freshwater Molasse** (UFM), but the distal portion of the basin now drained towards the west but no longer towards the east, which is revealed by a narrow complex of sandstones containing mica, the so-called 'Glimmersandrinne'. The 'Glimmersandrinne' can be traced from far in the east, the Bohemian massif, to the western end of the Molasse Basin and indicates west-directed palaeo-currents. The pebble spectrum in the conglomerates of the UFM is similar to that in the LFM (Kempf et al. 1999). Crystalline components from the Austroalpine and Penninic nappe systems make up less than 20% of the spectrum and decrease towards the top, and pebbles from the sedimentary rocks stem from the Austroalpine, Penninic and Helvetic nappe systems. Similar to the case for the LFM, this pebble spectrum also reflects the geology of the source area, with the lowest unit in the Alpine nappe pile, the Helvetic nappes, being the last to be exposed to erosion.

The 'Appenzell granite', a sedimentary complex in the lower portion of the Hörnli fan in eastern Switzerland formed due to a unique event. The entire suite of 'Appenzell granite' is interpreted as a deposit from numerous mud flows, which emanated from an enormous rock avalanche in the Austroalpine realm of the Alps or at their external edge (Bürgisser et al. 1981): the deposits from the rock avalanche were transported into the foreland by mud flows, but were then reworked and transported as far as the distal portion of the Hörnli fan in the form of fluviatile gully deposits, as mass flow or through additional fluviatile erosion.

The sedimentary suites in the eastern part of the Molasse Basin (see Fig. 4.7), in Bavaria and Austria, exhibit more marine conditions by comparison with the western portion. A detailed description of the sediments in the eastern Molasse Basin is given in Oberhauser (1980). In the Late Eocene and Early Oligocene, that is, contemporaneous with the sedimentation of the Northhelvetic Flysch, borehole records indicate that *Nummulite* limestones, marl and sandstone were deposited (Steininger & Wessely 2000): the thick Early Oligocene argillaceous marl suite is comparable to the LMM in the west; in contrast to the west, there is a marine suite of turbidite deposits (known as the Puchkirchen Group) above this, which at its base has a submarine erosional surface; pockets of conglomerates are found within this sandstone fan, the so-called 'Augensteine', that were shed from the south, from the Alps. At this time, there was a marine connection extending from the eastern Molasse Basin, towards the south from Salzburg, into the future Alps in the direction of the Inn Valley and, according to Wagner (1996), possibly into the Po Basin.

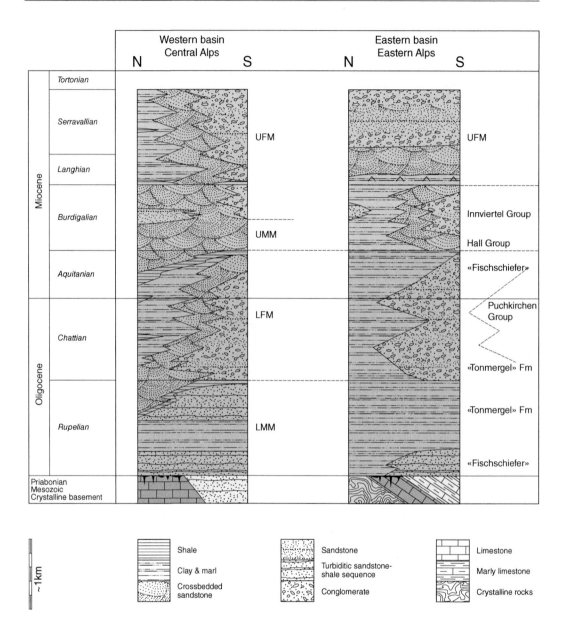

▲ **Figure 4.7** Schematic
overview of the sedimentary
sequence of the Molasse
Basin. Source: Adapted from
Schlunegger et al. (1996),
Kempf et al. (1999),
Steininger & Wessely (2000)
and Oberhauser (1980).

Marine pelagic marls characterize the sediments of the Early Miocene (Burdigalian) in the eastern part of the basin, when the UMM in the west exhibited a marine connection at this time towards the west via the Burdigalian Strait. At the southern edge of the eastern Molasse Basin: fans containing conglomerates from the Alps were also being built up; coastal sands were deposited at its northern transition to the pelagic marl; sands were also deposited to the north of the marls, which, however, were shed from a source in the north, from the Bohemian massif. From the Langhian onwards, contemporaneous with the UFM, terrestrial clastics were also deposited in the eastern part of the basin, as were coal-bearing freshwater sediments. Even so, there were several short periods (17, 15, 13 and 11 million years before present) of marine ingression from the east. In contrast to the western Molasse Basin, in the extreme east, drainage occurred in an easterly direction into the Danube system (Steininger & Wessely 2000), when the watershed was to the south of the spur of the Bohemian massif.

4.5 Oligocene–Pliocene Sediments in The Po Basin

A suite of sediments several kilometres thick was deposited in the Po Basin on the south side of the Alps. The edges of this basin were affected by the tectonic events taking place in the adjacent mountain chains, the Alps, Dinarides and Apennines, and the basin content reflects these events.

There is a thick suite of Eocene to the Late Miocene marine pelites and siltites in the centre of the basin, known only from boreholes (Pieri & Groppi 1981, Gunzenhauser 1985, Garzanti & Malusà 2008) (see Fig. 4.8), including pockets of Late Eocene neritic limestone (Ternate Formation in Fig. 4.8) and isolated thin beds of sandstone and conglomerate in the fine clastic sequence of the Miocene.

A completely different spectrum of sediments appears towards the Alps, in the form of the Gonfolite Lombarda Group. In the Early Oligocene, marine marls, pelites and turbiditic sandstones (Chiasso Formation in Fig. 4.8) are still present, but in the Late Oligocene, in the Chattian, thick suites of conglomerates (Como and Lucino formations in Fig. 4.8) suddenly appear. These conglomerates are a total of about 3.6 kilometres

thick and can be followed along the edge of the Alps as a belt that is 30 kilometres wide.

According to Gunzenhauser (1985), the conglomerates were deposited from the north, the Alps, in a system of submarine fans. The sudden onset of deposition in the Chattian can be explained in two ways, argues Gunzenhauser (1985): (i) through the dramatic drop in sea level that occurred at this time and subjected the debris on the now exposed coastal strips to erosion, and (ii) through the uplift in the hinterland, where an intensive phase of nappe stacking and thrusting along a steeply dipping thrust fault (the Insubric Fault) was taking place at this time. The Chiasso Formation underlying the Gonfolite Lombarda Group rests discordantly at angles of 10–20° on its substratum, which varies in composition laterally. These circumstances indicate a pre-Late Eocene deformation phase, which also can be assumed in the Southern Alps, where the Adamello intrusion cuts through a thrust fault (Brack 1981). In this context, it is also interesting to note that pebbles from the Bregaglia intrusion can be found in the conglomerates of the Como Formation. This intrusion occurred around 30 million years ago at a depth of approximately 5–6 kilometres, while the age of the conglomerates is about 25 million years, which indicates an intensive phase of exhumation at rates of over one millimetre per year.

Younger sediments, including the Pontegana conglomerate of the earliest Miocene (Messinian) and the Pliocene Balerna Shale, discordantly overlie the Miocene deposits. In the basin itself, Pliocene and the Pleistocene deposits are present in the form of a fine clastic suite.

Po Basin: Foreland basin shared by Alps, Dinarides and Apennines

Deep water clastics follow marine pelites and siltites

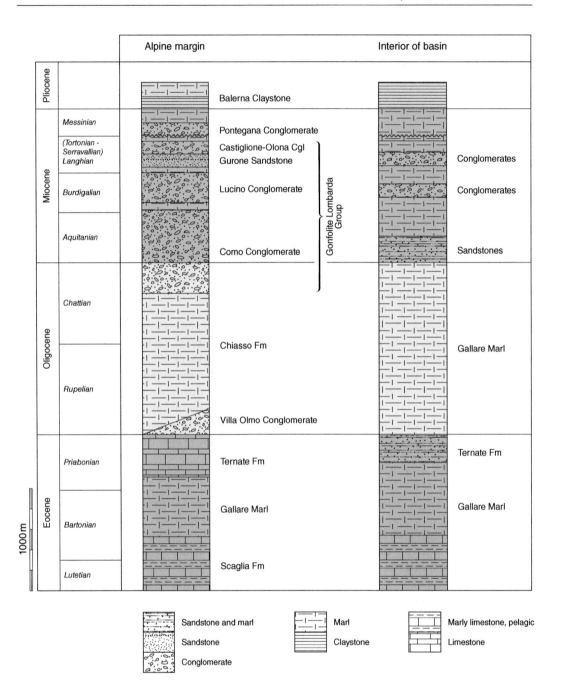

▲ Figure 4.8 The sedimentary sequence of the Po Basin. Source: Adapted from Gunzenhauser (1985) and Garzanti & Malusà (2008).

4.6 The Jura Mountains

The Cenozoic rocks in the Jura Mountains are often exposed in only small, isolated outcrops. Three types of sediments can be distinguished: the first comprises deposits in the distal region of the Molasse Basin, including, for example, UMM sandstones and fresh-water limestones and marls of the LFM and UFM; the second are local deposits of conglomerates with pebbles of limestone (the so-called gompholites); the third includes the 'Nagelfluh' and sands that were shed from the north, from the Black Forest and Vosges. The clastic deposits shed from the Vosges are mainly found in the Ajoie and Laufen basins. The clastic deposits shed from the Black Forest include the conglomerates in the Delémont Basin and in the region of Basel, Aargau and Randen, and also include the Wanderblock Formation (see Kemna & Becker-Haumann 2003). According to Kälin (1997) all these conglomerates were deposited in the Middle Miocene; the conglomerates in the Ajoie are the exception, for which a Late Miocene age has been determined. As the Cenozoic sediments are directly related to the folding of the Jura Mountains, they will be discussed in more detail in Chapter 6.

4.7   Intramontane Basins

From the Miocene onwards, two intramontane basins formed at the eastern end of the Eastern Alps, the Vienna Basin in the north and the Styrian Basin in the south (see Steininger & Wessley 2000 and Hamilton et al. 2000 for a detailed discussion).

In the case of the **Vienna Basin**, subsidence started in the Early Miocene (Burdigalian) in association with syn-sedimentary normal faulting. Subsidence was complete by the Late Badenian (14–13 million years ago) and amounted to 2000 metres. Overall, however, the basin had subsided by over 5000 metres. In the Early Miocene, marine conditions prevailed in the northern part of the basin, with deposits of sandstone and marl. In the southernmost part, fluviatile conglomerates

accumulated that were shed from the southwest, from the Alps. Marine sedimentation conditions predominated in the Middle Miocene, ranging from deep marine (with marl and sandstone) to neritic (with Lithothamnia limestones and localized conglomerates). Conditions increasingly turned brackish from the end of the Middle Miocene and the basin was filled up with freshwater deposits in the Late Miocene.

In the **Styrian Basin**, subsidence did not start until the early Middle Miocene. A syn-rift sequence was deposited that contained lacustrine sediments, thick fluviatile clastics and marine sediments, and a suite of andesitic volcanic rocks that erupted 17–13 million years before present. Marine post-rift sedimentation characterizes the Middle and Late Miocene, and limnic–fluviatile sediments mark the completion of the basin fill at the end of the Miocene. Overall, about 4000 metres of sediments were deposited in the Styrian Basin prior to inversion in the Pliocene. In addition to subsequent erosion, an eruption of a second suite of basaltic volcanic rocks occurred.

4.8   Plutonic and Volcanic Rocks

Along the Peri-Adriatic fault system, an entire series of larger and smaller plutons have intruded the rocks immediately next to the faults. In addition, numerous dyke swarms are exposed in a broad belt along this fault system. Rosenberg (2004) reviews this magmatism and, in particular, also discusses the mechanisms of emplacement of the plutons. The following description is based on this publication and the references cited therein. The regional expansion of these Cenozoic igneous

▶ **Figure 4.9** Tectonic map of the Alps showing the distribution of the Peri-Adriatic plutons, the associated dyke swarms and the volcanic rocks in the Northalpine Foreland. Source: Adapted from Rosenberg (2004).

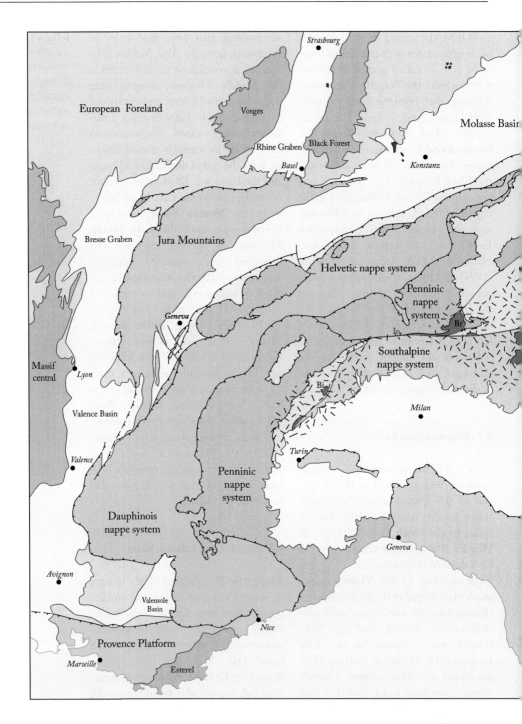

Strasbourg

European Foreland

Vosges

Molasse Basin

Rhine Graben   Black Forest

Basel

Konstanz

Bresse Graben   Jura Mountains

Helvetic nappe system

Penninic
nappe
system

Geneva

In.

Br

Massif
central

Lyon

Southalpine
nappe system

Bi

Milan

Valence Basin

Valence

Turin

Penninic
nappe
system

Dauphinois
nappe system

Genova

Avignon

Valensole
Basin

Nice

Provence Platform

Marseille

Esterel

rocks and their relationship to the Peri-Adriatic fault system is revealed in Fig. 4.9, which is based on Rosenberg (2004). According to Rosenberg (2004), emplacement of the **plutons** occurred in a transpressive regime along the Peri-Adriatic fault system and during a short period between 42 million years

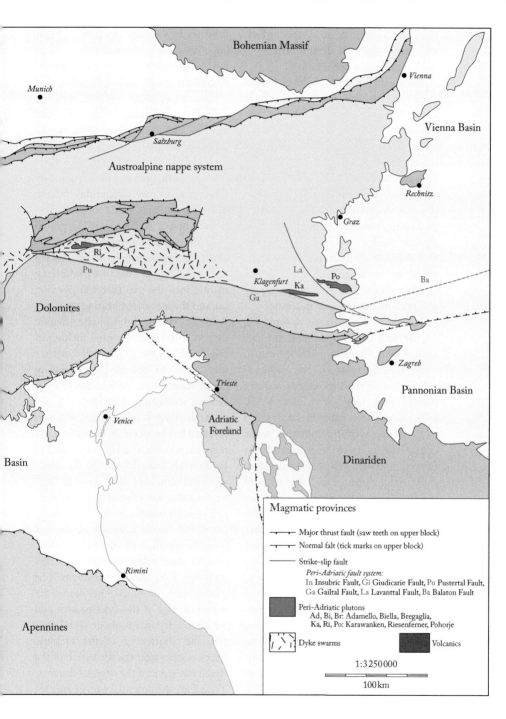

Magmatic provinces

- ⊥⊥⊥ Major thrust fault (saw teeth on upper block)
- ┬┬┬ Normal falt (tick marks on upper block)
- —— Strike-slip fault
  Peri-Adriatic fault system:
  In Insubric Fault, Gi Giudicarie Fault, Pu Pustertal Fault,
  Ga Gailtal Fault, La Lavanttal Fault, Ba Balaton Fault
- ▬ Peri-Adriatic plutons
  Ad, Bi, Br: Adamello, Biella, Bregaglia,
  Ka, Ri, Po: Karawanken, Riesenferner, Pohorje
- ⬚ Dyke swarms   ▬ Volcanics

1:3 250 000

100 km

(Middle Eocene) and 28 million years (Early Oligocene) before present, but is delimited to 34–28 million years before present (Early Oligocene) if the southern portion of the Adamello Pluton is excluded. This implies almost synchronous intrusion over a distance of 700 kilometres along the entire length of

▶ **Figure 4.10**

Cross-section through the
Bregaglia Pluton. Source:
Rosenberg (2004).
Reproduced with permission
of John Wiley & Sons.

*Peri-Adriatic
plutons*

the Peri-Adriatic fault system, which was active through dextral strike-slip motion and steep upthrusting.

All of the plutons belong to the calcareous alkaline series and are of I-type (Rosenberg 2004), that is, the product of intrusive melts. The isotope ratios exhibit a mixture of mantle and crustal signatures: mantle melts mixed with partially melted mafic lower crust, completed by fractionated crystallization, are responsible for the formation of dioritic, tonalitic, granodioritic and granitic melts, and accordingly, the Peri-Adriatic plutons are mainly composed of tonalite and granodiorite. Gabbro, diorite and granite are usually also present, but in lesser quantities.

Based on the extreme isotopic Cr, Ni, Sr and Nd enrichment, we can assume that the melts did not form in convection currents of the asthenosphere, but in the lithospheric mantle, namely, at a depth of at least 40–50 kilometres. The magma can be assumed to have risen over 30 kilometres, as the uppermost portions of some of the plutons solidified at a shallow depth of 5–10 kilometres. The emplacement mechanisms for the plutons vary from one location to another (see Rosenberg 2004 and references cited therein): caldera subsidence (Biella Pluton), ballooning (Adamello and Bregaglia plutons), stoping (Adamello Pluton) and updoming (Riesenferner and Karawanken plutons) have been proposed as emplacement mechanisms. Using the Bregaglia intrusion as an example, Fig. 4.10 shows the geometric relationships between the plutons and the country rock.

Dykes are similar in composition to the plutons, and their distribution in a belt of up to 50 kilometres or more wide on both sides of the plutons leads us to the assumption that the melts in the

lithospheric mantle covered a fairly large area and were not simply located immediately below the plutons. According to Rosenberg (2004), this circumstance, taken together with the contact relationships between the plutons and the country rock, can be explained only by a concentrated rise of the magma along the active Peri-Adriatic fault system, but the question remains as to why differential melting occurred in the first place. Von Blanckenburg & Davis (1995) postulated that when the subducted portion of the European plate broke off, this led to upwelling of the asthenospheric mantle into the gap that was opening up, and the associated heating up of the Adriatic plate's lithospheric mantle then resulted in differential melting in the vicinity of the gap, but this mechanism fails to provide a convincing explanation for the distribution of dyke swarms across large distances.

Andesitic volcanic activity, which is revealed by the spectrum of clastic components, was discussed in the context of the Northhelvetic Flysch. As the flysch sedimentation and dyke swarms may be the same age (due to their similarities in composition, many of the dykes are dated to the Oligocene), we can pose the question of whether the dyke swarms were volcanic vents associated with the **Taveyannaz volcanism**. The present-day distance between the northern edge of the dyke swarms and the flysch deposits is about 70 kilometres, but shortening of the Helvetic nappe system after the formation of the flysch has reduced the original distance. Therefore, the potential supply area for the andesitic material lies at a significant distance from the flysch basin, and we must continue to assume that the Taveyannaz volcanism occurred to the southwest of the flysch basin, but to

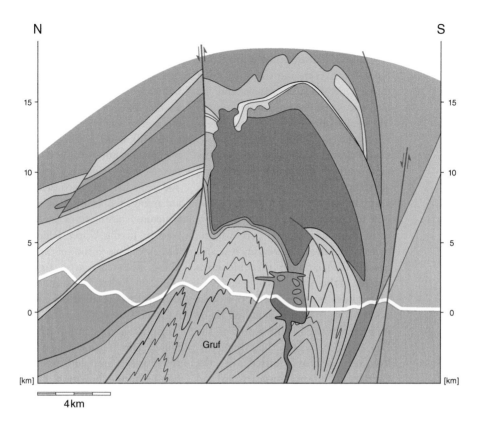

4 km

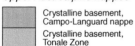

Cenozoic intrusives

Granodiorite
Tonalite   } Bregaglia
Granite   Novate

Upper Austroalpine nappes

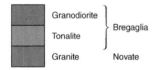

Crystalline basement,
Campo-Languard nappe
Crystalline basement,
Tonale Zone

Lower Austroalpine nappes

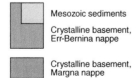

Mesozoic sediments
Crystalline basement,
Err-Bernina nappe

Crystalline basement,
Margna nappe

Southalpine nappes

Crystalline basement,
Upper Orobic nappe

Upper Penninic nappes/Piemont Ocean

«Bündnerschiefer», Avers nappe

Ophiolites, Malenco-Forno-Lizun nappe

Ophiolites, Platta nappe

Middle Penninic nappes/Briançon Rise

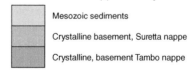

Mesozoic sediments

Crystalline basement, Suretta nappe

Crystalline, basement Tambo nappe

Lower Penninic nappes/Valais Trough

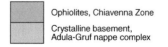

Ophiolites, Chiavenna Zone

Crystalline basement,
Adula-Gruf nappe complex

the northwest of the dyke swarms. These volcanoes were located on the now eroded Austroalpine nappes, but the volcanic vents must also have cut through the Penninic nappes and the Helvetic nappe system that (even then) lay below. Based on cross-sections reconstructed for the time (see Fig. 6.7), these volcanic vents could occur in the most southerly region of the Aar massif, an area that is now beneath the Gotthard massif and is thus not accessible for direct observation.

In addition to the peri-Adriatic plutons and the dyke swarms it is also necessary to consider the **volcanism outside of the Alps**, which forms a volcanic belt in central Europe that extends over several hundred kilometres from the Massif central to the Eger Graben outside the Alpine arc. Volcanic activity started in the Cretaceous (110 million years ago) and continued into the Holocene (11 000 years ago; see Franke 1992). Although some volcanoes, for example, Kaiserstuhl in the Upper Rhine Graben, occur in rift structures, Franke (1992) states there is no correlation between rifting and volcanism if the entire volcanic belt is considered, and this volcanism therefore should be interpreted geochemically as intraplate volcanism. The magmatic activity can probably be attributed to the process of asthenosperic upwelling, and the diatremes interpreted as explosive eruptions that originated at depths of over 100 kilometres and cross-cut the entire crust.

According to Keller (1984), volcanic activity peaked between 15 and 7 million years before present (Middle to Late Miocene) in the Swabian Alb (southern Germany), just adjacent to the western Molasse Basin in the north. The olivine melilitites found there can be interpreted as undifferentiated primary magmas (Keller 1984 and references therein) that formed at great depth. In contrast, the widespread phonolites (9.5 to 7 million years old) should be regarded as fractionated melts. The nappe tuffs (15–12 million years old), which include the vent fills, contain mantle xenoliths and also formed at great depth. A total of over 300 vents has been recorded and these are arranged as a band in a 'graben-like structure', starting from the Kaiserstuhl volcano in the Upper Rhine Graben and extending to the southeast (Keller 1984). Kaiserstuhl is a large volcanic complex (Keller 1984 and references therein) that is situated approximately over the Upper Rhine Graben mantle bulge. Primary mantle melts are also found here (olivine nephelinites and tephritic essexites), although fractionation led to the formation of phonolites (differentiated tephrite melts) and, finally, to bergalites and carbonatites.

4.9 Tectonic and Palaeogeographical Evolution

The Cenozoic sediments in and around the Alps were deposited in localized sedimentary basins, the location, shape and depth of which changed dependent on the tectonic processes during the formation of the Alps. The evolution of these sedimentary basins is best explained based on a chronological sequence of palaeogeographical maps, which are associated with a degree of uncertainty becuase the original relative position of individual crustal blocks is known only approximately, and in some cases must be estimated. The present-day outlines of the more important crystalline blocks, that is, external massifs and Alpine crystalline nappes, are indicated as points of reference in all maps and the movements that occurred can be deduced from their relative positions. These palaeogeographical maps were reconstructed for 90 (Late Cretaceous/Turonian), 55 (Early Eocene/ Ypresian), 35 (Eocene–Oligocene/ Priabonian–Rupelian boundary) and 25 million years ago (Late Oligocene/ Chattian).

Figure 4.11 shows the palaeogeographical situation in the **Turonian** (Late Cretaceous), 90 million years ago. At this time, pelagic limestones were deposited on the European shelf in the Northalpine foreland and the Helvetic and Dauphinois realms. In the southwest, the Mesozoic sediments were deformed by a north–south directed compression that was linked to the formation of the Pyrenees. This is indicated symbolically in Fig. 4.11 by a north-vergent thrust fault at the northern margin of the Provence Platform. The earlier subsidence of the Vocontian Trough was probably no longer active at this time, but the structure of the Penninic realm persisted in the types of sediments. The thick, sometimes sandy and marly limestones and marls of the Prättigau Flysch (or the 'pre-flysch') and of the Rhenodanubian Flysch were deposited in the Valais Trough. As mentioned previously, these sediments constitute a transition between the Jurassic–Cretaceous Bündnerschiefer Group and the Paleogene Prättigau Flysch. In contrast, pelagic marly limestones were deposited on the Briançon Rise. The older Southpenninic flysch basins formed in the region of the Piemont Ocean in front of the advancing Alpine nappes. The palaeocurrents in the Gets, Simme and Arblatsch flyschs came from the southeast, from the uplifting Alps. In the case of the Northpenninic Rhenodanubian Flysch,

palaeocurrents came from the northeast. Clearly, the Briançon Rise had subsided at this time and was completely flooded, such that it could not act as a source area for detritus. A series of small intramontane basins formed within the uplifting orogen at this time, the Gosau Basins. Debris from immediately adjacent areas accumulated in these basins. The structural relationships make clear that this sedimentation occurred in parallel with active thrust tectonics (Ortner 2001), a point that will be considered later. The Lombardian Flysch was shed from the south side of the orogen. In this case, the Alps were the source area. The Austroalpine region in Fig. 4.11 had been more or less entirely exhumed in the Cretaceous by the orogeny and subjected to subaerial erosion. This orogen was a result of convergent plate movement between the Eurasian and African or Adriatic plates, as will be discussed later. During its uplift, the Piemont Ocean was subducted under the Adriatic continental margin. Dating of high-pressure metamorphism in Austroalpine rocks provides evidence that, 90 million years ago, rock units had already sunk to depths of over 20 kilometres in the subduction zone (Oberhänsli et al. 2004, Berger & Bousquet 2008). The subduction zone is indicated symbolically in Fig. 4.11 by the thrust fault at the boundary between the (remainder of) Piemont Ocean and the Austroalpine nappe system.

By the Early Eocene, 55 million years ago, the distribution of land and water in the Alpine region had changed completely. Only one isolated basin remained between the northern foreland and the uplifting Alps, as is indicated in Fig. 4.12, which comprised parts of the past Northpenninic Valais Trough and the

▶ **Figure 4.11**
Palaeogeographical map of the future Alpine realm in the Turonian (90 Ma). The present-day outcrop shapes of the various basement units are given as a frame of reference. The Southpenninic flysch basins were narrow troughs straddling the active plate margin where the Piemont Ocean plunged beneath the Adriatic plate. Evidence from the Gosau basins and the Lombardian Flysch indicates the presence of a mountain range consisting of Austroalpine units: arrows indicate current directions.

Closure of the Piemont Ocean

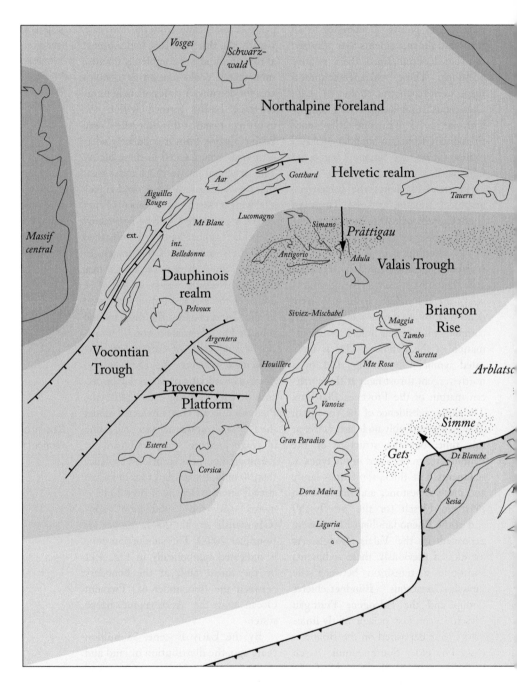

Helvetic and Dauphinois nappe systems. Two Northpenninic flyschs, the Prättigau and the Niesen flyschs, were deposited in the part of this basin closest to the Alps. The limestones of the Einsiedeln Formation were deposited on the Helvetic shelf in the northwestern part of the basin (Kempf & Pfiffner 2004). Both parts of the basin were bounded by a rises, from

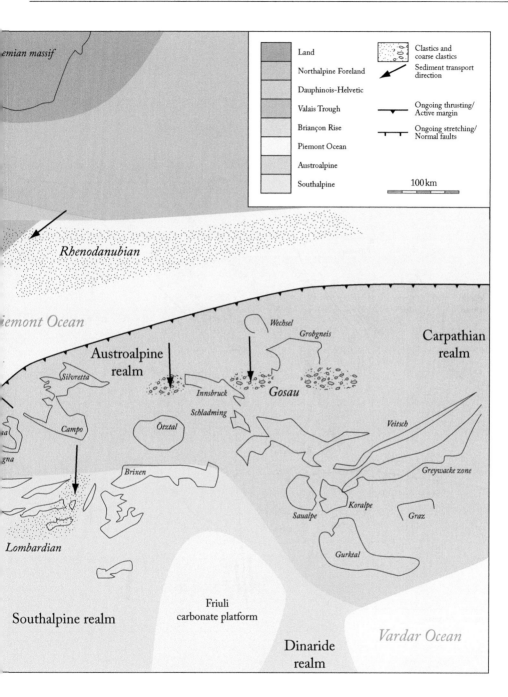

where the Niesen Flysch (Ackermann 1986) and Prättigau Flysch (Nänny 1948) were shed. Conglomerates (Argens Conglomerates) that represent palaeovalley fills are observed at the southwestern end of the basin (Sinclair 1997, Lickorish & Ford 1998). According to Lickorish & Ford (1998), the clastics were shed from the south, in the ancestral

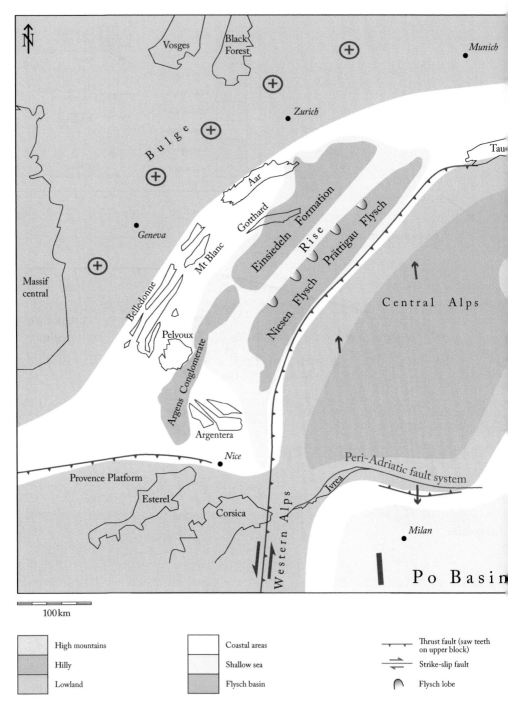

Western Alps of Corsica. The neighbouring extension to the Pyrenees,
the Provence Platform, is also a
potential source area. In the northeast,
the flysch basin must be underneath
the present-day Austroalpine nappes,

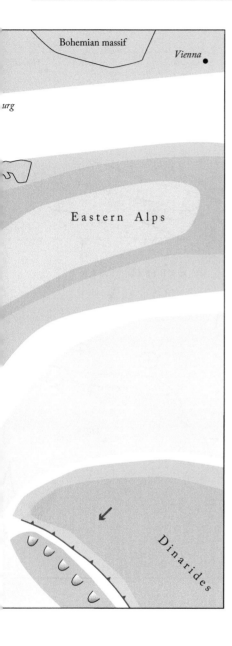

Relative motion between Adria and Europe

Local nappe-transport direction

⊕ Uplift/erosion

the Po Basin, as is indicated by the marly limestones of the Scaglia. Flysch sediments were deposited in the Po Basin by palaeocurrents originating in the Dinarides (Doglioni & Bosellini 1987). South-directed thrust faults were possibly still active in the Early Eocene in the southalpine realm of the Alps (Schönborn 1992). Sinistral strike-slip motion must be assumed in the ancestral Western Alps, which compensated for the movement of the Adriatic plate relative to Eurasia.

The Northpenninic flysch basin should be interpreted as a foreland basin of the Alpine orogen, namely, as an underfilled basin. The basin reflects down-flexing of the European plate margin under the weight of the plate on top, as it was being subducted under the Adriatic plate, and the already existing mountain chain. Due to the elastic stiffness of the European plate, this flexing process produced a bulge in the foreland in the northwest. This exposed the Mesozoic limestones in the foreland to subaerial erosion and, thus, karstification. The Cretaceous sediments were fully eroded at many locations, such that the younger Cenozoic sediments were deposited directly on the Late Jurassic limestones. On the other hand, the rocks in the subducting European plate were being transported to ever greater depths in the subduction zone. The resultant metamorphism now gradually also started to affect the rocks of the distal European margin, for example, the Dora Maira nappe (see Schwartz et al. 2007).

The change in the land–water distribution in the Alpine region and its foreland continued. Figure 4.13 shows the situation at the Eocene–Oligocene

because the area between Salzburg and Vienna was above sea level at that time. Pelagic conditions prevailed in

◄ **Figure 4.12**

Palaeogeographical map of the future Alpine realm in the Early Eocene (55 Ma). The present-day outcrop shapes of the various basement units are given as a frame of reference. The Northpenninic flysch basins were narrow troughs in the northwestern Alpine foreland. A rise separated two flysch basins and acted as the source of detritus. Continental conditions existed on a regional bulge to the northwest of the flysch basins.

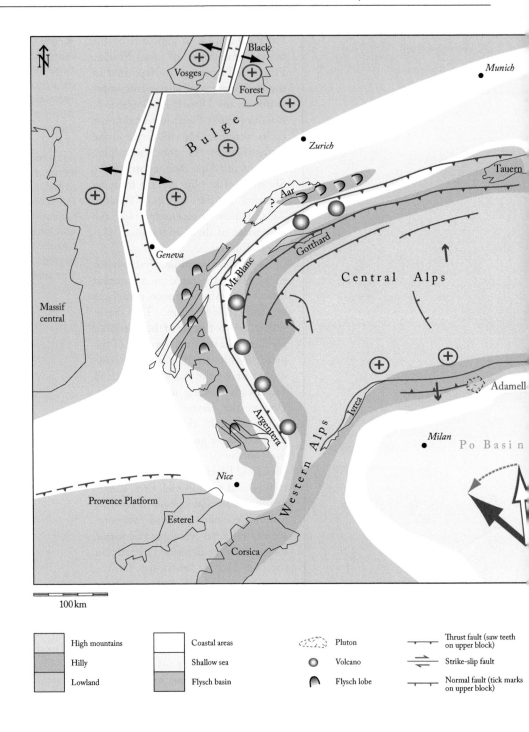

100 km

	High mountains		Coastal areas		Pluton		Thrust fault (saw teeth on upper block)
	Hilly		Shallow sea		Volcano		Strike-slip fault
	Lowland		Flysch basin		Flysch lobe		Normal fault (tick marks on upper block)

boundary, 35 million years ago. The foreland basin in the north had shifted further to the north and west and extended as far as the Salzburg area in the east. The sediments that were deposited in the foreland basin are

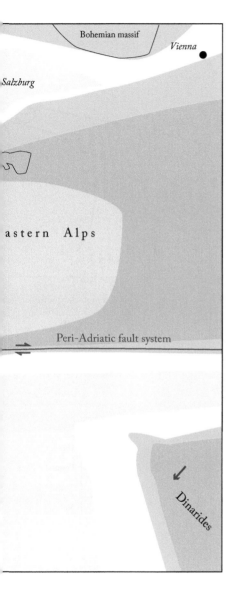

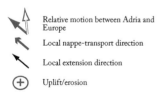

Alps, and to the Annot Sandstone in the Western Alps. The turbidites were sourced from the basin margin closest to the Alps, but the direction of flow was parallel to the axis of the basin. It is unclear whether the flysch basin in the east was directly connected to that in the west, or whether they were arranged en échelon in the region above the future Aar massif. The palaeogeography of the southern part of the basin (Argentera) is based on Sinclair's (1997) reconstruction.

Figure 4.13 indicates the presence of a series of volcanoes to explain the origin of the andesitic components and pebbles in the flysch. Their exact location is debatable, but their presence is necessary in order to explain the proportionally important volcanic detritus. The hypothetical southern volcanoes would have been located in the westward prolongation of the Peri-Adriatic fault system. As discussed above, this fault system would have opened up the passages for the rise of the melts. No evidence has been found for a western continuation of the fault system in the direction of Argentera for this period. A connection with this fault system is unlikely for the northern series of volcanoes in Fig. 4.13. The Oligocene dyke swarms (see Fig. 4.9) are also located at some distance from the flysch basin. Finally, these volcanoes can also not be explained through classic andesitic volcanism at a subduction zone, as the Eocene subduction zone that dipped towards the west only reached the critical depth of 120 kilometres at a substantial distance from the internal margin of the flysch basin.

South-directed overthrusting occurred in the Southalpine region at the southern margin of the Alpine orogen and a steep, south-directed upthrust started in the Peri-Adriatic fault

◄ Figure 4.13

Palaeogeographical map of the future Alpine realm at the transition from the Eocene to the Oligocene (35 Ma). The present-day outcrop shapes of the various basement units are given as a frame of reference. The Northhelvetic Flysch was deposited in a narrow curved trough that extended from the Western Alps to the Central Alps. Marine conditions in the Northalpine Foreland of the Eastern Alps reached as far as Salzburg.

thick turbiditic sequences, which are assigned to the Northhelvetic Flysch (Taveyannaz Formation) in the Central

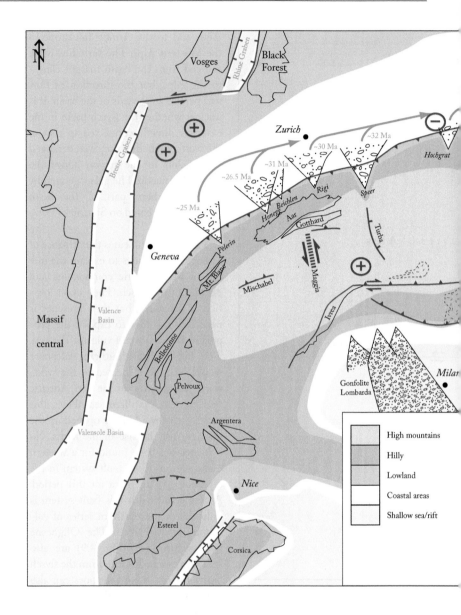

system, associated with dextral strike-slip motion. It therefore can be assumed that this southern portion of the Central Alps experienced high rates of uplift at this time. At the time, the surface rocks in the Eastern and Central Alps were mainly made up of Austroalpine rocks, although the Penninic nappe pile had generally already formed at the subsurface. In the Western Alps, rocks that were exposed at the surface were probably mainly Southalpine rock units. The Penninic nappe pile was, however, also generally already present at depth.

In the northern foreland, east–west-directed, distensive movements started. As a result of this, the European continental margin broke up and a north–south-oriented rift system developed. This was the Rhine Graben (see

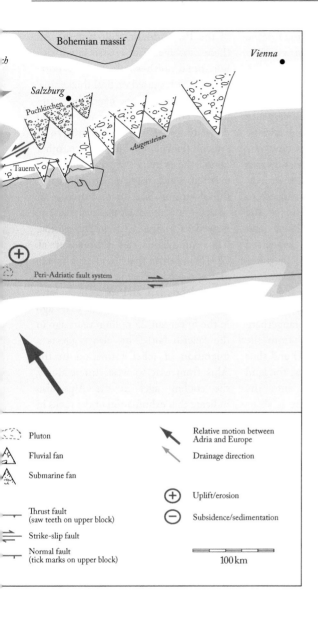

Figure 4.14

Palaeogeographical map of the future Alpine realm in the Late Oligocene (25 Ma). The present-day outcrop shapes of the various basement units are given as a frame of reference. Continental conditions are indicated by fluvial fans in the Northalpine Foreland of the Central and Eastern Alps, which had filled the foreland basin completely. Marine conditions prevailed from Salzburg to Vienna, where, as well as in the Po Basin, submarine fan deposits accumulated. Marine conditions also reigned in a narrow north–south oriented basin to the west of the Western Alps (Valensole and Valence basins) that connected to the rift structures of the Bresse and Rhine grabens. See text for discussion.

Schumacher 2002) in the north, which lead into the, laterally displaced, Bresse Graben in a transfer zone to the south of the Vosges.

In the Early Oligocene, the foreland basin was entirely filled and became dry land to the north of the Central Alps: the underfilled flysch basin had turned into an overfilled molasse basin. Thick fluvial fans had built up, as shown in Fig. 4.14. The ancestral Alpine rivers deposited masses of sediment and the fans prograded, that is, they slowly shifted in a northerly direction. However, the transformation of the foreland basin into dry land was not purely due to being filled up by fluvial deposits, but was also associated with a marked worldwide fall in sea level at the start of the Chattian.

Since the transition from the Eocene to the Oligocene, the basin axis had shifted about 50 kilometres to the north over a period of 10 million years, which equates to a rate of 5 millimetres per year, corresponding to the convergence rate between the European and Adriatic plates at the time. Also at this time progradation of the fluvial fans occurred towards the east along the basin axis, and marine conditions continued to prevail in the foreland basin of the Eastern Alps, between Vienna and Salzburg. Wagner (1996) suggested that a strait existed between the foreland basin and the Po basin at the longitude of Salzburg: Cenozoic sediments in the lower Inn Valley contain conglomerates (Chattian Oberangerberg Formation) that were shed from the west (Ortner & Stingl 2001) and thus indicate an embayment in the foreland basin towards the south, into the Eastern Alps. However, there is absolutely no sign of Cenozoic deposits in the sedimentary suites in the Tauern region to the south of this bay. Interestingly, the conglomerates in the Oberangerberg Formation also contain volcanic debris. Around this time, the intrusion phase along the Peri-Adriatic fault system had just concluded (Rosenberg 2004). The location of the isolated volcano marked in Fig. 4.14 could represent the source area for the volcanic debris.

In the Central Alps, nappe-stacking was in full flow, with north-vergent overthrusting in the Helvetic nappe system and south-vergent overthrusting in the Southalpine nappe system. At this time, south-vergent back-thrusting and back-folding were taking place in the Penninic nappe complex (Mischabel back fold, vertical folding in the Gotthard massif and back-folding and back-thrusting in the Insubric segment

of the Peri-Adriatic fault system). All these movements were associated with horizontal, north–south-directed shortening, which caused vertical thickening. As a result, on the one hand, the crust–mantle boundary (Moho) was pushed down to greater depths, referred to as crustal thickening, but on the other hand, the surface of the land was uplifted, which triggered an increase in fluvial erosion. The first occurrence of coarse-grained conglomerates can be viewed as a sign of intensive erosion. The compilation by Pfiffner et al. (2002) indicates that these conglomerates appeared earlier in the east, in the Speer debris fan, and successively later towards the west (32 million years ago in the Speer fan, 25 million years ago in the Pèlerin fan). This also suggests a migration of relief formation in the Alps from east to west. Interestingly, the cooling ages in the Alps also indicate that exhumation of the rocks in the core of the Alps progressed based on the same pattern. In the Chattian, 25 million years ago, there is rapid uplift (and denudation) in the Lepontine region, to the east of the Maggia transverse zone (marked with a + in Fig. 4.14). The Maggia transverse zone, itself, played a role in the form of a dextral strike-slip fault in the stack of Penninic crystalline nappes. Later, in the Miocene, maximum uplift (and denudation) had shifted to the Toce region, to the west of the Maggia transverse zone. Pfiffner et al. (2002) attribute this shift in uplift to the effects of the Ivrea Body in the northwestern corner of the Adriatic plate. The Ivrea Body functioned as a rigid block, which was pressed or indented into the European plate. At the same time, the Adriatic plate was moving in a westerly direction along the Peri-Adriatic fault system. The westward shift of the

Adriatic indenter was remarkably parallel in time and space to the first appearance of the conglomerates in the Molasse Basin.

Conglomerates, the so-called 'Augensteine', were also shed from the Eastern Alps. These were deposited on the southern shore of the marine foreland basin.

On the south side of the Central Alps, coarse clastic deposits also suddenly appear in the Chattian, with the conglomerates of the Gonfolite Lombarda. As discussed in Section 4.5, both the fall in sea level, which exposed large areas of sediment that had previously been deposited near the shore of the Po Basin, and the intensive uplift in the region of the Insubric segment of the Peri-Adriatic fault system, contributed towards this sudden appearance of conglomerates. The clasts were transported in deep marine canyons into the Po Basin and deposited in deeper water. The term 'Molasse' therefore should not be used for the Gonfolite Lombarda. The resedimen-tation of pebbles from the Bregaglia Pluton indicates an extremely high rate of exhumation of 1 millimetre per year, a rate that is comparable to the convergence rate of 5 millimetres per year mentioned above.

A larger rift system cross-cut the European plate in the foreland of the Alps and Pyrenees. This rift system had already formed in the Eocene, as a reaction to the tensile stresses caused within the European plate by the collision with the Adriatic plate. Ziegler & Dèzes (2007) provide an in-depth discussion on this. The Rhine and Bresse grabens in the foreland of the Alps developed as a result of this rift system. The two grabens are offset from each other by a transfer zone to the south of the Vosges. In the south, the Bresse

Graben leads into the Valence Basin, which, in turn, leads into the Valensole Basin. The Valence Basin is asymmetric, exhibiting numerous normal faults that caused the eastern block to subside and produced local basins with enormously thick sediments (Séranne 1999).

In the south of the Alps, there was slow development of a rift between Corsica and the Provence Platform (Esterel). Later, this rift spread rapidly: Corsica rotated away anti-clockwise from Esterel. New oceanic crust was formed in the opening between, which is now the sea floor in the Ligurian Sea (Séranne 1999).

At this time, there was a high in the area of the future Jura Mountains between the Bresse Graben and the Molasse Basin. Initial folding was probably already taking place in the southern portion of this high, as indicated by an angular undonformity at the base of the Miocene sediments, as will be discussed later.

References

Ackermann, A., 1986, Le Flysch de la nappe du Niesen. Eclogae geologicae Helvetiae, 79/3, 641–684.

Allen, P. A., Mange-Rajetzky, M., Matter, A. & Homewood, P., 1985, Dynamic palaeogeography of the open Burdgalian seaway, Swiss Molasse basin. Eclogae geologicae Helvetiae, 78/1&2, 351–381.

Berger, A. & Bousquet, R., 2008, Subduction-related metamorphism in the Alps: a review of isotopic ages based on petrology and their geodynamic consequences. In: Siegesmund, S., Fügenschuh, B. & Froitzheim, N. (eds), Tectonic Aspects of the Alpine–Dinaride–Carpathian System. Geological Society London, Special Publications, 298, 117–144.

Berger, J.-P., 1996, Cartes paléo-géographiques-palinspastiques du bassin molassique suisse (Oligocène inférieur–Miocène moyen). Neues Jahrbuch fur Geologie und Palaontologie-Abhandlungen, 202, 1–44.

Bichsel, M. & Häring, M. O., 1981, Facies evolution of Late Cretaceous Flysch in Lombardy (northern Italy). Eclogae geologicae Helvetiae, 74/2, 383–420.

Brack, P., 1981, Structures in the southwestern border of the Adamello intrusion (Alpi Bresciane, Italien). Schweizerische mineralogisch-petrographische Mitteilungen, 61/1, 37–50.

Bürgisser, H., Frei, H.-P. & Resch, W., 1981, Bericht über die Exkursion der Schweizerischen Geologischen Gesellschaft in die Molasse der Nordostschweiz und des Vorarlbergs vom 19. bis 21. Oktober 1980. Eclogae geologicae Helvetiae, 74/1, 297–310.

Caron, C., 1972, La Nappe Supérieure des Préalpes: subdivisions et principaux caractères du sommet de l'édifice préalpin. Eclogae geologicae Helvetiae, 65/1, 57–73.

Diem, B., 1986, Die Untere Meeresmolasse zwischen der Saane (Westschweiz) und der Ammer (Oberbayern). Eclogae geologicae Helvetiae, 79/2, 493–559.

Doglioni, C. & Bosellini, A., 1987, Eoalpine and mesoalpine tectonics in the Southern Alps. Geologische Rundschau, 76/3, 735–754.

Franke, W., 1992, Phanerozoic structures and events in Central Europe. In: Blundell, D., Freeman, R. & Mueller, St. (eds), A Continent Revealed: The European Geotraverse, Cambridge University Press, 164–180.

Gunzenhauser, B. A., 1985, Zur Sedimentologie und Paläogeografie der oligo-miocaenen Gonfolite Lombarda zwischen Lago Maggiore und der Brianza (Südtessin, Lombardei). Beiträge zur geologischen Karte der Schweiz, N.F. 159, 114 pp.

Hamilton, W., Wagner, L. & Wessely, G., 2000, Oil and Gas in Austria. In: Neubauer, F. & Höck, V. (eds), Aspects of Geology in Austria. Österreichische Geologische Gesellschaft, Wien, 235–262. Also published in Mitteilungen der Österreichischen Geologischen Gesellschaft, 92. Band (1999).

Herb, R., 1988, Eocaene Paläogeografie und Paläotektonik des Helvetikums. Eclogae geologicae Helvetiae, 81, 611–657.

Hsü, K. J. & Briegel, U., 1991, Geologie der Schweiz: Ein Lehrbuch für den Einstieg, und eine Auseinandersetzung mit den Experten. Birkhäuser, Basel, 219 pp.

Kälin, D., 1997, Litho- und Biostratigraphie der mittel- bis obermiozänen Bois de Raube-Formation (Nordwestschweiz). Eclogae geologicae Helvetiae, 90/1, 97–114.

Kaufmann, F. J., 1877, Geologische Beschreibung der Kalkstein- und Schiefergebiete der Kantone Schwyz, Zug und des Bürgenstocks bei Stanz. Beiträge geologische Karte Schweiz, 14/2. Abteilung, 180 pp.

Kaufmann, F. J., 1886, Centralgebiet der Schweiz: Emmen- und Schlierengegenden nebst Umgebungen Brünigstrasse und Linie Lungern-Grafenort. Beiträge geologische Karte Schweiz, 24/1. Theil, 608 pp.

Kemna, H. A. & Becker-Haumann, R., 2003, Die Wanderblock-Bildungen im Schweizer Juragebirge südlich von Basel: Neue Daten zu einem alten Problem. Eclogae geologicae Helvetiae, 96/1, 71–83.

Kempf, O. & Pfiffner, O. A., 2004, Early Tertiary evolution of the North Alpine Foreland Basin of the Swiss Alps and adjoining areas. Basin Research, 16, 549–567, doi: 10.1111/j.1365-2117.2004. 00246.x.

Kempf, O. & Pross, J., 2005, The Lower Marine to Lower Freshwater Molasse transition in the northern Alpine foreland basin (Oligocene; central Switzerland–south Germany): age and geodynamic implications. International Journal of Earth Sciences, 94, 160–171.

Kempf, O., Matter, A. Burbank, D. W. & Mange, M., 1999, Depositional and structural evolution of a foreland basin margin in a magnetostratigraphic framework: The eastern Swiss Molasse Basin. International Journal Earth Sciences, 88, 253–275.

Lickorish, W. H. & Ford, M., 1998, Sequential restoration of the external Alpine Digne thrust system, SE France, constrained by kinematic data and synorogenic sediments. In: Mascle, A., Puigdefabregas, C., Luterbacher, H. P. & Fernandez, M. (eds), Cenozoic Foreland Basins of Western Europe, Geological Society, London, Special Publications, 134, 189–211.

Marthaler, M., 1984, Géologie des unités penniques entre le Val d'Anniviers et le Val de Tourtemagne. Eclogae geologicae Helvetiae, 77/2, 295–448.

Menkveld-Gfeller, U., 1995, Stratigraphie, Fazies und Paläogeografie des Eocaens der helvetischen Decken der Westschweiz. Eclogae geologicae Helvetiae, 88, 115–134.

Nänny, P., 1948, Zur Geologie der Prätigauschiefer zwischen Rhätikon und Plessur. Dissertation University Zürich und Mitt. Geol. Inst. ETH u. University Zürich, Serie C, No. 30, 127 pp.

Oberhänsli, R., Bousquet, R., Engi, M., Goffé, B., Gosso, G., Handy, M., Höck, V., Koller, F., Lardeaux, J.-M., Polino, R., Rossi, Ph., Schuster, R., Schwartz, S. & Spalla, I., 2004, Metamorphic structure of the Alps (Karte 1:1 000 000). Commission for the Geological Map of the World, SGMW, Paris.

Oberhauser, R. (ed.), 1980, Der geologische Aufbau Österreichs, Springer Verlag, 699 pp.

Ortner, H., 2001, Cretaceous thrusting in the western part of the Northeren Calcareous Alps (Austria) – evidences from synorogenic sedimentation and structural data. Mitteilungen der Österreichischen Geologischen Gesellschaft, 94, 66–77.

Ortner, H. & Stingle, V., 2001, Facies and basin development of the Oligocene in the Lower Inn Valley, Tyrol/Bavaria. In: Piller, W. E. & Rasser, M. W. (eds), Paleogene of the Eastern Alps. Österreichische Akademie der Wissenschaften, Schriftenreihe Erdwissenschaftliche Kommission, 14, 153–196.

Pfiffner, O. A., Schlunegger, F. & Buiter, S. J. H., 2002, The Swiss Alps and their

peripheral foreland basin: Stratigraphic response to deep crustal processes. Tectonics, 21/2, 3-1–3-16, doi: 10.1029/2000TC900030.

Pieri, M. & Groppi, G., 1981, Subsurface geological structure of the Po Plain. Progetto finalizzato geodinamica/Sottoprogetto 'Modello strutturale'. Pubblicazione Consiglio nazionale di Ricerca, 414.

Rosenberg, C. L., 2004, Shear zones and magma ascent: A model based on a review of the Tertiary magmatism in the Alps. Tectonics, 23, TC3002, doi 10.1029/2003TC001526.

Schlunegger, F., Burbank, D. W., Matter, A., Engesser, B. & Mödden, C., 1996, Magnetostratigraphic calibration of the Oligocene to middle Miocene (30–15 Ma) mammal biozones and depositional sequences of the Swiss Molasse Basin. Eclogae geologicae Helvetiae, 89, 753–788.

Schönborn, G., 1992, Alpine tectonics and kinematic models of the central Southern Alps. Memorie di scienze geologiche (Memorie degli Istituti di Geologia e Mineralogia dell'Università di Padova), XLIV, 229–393.

Schumacher, M. E., 2002, Upper Rhine Graben: Role of preexisting structures during rift evolution. Tectonics, 21/1, 6-1–6-17, doi: 10.1029/2001TC900022.

Schwartz, S., Lardeaux, J. M., Tricart, P., Guillot, S. & Labrin, E., 2007, Diachronous exhumation of HP–LT metamorphic rocks from south-western Alps: evidence from fission-track analysis. Terra Nova, 19, 133–140.

Séranne, M., 1999, The Gulf of Lion continental margin (NW Mediterranean) revisited by IBS: an overview. In: Durand, B., Jolivet, L., Horvath, F. & Seranne, M. (eds), The Mediterranean Basins: Tertiary Extension within the Alpine Orogen. Geological Society London, Special Publications, 156, 15–36.

Siegenthaler, Ch., 1974, Die Nordhelvetische Flysch-Gruppe im Sernftal (Kt. Glarus). Dissertation University Zürich, 83 pp.

Sinclair, H. D., 1997, Tectonostratigraphic model for underfilled peripheral foreland basins: An Alpine perspective. Bulletin Geological Societey of America, 109, 324–346.

Steininger, F. F. & Wessely, G., 2000, From the Tethyan Ocean to the Paratethys Sea: Oligocene to Neogene Stratigraphy, Paleogeography and Paleobiogeography of the circum-Mediterranean region and Oligocene to Neogene Basin evolution in Austria. In: Neubauer, F. & Höck, V. (eds), Aspects of Geology in Austria. Österreichische Geologische Gesellschaft, Wien, 95–116. Also published in Mitteilungen der Österreichischen Geologischen Gesellschaft, 92. Band (1999).

Studer, B. 1827, Remarques géognostiques sur quelques parties de la chaîne septentrionale des Alpes. Annales des Sciences naturelles, Paris, 11, 1–47.

Studer, B., 1853, Geologie der Schweiz. Stämpfli, Bern, 2, 497 pp.

Thum, I. & Nabholz, W., 1972, Zur Sedimentologie und Metamorphose der penninischen Flysch- und Schieferabfolgen im Gebiet Prättigau-Lenzerheide-Oberhalbstein. Beiträge geologische Karte Schweiz, neue Folge 144, 112 pp.

Trümpy, R., Fumasoli, M., Hänny, R., Klemenz, W., Neher, J. & Streiff, V., 1970, Aperçu général sur la géologie des Grisons. C. R. des Séances de la Société géologique de France, Fasc 9, 1969/9, 330–364 and 391–194.

Van Stuijvenberg, J., 1979, Geology of the Gurnigel area (Prealps, Switzerland). Beiträge geologische Karte Schweiz, neue Folge 14, 55 pp.

Von Blanckenburg, F. & Davis, J. H., 1995, Slab breakoff: A model for syncollisional magmatism and tectonics in the Alps. Tectonics, 14, 120–131.

Vuagnat, M., 1952, Pétrographie, répartition et origine des microbrèches du Flysch nordhelvétique. Beiträge zur geologischen Karte der Schweiz, N.F. 127, 103 pp.

Wagner, L. R., 1996, Stratigraphy and hydrocarbons in he Upper Austrian Molasse Foredeep (active margin). In: Wessely, G. & Liebl, W. (eds), Oil and Gas in Alpidic Thrust belts and Basins of Central and Eastern Europe. EAGE Special Publication, 5, 217–235.

Wildi, W., 1985, Heavy mineral distribution and dispersal pattern in penninic and ligurian flysch basins (Alps, northern Apennines). Giornale di Geologia, seria 3a, 47/1–2, 77–99.

Winkler, W., Wildi, W., van Stuijvenberg, J. & Caron, C., 1985, Wägital-Flysch et autres flyschs penniques en Suisse Centrale: Stratigraphie, sédimentologie et comparaisons. Eclogae geologicae Helvetiae, 78/1, 1–22.

Ziegler, P. A. & Dèzes, P., 2007, Cenozoic uplift of Variscan Massifs in the Alpine foreland: timing and controlling mechanisms. Global and Planetary Change, 58, 237–269.

Ziegler, W. H., 1956, Geologische Studien in den Flyschgebieten des Oberhalbsteins (Graubünden). Eclogae geologicae Helvetiae, 49/1, 1–78.

5 Tectonic Structure of the Alps

We are, unavoidably, left with the impression of a chaotic pattern of coloured spots when we look at a geological map of the Alps. Sedimentary rocks of varying ages lie next to each other at one location, then over each other at the next location and, in many cases, rocks of different ages are in direct contact with each other. These chaotic patterns reflect the tectonic structure of the Alps, which is based on vast movements and flexing of crustal rocks.

Humans have used rock as a building material since the dawn of history for constructions of all sizes. The selection of rocks for urban buildings usually focused on particularly attractive colouring and material that was easy to process and resistant to weathering. A sustainable supply of rock as a raw material requires thorough knowledge on where it can be exploited. Rock was also used as the raw material for smaller constructions in rural areas, for example, for dry stone walls, building walls and roofs, and for many other purposes. We rapidly realize that local rocks are used when we look at their composition in such rural constructions, but also see that the ease with which the rock can be processed played a role in its selection. The inhabitants clearly had a very good knowledge of their 'local geology', which is also apparent in the agricultural use of the mountain landscape, as woodland was cleared only in those locations where the soil could be utilized for pasture and crops.

The Alpine mountain chain has, however, also fascinated researchers for centuries, and the early 'geognostics' soon realized that the distribution of individual types of rock was governed by certain laws. They also came to

The birthplace of nappes

understand that there were exceptions and contradictions to these laws. An overall picture of the distribution of layers of rock was obtained from the many observations that had been carefully recorded on to maps, and this picture was then fitted into the contemporaneous view of the world. However, the outcome was not entirely satisfactory and many decades passed before apparent contradictions were explained (see Franks et al. 2000).

For example, in 1809, Hans Conrad Escher noted that apparently older rocks (greywacke) overlie younger rocks (Alpine limestone) in the Alps of the Glarus area in eastern Switzerland and in central Switzerland, which contradicted all ideas at the time. He invited the eminent Prussian geologist, Leopold von Buch, to the Glarus area, who in 1810 stated that the apparently older 'greywacke' must be slightly younger, as there were never any exceptions. Figure 5.1 is a watercolour of the Tschingelhoren with the Martin's Hole, which Hans Conrad Escher painted in 1812, in which two different rock formations separated by a sharp line are clearly visible. The photograph in Fig. 5.2, taken from about the same location, shows how meticulous Escher was in his observation. The summit of the Tschingelhoren is composed of Permian sediments (Verrucano Group) and overlies grey, Late Jurassic limestone (Quinten Limestone), within which the Martin's Hole is located. The Tschingelhoren lies within the Tectonic Arena Sardona, which was included in the UNESCO list of World Natural Heritage Sites in 2008. The enormous Glarus Thrust, which is impressively easy to see in a number of locations and

Geology of the Alps: Revised and updated translation of Geologie der Alpen, Second Edition. O. Adrian Pfiffner.
© 2014 John Wiley & Sons, Ltd. Published 2014 by John Wiley & Sons, Ltd.

Figure 5.1 Glarus Thrust and Martin's Hole in the Tschingelhoren. Source: Aquarell by H.C. Escher (1812).

Figure 5.2 Glarus Thrust and Martin's Hole in the Tschingelhoren.

Fig. 4. Profil an der Lochseite bei Schwanden
(nach Escher)

clearly visualizes the process of mountain building, was the main reason for the inclusion of this area as a World Heritage Site.

Hans Conrad's son, Arnold Escher, pursued the issue further and invited another eminent geologist, Murchison, to the Glarus area, who confirmed to Escher that this was clearly the case of an enormous overthrust at the Panix pass. However, Arnold Escher did not have the confidence to announce this 'monstrosity' to the specialists – he had no wish to be regarded as a fool. He therefore conceived of an enormous recumbent double fold, in the reverse limb of which old rocks lay over young rocks, and presented this idea in 1866. Albert Heim, who had just finished a detailed study on the mechanisms of mountain building (Heim 1878), adopted the idea. Figure 5.3 shows a

◀ Figure 5.3 Glarus Thrust at the Lochsite locality. Source: Drawing by A. Escher (in Alb. Heim, 1878).

sketch of the outcrop at the famous location of 'Lochsiten' in the Glarus area, produced by Heim (Heim 1878) and based on A. Escher. A thin band of limestone, the 'Lochsitenkalk' lies between the Permian Verrucano above and the Cenozoic flysch below. It was assumed to date from the Late Jurassic ('Malm Limestone'), which resulted in the interpretation of an apparently simple inverted fold limb. Yellow portions in the upper part of the 'Lochsitenkalk', as are found in many locations, were interpreted to be Triassic dolostones. Trümpy & Westermann (2008) provide a fitting account of the long path that Albert Heim had to tread before he finally accepted the existence of thrust faults in 1901.

The Mythen, in central Switzerland, is a good example to use to illustrate the problem of 'old on top of young'.

▼ **Figure 5.4** The Penninic klippen in the Mythen (canton Schwyz, Central Switzerland).

Figure 5.5 The Penninic klippen in the Mythen (canton Schwyz, Central Switzerland). Source: Drawing by J. Kaufmann (1877).

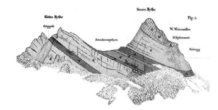

Figure 5.4 shows the Mythen as seen from the west. Of note from a distance is the fact that the summit is composed mainly of limestones that appear to float on a layer of marls. In 1877, Kaufmann, who travelled through the region carrying notes from Hans Conrad Escher's legacy, recognized that this was a special case of different rocks lying next to and on top of each other. His original sketch (from Kaufmann 1877) is reproduced in Fig. 5.5. Even today, it would be difficult to produce a better illustration of the profile view of the Mythen. However, Kaufmann remained silent on the implications of his observations.

In 1884, Bertrand, a Belgian mining geologist, who had never visited the Alps, stated that the field finds of Escher and Albert Heim were to be interpreted as an overthrust. In 1893, Schardt, who was working on the Prealps in western Switzerland, stated that an enormous overthrust must also be present there. Finally, it was Maurice Lugeon who persuaded the geological experts of the existence of overthrusts in 1896–1902. Albert Heim found it very difficult to accept the new idea, in spite of his in-depth knowledge of the Alps and the deformation processes affecting rocks. After numerous fierce debates, he was finally won over in 1902. The idea of nappe theory was borne out of this statement, whereby larger crustal blocks are thrust over each other and compressed during an orogeny. It is easy to understand how old strata can come to lie on top of young strata as a result of this process, while the 'normal' sequence with the younger stratum on top is found in other locations. This insight now made it possible to understand the tectonic structure of the Alps and to investigate it further.

There are substantial differences between the tectonic structures of the Western Alps, the Central Alps and the Eastern Alps. Therefore, these regions will be discussed separately below, even though certain large-scale tectonic units extend across more than just one region. The discussion of these regions always starts with the external portions (European plate margin), and works towards the internal portions (Adriatic plate margin).

5.1 The Western Alps

The Western Alps, a mountain chain that runs north–south, contain the entire spectrum of tectonic Alpine building blocks. The tectonic map in Fig. 5.1-1 shows the Jura Mountains in the northwest, which represent the outermost margin of the deformed European continental margin, and the Ivrea Zone in the east, which represents the edge of the Adriatic continental margin. In between, the nappe piles formed by the Penninic nappes and the Subalpine chains of the Dauphinois run southwards in two separate bands from the Central Alps. In the Subalpine chains, a distinction should be made between an allochthonous, that is, thrust over a long distance, nappe complex and a parautochthonous fold belt that has not been displaced to any great extent. Larger crystalline uplifts extend along the internal margin of this zone (Belledonne, Pelvoux and Argentera massifs). These are also called the 'external massifs' and represent the continuation of the Aiguilles Rouges and Mont Blanc massifs in the Central Alps. In a southerly direction, the allochthonous nappe complex of the Subalpine chains ends at about the southern end of the crystalline uplifts. In the parautochthonous

fold belt, the axes of the large-scale folds trend in two directions, east–west and north–south. In the extreme south, the folds with an east–west trend are associated with north-vergent thrust faults. These structures can be interpreted as extensions of the Pyrenees and they are older than folds with a north–south trend. To the south of the Argentera massif, at the Digne thrust fault, the parautochthonous fold belt has been overthrust by several kilometres in a southwesterly direction. The frontal folds of this thrust system form an arc that can be followed all the way to the area around Nice. Two larger crystalline uplifts are also discernible within the Penninic nappes, the Gran Paradiso and the Dora Maira massifs, which are also known as 'internal massifs'. The crystalline basement of both these units is part of the Briançon microcontinent and was overthrust in a westerly direction by more than 100 kilometres. The Western Alps are surrounded by Cenozoic sedimentary basins: the Po Basin in the east, the Bresse Graben in the northwest and the Valence and Valensole basins in the southwest.

Figure 5.1-1 also shows the locations of geological cross-sections based on which the structure of the Western Alps is explained. The first cross-section (Fig. 5.1-2) is through the northern Western Alps (adapted from Guellec et al. 1990, Mugnier et al. 1990, Schmid & Kissling 2002, Bucher et al. 2004). In addition to the structural observations, this cross-section also gives information on the deep structure that was obtained from geophysical investigations conducted by ECORS-CROP (Nicolas et al. 1990, Roure et al. 1996) and by teleseimic studies carried out by Waldhauser et al. (2002), Diel et al. (2009) and Wagner et al. (2012). In this cross-section,

▶ **Figure 5.1-1** Simplified geological-tectonic map of the Western Alps with traces of cross-sections (green lines–numbers are figure numbers).

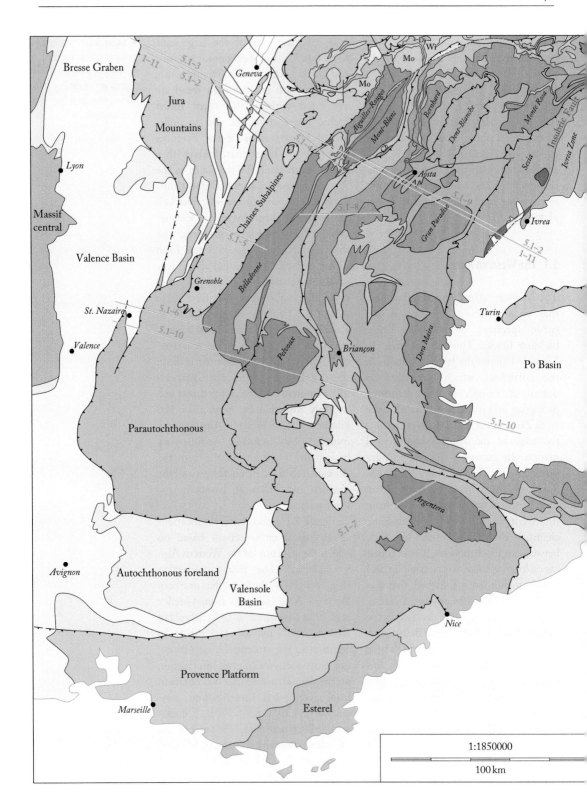

Bresse Graben

Jura

Mountains

Geneva

Lyon

Massif
central

Valence Basin

Chaînes Subalpines

Grenoble

Belledonne

St. Nazaire

Valence

Parautochthonous

Pelvoux

Briançon

Dora Maira

Po Basin

Argentera

Avignon

Autochthonous foreland

Valensole
Basin

Nice

Provence Platform

Marseille

Esterel

Aiguilles Rouges

Mont-Blanc

Bernhard

Dent-Blanche

Monte Rosa

Insubric Fault

Ivrea Zone

Ch.

Aosta

Gran Paradiso

Sesia

Ivrea

Mo

Mo

Wi

Turin

1:1850000

100 km

ropean (Northalpine) Foreland

zoic sediments

Subalpine Molasse

Molasse Basin, Bresse Graben,
Valence- and Valensole Basin

zoic sediments

Autochthonous Mesozoic Foreland

Jura Mountains

Triassic basement

Crystalline basement (Massif central)

Adriatic Foreland

Cenozoic rocks

Sediments of Po Basin

Peri-Adriatic plutons

Nappe systems of the Adriatic continental margin

Austroalpine nappe complex

Sesia Zone, Dent Blanche nappe

Southalpine crystalline basement blocks

Strona-Ceneri Zone (upper crust)

Ivrea Zone (lower crust)

ppe sytems of the European continental margin

vetic/Dauphinois nappe complex

Ultrahelvetic cover nappes

Ultrahelvetic basement nappe (Verampio)

Chaînes Subalpines, Wildhorn nappe (Wi)

Diablerets nappe (Di)

ahelvetic/Infradauphinois complex

zoic sediments

Morcles nappe (Mo)

Parautochthonous cover

Triassic basement (mainly crystalline rocks)

Chétif (Ch)

Argentera, Pelvoux, Mont Blanc

Belledonne, Aiguilles Rouges

raalpine units

Mesozoic sediments (Provence Platform)

Crystalline basement (Esterel)

Penninic nappe complex

Upper Penninic nappes / Piemont Ocean

Ophiolite-bearing nappes

Middle Penninic nappes / Briançon Rise

Cover nappes

Palaeozoic sediments (Zone Houillère)

Crystalline nappes:
Dora Maira, Gran Paradiso,
Bernhard, Monte-Rosa

Lower Penninic nappes / Valais Trough

Flysch nappes

Cover nappes

Crystalline nappes

Apennine (Monferrato unit)

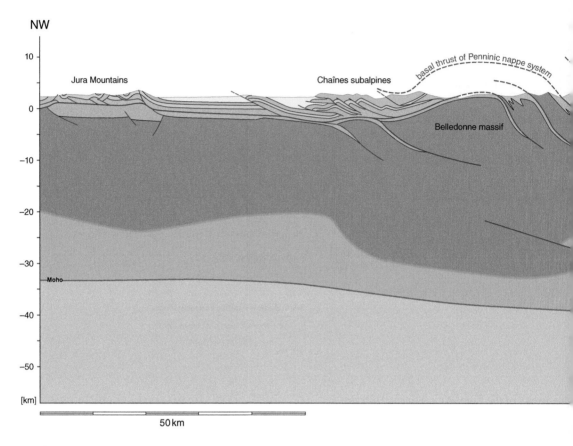

NW

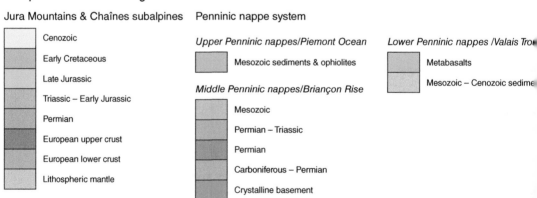

European continental margin

Jura Mountains & Chaînes subalpines

- Cenozoic
- Early Cretaceous
- Late Jurassic
- Triassic – Early Jurassic
- Permian
- European upper crust
- European lower crust
- Lithospheric mantle

Penninic nappe system

Upper Penninic nappes/Piemont Ocean

- Mesozoic sediments & ophiolites

Middle Penninic nappes/Briançon Rise

- Mesozoic
- Permian – Triassic
- Permian
- Carboniferous – Permian
- Crystalline basement

Lower Penninic nappes /Valais Tro

- Metabasalts
- Mesozoic – Cenozoic sedime

the European crust dips below the Adriatic crust towards the east. In the northwest, in the Jura Mountains and the Subalpine chains, the Mesozoic–Cenozoic sedimentary cover has been peeled off and transported to the northwest. During this process, the displaced sediments were greatly shortened internally, but to a greater extent in the Subalpine chains than in the Jura

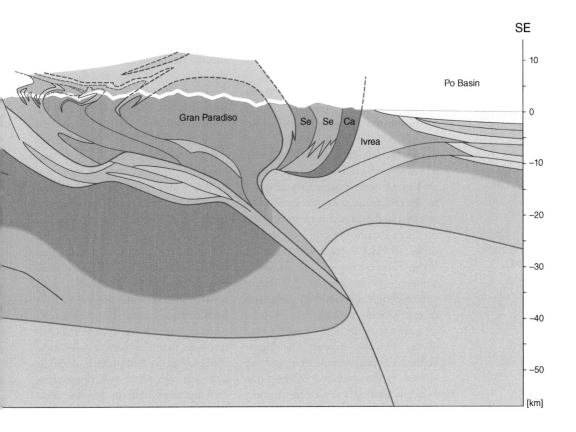

SE

Po Basin

Gran Paradiso

Se Se Ca

Ivrea

10
0
−10
−20
−30
−40
−50
[km]

Adriatic continental margin

Austroalpine nappe system (Sesia Zone, Se)

- Eclogitic mica-schist
- Gneiss Minuti

Southalpine nappe system

- Pliocene – Quaternary
- Burdigalian – Tortonian
- Late Cretaceous – Aquitanian
- Triassic – Early Cretaceous
- Canavese Zone (Ca)
- Adriatic upper crust
- Adriatic lower crust
- Lithospheric mantle

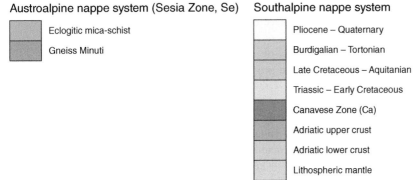

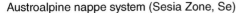

Mountains. The pre-Triassic basement beneath the Jura Mountains is only slightly deformed and forms a local high, whereas the crystalline basement beneath the Subalpine chains was shortened internally by thrusting. A dome-like bulge raised the top of the crystalline basement up to a much higher level in the Belledonne and Mont Blanc massifs than in the

◀ **Figure 5.1-2**
Geological cross-section
through the Western Alps
along the seismic lines of
ECORS/CROP (redrawn after
Schmid & Kissling 2000).
Source: Adapted from
Guellec et al. (1990),
Mugnier et al. (1990),
Schmid & Kissling (2000)
and Bucher et al. (2004).

*Western Alps:
Earth's mantle
involved in
thrusting*

foreland. The crystalline basement in the internal portion of the orogen, which corresponds to the distal edge of the European continental margin, is even more strongly deformed internally, and this formed the substratum of the sediments now present in the Lower and Middle Penninic nappes. Even though the different crystalline basement blocks were thrust on top of each other, the tectonic contacts were later subjected to ductile folding, as for example is revealed in the largest crystalline block, the Gran Paradiso massif – the domed surface of which forms the contact to the Upper Penninic nappes. The Mesozoic sediments in the Lower and Middle Penninic nappes have been peeled off their crystalline susbstrate and form a separate nappe pile in the frontal region of the Penninic nappes. The Upper Penninic nappes overlie this and are located to the southeast of the Gran Paradiso massif: these are oceanic rocks that underwent high-pressure metamorphism during the Alpine orogeny, but most of these rock units depicted in Fig. 5.1-2 have been removed by erosion. Figure 5.1-1, however, shows how these rock units remain in place around the Gran Paradiso massif, such that there is no doubt that they once also overlay the massif. Based on geophysical data it can be assumed that the European lower crust was thickened by tectonic shortening beneath the Penninic nappes, as the boundary between the crust and the mantle, the Moho, is at a depth of over 50 kilometres below the Gran Paradiso massif. The shortening of the lower crust is indicated by the wavy nature of the contact between lower and upper crust, the so-called Conrad disconinuitiy.

The situation is completely different in the southeast, on the Adriatic side, although the upper crust has also been

thickened by thrusting. The Mesozoic–Cenozoic sediments above it are also affected by thrusting, but without actually being sheared off, as was the case for the European crust. What is interesting, is the fact that the lower crust and the Adriatic mantle rise up towards the northwest. The lower crust thereby reaches the Earth surface and is now exposed in the Ivrea Zone, and the mantle below, also called the Ivrea Body or the Ivrea Mantle, also rises up and almost reaches the Earth surface. Ultramafic rocks are exposed near Baldissero that probably stem from the mantle, but they could also be intrusions of mantle melts in the lowest crust. Tectonic contacts between the Penninic nappes and the Ivrea Body are steep and Upper Penninic oceanic rock units and continental crustal blocks from the Adriatic continental margin (e.g. the Sesia-Lanzo Zone) lie alongside each other. Two larger, steep, shearing zones indicate how the rocks were overthrust onto the Ivrea Body (Insubric Fault) and how the rocks of the Gran Paradiso massif were overthrust onto the steep zone. Simultaneously, the highly metamorphic rocks moved closer to the Earth surface due to erosion, a process known as exhumation.

The Jura Mountains

At the outermost northwestern margin of the Western Alps, the Jura Mountains became a range in their own right, separate from the Alps or, more precisely, the Subalpine chains (see Fig. 5.1-1). The Jura Mountains were built up from numerous folds that became detached from their pre-Triassic basement and were shifted to the northwest on a basal decollement horizon in Triassic evaporites. The cross-section in Fig. 5.1-3 is based on Guellec et al. (1990) and Mugnier et al. (1990) but has been

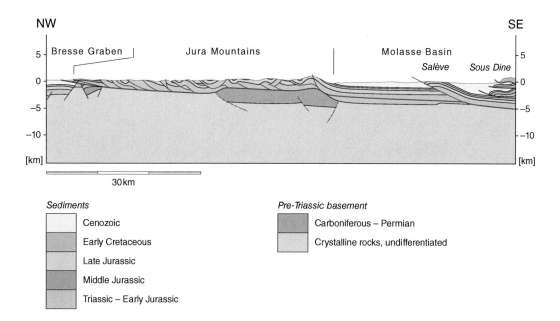

NW   SE

Bresse Graben   Jura Mountains   Molasse Basin
                                                    Salève   Sous Dine

30 km

Sediments

Cenozoic

Early Cretaceous

Late Jurassic

Middle Jurassic

Triassic – Early Jurassic

Pre-Triassic basement

Carboniferous – Permian

Crystalline rocks, undifferentiated

slightly modified. In the extreme northwest, the Jura Mountains have been thrust onto the Cenozoic fill of the Bresse Graben, and in the southeast, the Mesozoic sediments plunge beneath the Cenozoic deposits of the Molasse Basin. The internal structure of the Jura Mountains is mostly characterized by northwest-vergent thrust faults that rise up in the form of ramps from the decollement horizon, which lies parallel to the strata and either come to a dead end in the core of anticlinal structures in the Triassic, or transport the Triassic in the hanging wall over large distances onto the younger rocks. There was also local development of southeast-directed thrust faults. Interestingly, the basal decollement horizon in the southeastern portion of the Jura Mountains is structurally higher up than in the northwest. It therefore can be assumed that the decollement horizon in the southeastern segment was uplifted into its higher position after the sediments had been detached and sheared off, as the decollement horizon could not have

been active on such an uneven surface. In Fig. 5.1-2 the interpretation of this situation is that the inversion of the Permo-Carboniferous trough is responsible for this structure.

The Subalpine Chains of the Dauphinois

The Subalpine chains form the external portion of the Western Alps and its structure changes along the strike. In the north, in the Haute Savoie, Subalpine chains are present in the form of one allochthonous unit, which ends in the south, approximately at the southern end of the Belledonne massif (see Fig. 5.1-1). In the south, in the Vercors, Mesozoic sediments of the Dauphinois are cross-cut by a series of thrust faults with a small transport distance. This change in the internal structure is explained in more detail based on three cross-sections.

The lower portion of the cross-section in Fig. 5.1-4 is based on the reflection seismic data produced by ECORS-CROP (Guellec et al. 1990,

▲ **Figure 5.1-3** Geological cross-section through the Jura Mountains along the seismic lines of ECORS. Source: Adapted from Guellec et al. (1990) and Mugnier et al. (1990).

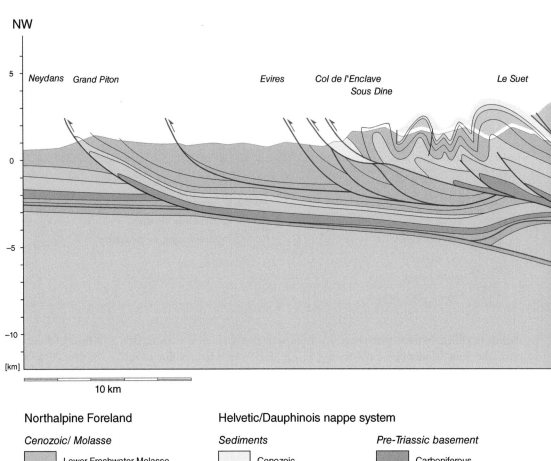

NW

Neydans   Grand Piton   Evires   Col de l'Enclave   Le Suet
                                                      Sous Dine

10 km

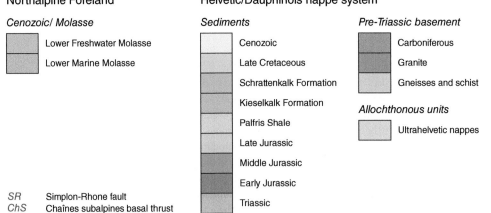

Northalpine Foreland

Cenozoic/ Molasse

- Lower Freshwater Molasse
- Lower Marine Molasse

Helvetic/Dauphinois nappe system

Sediments

- Cenozoic
- Late Cretaceous
- Schrattenkalk Formation
- Kieselkalk Formation
- Palfris Shale
- Late Jurassic
- Middle Jurassic
- Early Jurassic
- Triassic

Pre-Triassic basement

- Carboniferous
- Granite
- Gneisses and schist

Allochthonous units

- Ultrahelvetic nappes

SR   Simplon-Rhone fault
ChS   Chaînes subalpines basal thrust

▲ **Figure 5.1-4**

Geological cross-section through the Chaînes subalpines of Haute Savoie (France). Source: Adapted from Guellec et al. (1990) and Mugnier et al. (1990).

Mugnier et al. 1990). However, the internal structure of the Subalpine chains has been newly reconstructed and is based on a recent structural geological analysis of the surface data (Pfiffner et al. 2010, Pfiffner 2011), which shows that the large-scale folds in the Haute Savoie are not simply the result of a northwest-vergent imbricate thrust structure, but that many of the folds should be interpreted as detachment folds with a more symmetrical structure that were sheared off in the evaporites of the Triassic. The anticlinal

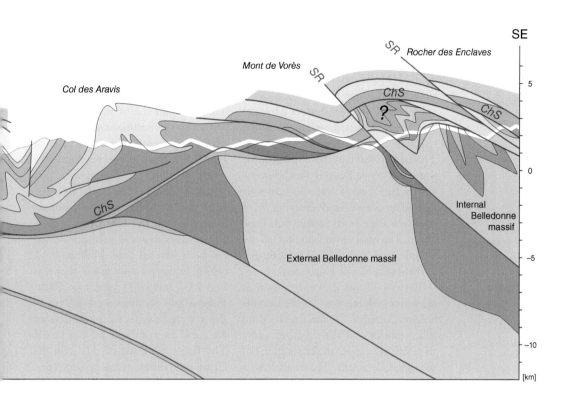

SE

Rocher des Enclaves

Mont de Vorès

Col des Aravis

ChS

ChS

ChS

?

SR

SR

SR

Internal
Belledonne
massif

External Belledonne massif

5

0

−5

−10

[km]

Penninic nappe system

Penninic nappes undifferentiated

cores of the detachment folds, however, are filled with the thick argillaceous–marly rocks of the Early and Middle Jurassic, and the fold structure is very much influenced by the mechanically strong, thick limestones of the Late Jurassic and Early Cretaceous (Quinten Limestone and Urgonian Limestone or Schrattenkalk Formation). In the north, the Subalpine chains give way to the Jura Mountains, and . The upthrusted anticline of the Salève is shown at the transition between the two in Fig. 5.1-3. The Salève and neighbouring anticlines rise above the flat, Cenozoic strata of the Molasse Basin similar to monadnocks. Towards the Jura Mountains, the folds become more frequent and are closer together (see Fig. 5.1-3). Molasse sediments (Lower Freshwater Molasse) overlie the Mesozoic sediments in the southern limb of the Salève anticline. This makes clear that the basin fill of the western Molasse Basin is allochthonous ('piggy-back basin'). The allochthonous folds in the Sous Dine chain in Fig. 5.1-4 mark the edge of the actual Subalpine chains. The basal thrust fault first follows Middle Jurassic shales and then, further to the south, Triassic evaporites.

Dauphinois nappe system: Detached Mesozoic cover, but basement involved in thrusting, too

In the southeast, the basal thrust rises up, forms a dome over the Belledonne and/or Aiguilles Rouges and Mont Blanc massifs and then plunges under the Penninic nappes to the southeast of the external massifs. The frontal Sous Dine chain was thereby transported over 50 kilometres to the northwest relative to the crystalline basement. This basal thrust clearly runs over the top of the Morcles nappe in the northeastern part of the Aiguilles Rouges massif. The nappe associated with the Subalpine chains is therefore in the same tectonic position as the Diablerets nappe in the Central Alps, that is, the allochthonous Subalpine chains can be correlated to the Helvetic nappes. From a palaeogeographical perspective, its structural position over the Morcles nappe also indicates that the Subalpine chains originated southeast of the Mont Blanc massif and not in the Chamonix Zone, as was proposed by Affolter et al. (2008). Figure 5.1-4 shows that the basal thrust in the Subalpine chains is displaced by two steep thrust faults, which are extensions of the Rhone–Simplon Fault that exhibit a thrust fault component in this segment, in addition to the strike-slip component.

The fold trains of the Subalpine chains exhibit a pronounced arcuate shape in the Haute Savoie, which Affolter et al. (2008) attribute to the distribution of evaporites in the Triassic. The Subalpine chains were transported further to the northwest where evaporites are present, as, for example, in Fig. 5.1-4, than in the region to the south of this, where evaporites are absent. The fold trains can be followed from the Haute Savoie, in a southerly direction, into the Chartreuse region. Figure 5.1-5 shows a cross-section that has been adapted from Philippe et al. (1998). In this cross-section, as shown in Figs 5.1-1 and 5.1-5, there is an allochthonous unit of Mesozoic sediments in the west that overlies the autochthonous Mesozoic cover of the Belledonne massif. The basal overthrust of this unit, the Chartreuse thrust, forms a depression and rises over the Belledonne massif in an easterly direction. According to Philippe et al. (1998), the thrust fault is not located in Triassic evaporites, but in the argillaceous–marly rocks of the Middle Jurassic and, stratigraphically, towards the west also rises into the basal strata of the Late Jurassic (see Fig. 5.1-5). In the footwall of this Chartreuse thrust, the Jurassic–Cretaceous sediments exhibit an imbricate structure due to numerous thrust faults that are minimally offset from each other. The youngest sediments in the frontal thrust sheets are the Cenozoic sediments of the Valence Basin. Towards the east, the thrust faults that produced this imbricate structure merge deep down and cut into the crystalline basement of the Belledonne massif. A retro-deformation of the cross-section reveals that the units that make up the imbricate structure were originally the autochthonous cover of the (external) Belledonne massif, while the Subalpine chains in the

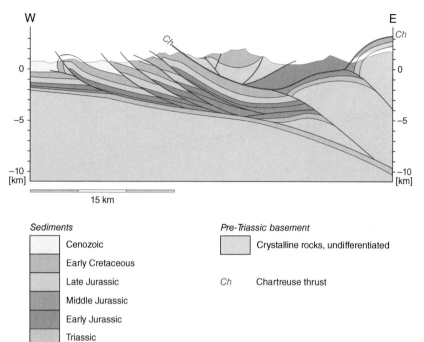

Figure 5.1-5 Geological cross-section through the Chaînes subalpines of Chartreuse (France). Source: Adapted from Philippe et al. (1998).

hangingwall of the Chartreuse thrust must stem from a more internal location. The frontal anticline above the Chartreuse therefore thrust must have been transported at least 35 kilometres to the west.

The structure of the Subalpine chains changes further to the south, in the Vercors. The Chartreuse thrust ends at the southern edge of the Belledonne massif. Figure 5.1-6 (adapted from Philippe et al. 1998) reveals four thrust faults that dip towards the east and that merge into the basal decollement horizon in the evaporites of the later Triassic or the clays of the Early Jurassic. The cross-section, constructed with the aid of boreholes and reflection seismic data, reveals how the thickness of the Mesozoic sediments increases towards the east (see Philippe et al. 1998). This is due to the fact that this section represents the western edge of the Vocontian Trough (see Chapter 3) in the Dauphinois. This trough opened in the Jurassic and synsedimentary normal faults caused abrupt changes in the thickness of the basin fill. The three smaller normal faults that dip towards the east into the crystalline basement in the western portion of the cross-section demonstrate the situation at the western edge of the trough. There are, however, also normal faults in many locations in the Mesozoic sediments that, for example, displace the sediments of the Middle Jurassic and indicate abrupt changes in thickness. At the eastern edge of the cross-section, the larger normal faults that dip towards the west reflect the eastern edge of the Vocontian Trough and an adjacent basement high (La Mure in Fig. 5.1-6). The Alpine east–west-directed compression resulted in the La Mure crystalline block being thrust westwards onto the crystalline basement of the Vocontian Trough, and at the same time

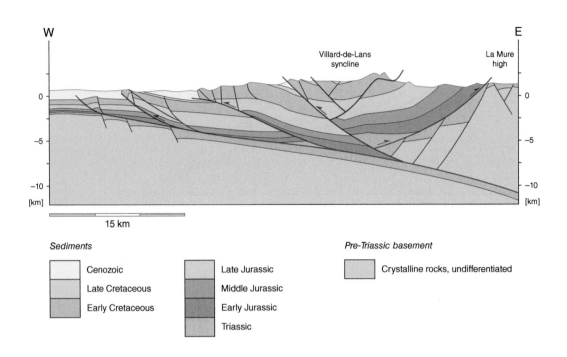

W   E

Villard-de-Lans
syncline

La Mure
high

15 km

Sediments

	Cenozoic		Late Jurassic
Late Cretaceous		Middle Jurassic
Early Cretaceous		Early Jurassic
	Triassic

Pre-Triassic basement

Crystalline rocks, undifferentiated

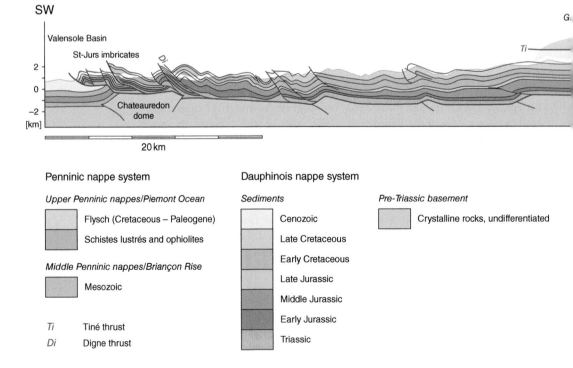

SW   G

Valensole Basin

St-Jurs imbricates

Di

Ti

Chateauredon
dome

20 km

Penninic nappe system

Upper Penninic nappes/Piemont Ocean

Flysch (Cretaceous – Paleogene)

Schistes lustrés and ophiolites

Middle Penninic nappes/Briançon Rise

Mesozoic

Ti   Tiné thrust
Di   Digne thrust

Dauphinois nappe system

Sediments

Cenozoic

Late Cretaceous

Early Cretaceous

Late Jurassic

Middle Jurassic

Early Jurassic

Triassic

Pre-Triassic basement

Crystalline rocks, undifferentiated

the sedimentary fill of the trough was thrust in an easterly direction towards the basement high. This basin inversion resulted in a certain symmetry in the structure of the Subalpine chains, as can be seen, for example, in the Villard-de-Lans synclines in Fig. 5.1-6. It should now be noted that, during the formation of the Pyrenees in the Cretaceous, a north–south-directed shortening resulted in the development of multiple fold structures running east–west in the Vercors region. These structures were subsequently over-printed during the Cenozoic Alpine east–west-directed shortening.

Alpine shortening in the Subalpine chains of Chartreuse and Vercors pro-gressed uniformly in a WNW–ESE direction (Philippe et al. 1998) and the nappes were transported towards the west-northwest. In contrast, further to the southeast, in the southern Subalpine chains in the Digne region, this shortening

occurred in a southwest–northeast direction and the nappes were trans-ported to the southwest. Lickorish & Ford (1998) produced a balanced cross-section extending from the Argentera massif into the Valensole Basin that is shown in Fig. 5.1-7. The Mesozoic (and Cenozoic) sediments have been sheared off their pre-Triassic basement. The Triassic evaporites formed the decollement horizon. In the southwest, we can see how the St Jurs imbricates have been pushed on top of the Cenozoic sediments of the Valensole Basin. The most important thrust fault is the Digne thrust immediately to the northeast. The thickness of the Mesozoic formations in the Digne nappe is far greater than that of the St Jurs imbricates because the Digne thrust represents a reactivated synsedi-mentary normal fault. This normal fault was one of the structures associated

◀ **Figure 5.1-6**
Geological cross-section through the Chaînes subalpines of Vercors (France). Source: Adapted from Philippe et al. (1998).

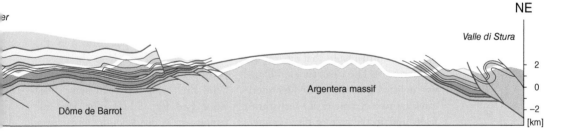

Figure 5.1-7
Geological cross-section through the Chaînes subalpines near Digne (France). Source: Adapted from Lickorish & Ford (1998).

with the formation of the Vocontian Trough in the Jurassic, and we therefore now find the sediments of the Vocontian Trough in the Digne nappe. The Digne nappe is internally shortened by further thrust faults that level off into the evaporitic decollement horizon towards the base, and according to this reconstruction by Lickorish & Ford (1998) the overall shortening in this cross-section is 21.5 kilometres. Relative to the crystalline basement, the frontal anticline in the Digne nappe was transported 11 kilometres to the southwest, the back end by 23 kilometres. This difference is explained by the internal shortening of the nappe, which shortened the original length of the nappe and thereby the back end was transported further along the thrust fault. The most important internal shortening occurred in the Tinée thrust, which on its own was responsible for a shortening of about 6.5 kilometres.

Figure 5.1-7 also shows that the crystalline basement was shortened as well and that it deformed the Digne thrust. The top of the crystalline basement was uplifted by thrust faults that dip towards the northeast. The Chateauredon Dome was uplifted beneath the St Jurs imbricates by two symmetrical, steep thrust faults and, in the case of the Argentera massif, a large-scale bulge developed. The activity of the Digne thrust must be older than this deformation, as shearing-off on such an irregular surface is not conceivable. On the other hand, the Digne thrust must be younger than the Miocene–Pliocene sediments of the Valensole basin that were overthrust which indicates an early Pliocene age for the Digne thrust and an even younger age for the uplift of the Argentera massif.

Comparison of the transport distances for the individual frontlines of the different allochthonous units in the Subalpine chains is also of interest. These distances are greatest, at more than 50 kilometres, in the north of the Haute Savoie and are directed west-northwest; they are a minimum of 35 kilometres in the Chartreuse and are west-directed; in the case of the Digne thrust they amount to only 11 kilometres and are southwest-directed. In all cases, the basal decollement horizon was curved by a later phase of deformation in the crystalline basement below, and partially subjected to small-scale deformation. The basement apparently remained unyielding over long periods and only became deformed under the substantial weight of the overlying nappe pile after the shearing-off of the Ssubalpine chains. This pattern is also widespread in the Helvetic and Penninic nappes.

The Penninic Nappes and their Contact with the Adriatic Continental Margin

The internal portion of the Western Alps includes the nappe pile formed by the Penninic nappes, which in the eastern portion are flanked by the Austroalpine Sesia Zone and the Southalpine Ivrea Zone (see Fig. 5.1-1). The Penninic nappe pile is divided into three parts: the Lower Penninic nappes stem from the Valais Trough, the Middle Penninic nappes from the Briançon Rise and the Upper Penninic nappes can be assigned to the Piemont Ocean.

Within these Penninic nappes, three types can be distinguished based on the composition of the associated rocks: Type I comprises Mesozoic–Cenozoic sediments that have been sheared off their crystalline basement; Type II comprises mainly crystalline basement and represents the original substratum for Type I; Type III comprises

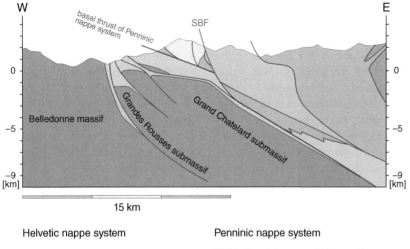

Figure 5.1-8 Geological cross-section through the external Penninic nappes and the internal Belledonne massif (France). Source: Ceriani & Schmid (2004). Reproduced with permission of Swiss Geological Society.

Helvetic nappe system

Mesozoic cover sediments

Crystalline basement

SBF   «Subbriançonnais» fault

Penninic nappe system

Middle Penninic nappes/Briançon Rise

Mesozoic sediments

Carboniferous – Permian (Zone Houillère)

Crystalline rocks/Permo-Carboniferous

Lower Penninic nappes/Valais Trough

Cenozoic (flysch)

Mesozoic (Bündnerschiefer/Schistes lustrés and ophiolites)

ophiolitic rocks and oceanic sediments. Types I and II are located in the Lower and Middle Penninic nappes, whereas Type III is in the Upper Penninic nappes.

The tectonic map in Fig. 5.1-1 reveals that the Lower Penninic nappes become narrower and peter out towards the south. This disappearance is associated with the fact that the Valais Trough is replaced laterally by the Vocontian Trough in the Dauphinois in the southern portion of the Western Alps. In contrast, the Middle Penninic nappes, which are made of crystalline basement derived from the Briançon microcontinent and sediments that were deposited on the Briançon Rise, form a broad band, within which the little town of Briançon is located, from where the name was taken. The Upper Penninic

nappes are mainly exposed in the eastern portion of the Western Alps, where two notable windows are formed by the dome-like bulges of Dora Maira and Gran Paradiso (see Fig. 5.1-1). Here, crystalline rocks from continental crust of the Briançon microcontinent are exposed in the centre of these windows, which are assigned to the Middle Penninic nappes.

The structure of the internal zone of the Western Alps and its relationship to the external zone is elucidated using three cross-sections. The first cross-section, Fig. 5.1-8, is based on work carried out by Ceriani & Schmid (2004). This shows the tectonic apposition of a flysch sequence onto the Mesozoic sediments of the Dauphinois in the roof of the internal Belledonne massif (Grandes Rousses and Grand

Chatelard submassifs). This flysch was deposited in the Valais Trough and is assigned to the Lower Penninic nappes, due to its present-day position at the bottom of the Penninic nappe pile. Another unit that is similar in origin and position is suspected to exist at great depth at the eastern end of the cross-section, and is composed of Mesozoic sediments (schistes lustrés or Bündnerschiefer) and includes ophiolitic material. The volume of these units from the Valais Trough is small because of their location near the southwestern end of the trough. In the east, the Lower Penninic nappes are overlain by Mesozoic or Permocarboniferous sediments of the Middle Penninic nappes, which are assigned to the Briançon Rise. A larger fault with a normal fault component (the 'Subbriançonnais' Fault, SBF) cross-cuts all these nappe contacts. The tectonic contact at the western delimitation of the Zone Houillère has a strike-slip component. In the extreme east, the Zone Houillère is in contact with crystalline basement that is also allocated to the Middle Penninic nappes, and we can see here how later deformation processes folded this nappe contact intensively.

The cross-section in Fig. 5.1-9 follows the ECORS-CROP seismic line and is based on research carried out by Bucher et al. (2004). The Lower Penninic nappes overlie sediments of the Dauphinois above the Mont Blanc massif in the form a nappe stack that is over 5 kilometres thick. The Versoyen nappe also contains prasinites, in addition to shales, while the Moûtier and Petit St Bernard nappes are composed of Mesozoic and Cenozoic clastic sediments (Bündnerschiefer or schistes lustrés). The Middle Penninic nappes that follow above this (Zone Houillère, Ruitor and Gran Paradiso) are mainly

Penninic nappes: Detached Mesozoic cover, detached, basement, folded thrust faults

composed of crystalline basement that represents the original upper crust of the Briançon Rise. The Zone Houillère has a core composed of crystalline basement that is overlain by a thick sequence of Carboniferous sediments. The name of this nappe stems from the coal seams in the Carboniferous sediments. The crystalline basement in the Ruitor nappe is overlain by a thin Permo-Triassic quartzite, which also applies to the internal zone (or the Vanoise nappe). The highest Middle Penninic unit is the Gran Paradiso massif, the dome-like surface of which is clearly recognizable in Fig. 5.1-9. During the Alpine orogeny, the nappe pile composed of crystalline basement was first stacked up in an imbricate fashion (Bucher et al. 2004) and the nappe contacts were then intensively folded, as shown in Fig. 5.1-9 – a process known as 'post-nappe folding'. The Ruitor fold that closes towards the west and the Valsavanche fold that closes towards the east are both particularly impressive. Their subhorizontal axial surfaces indicate a vertical shortening of the nappe pile, which, in turn, is contingent on horizontal stretching due to volume conservation. This phenomenon can be interpreted as a form of 'collapse structure': the nappe pile that had been compressed by plate collision partially escaped upwards to form a mountain chain, which disintegrated under its own weight and produced horizontal stretching in the direction of the plate convergence in the uppermost portion of the orogen – numerical models exhibit exactly this behaviour (see Selzer et al. 2008).

The Upper Penninic nappes form a cap over the Middle Penninic nappes, but they also plunge to depth in the east. They are composed of calcareous shales that contain metamafics and

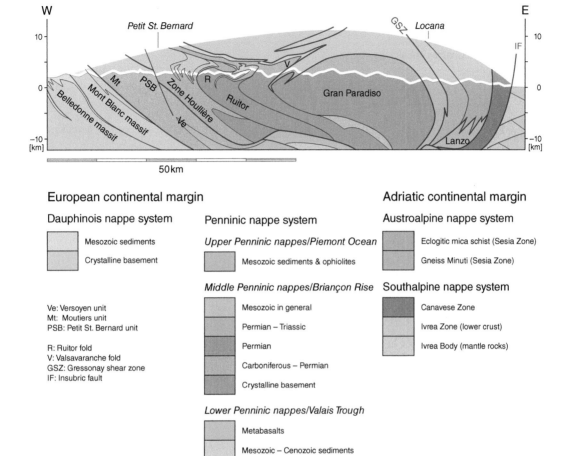

European continental margin

Dauphinois nappe system

	Mesozoic sediments
	Crystalline basement

Ve: Versoyen unit
Mt: Moutiers unit
PSB: Petit St. Bernard unit

R: Ruitor fold
V: Valsavaranche fold
GSZ: Gressonay shear zone
IF: Insubric fault

Penninic nappe system

Upper Penninic nappes/Piemont Ocean

	Mesozoic sediments & ophiolites

Middle Penninic nappes/Briançon Rise

	Mesozoic in general
	Permian – Triassic
	Permian
	Carboniferous – Permian
	Crystalline basement

Lower Penninic nappes/Valais Trough

	Metabasalts
	Mesozoic – Cenozoic sediments

Adriatic continental margin

Austroalpine nappe system

	Eclogitic mica schist (Sesia Zone)
	Gneiss Minuti (Sesia Zone)

Southalpine nappe system

	Canavese Zone
	Ivrea Zone (lower crust)
	Ivrea Body (mantle rocks)

have been overprinted by high-pressure metamorphism: a mixture of blueschist-facies and eclogitic metamafics, as well as high-pressure metasediments are now present. To the northwest of the Gran Paradiso, the basal thrust of the Upper Penninic nappes has been transformed into a complex shape by the thrust faults within the Middle Penninic nappes and by younger fold-ing processes, which allows the conclu-sion that this contact is the oldest tectonic contact within the entire nappe pile (see Bucher et al. 2004). The contact to the units of the Adriatic continental margin exhibits imbricate structure, as seen for example in the contact between the Lanzo unit (Piemont Ocean) and the Sesia Zone (Adriatic continental margin), which can be explained by the kinematic processes within the subduction zone. Numerical modelling (Pfiffner et al. 2000) has shown that a broad deformation zone develops at the contact between the plates in a subduc-tion zone, in which a mixture of rocks of different origins is produced, depend-ing on the changes in the subduction angle or in the mechanical properties of the subducted rocks – this deformation zone is also called a subduction channel. The Gneiss Minuti in the Sesia, Lanzo and Canavese zones (Fig. 5.1-9) would form part of this subduction channel.

▲ **Figure 5.1-9**
Geological cross-section through the Penninic nappe system along the seismic line ECORS/CROP. Source: Bucher et al. (2004). Reproduced with permission of Swiss Geological Society.

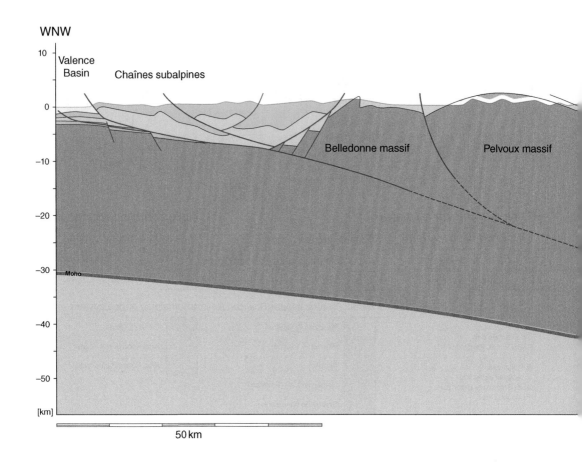

WNW

50 km

European continental margin

Dauphinois nappe system

- Cenozoic
- Cretaceous
- Jurassic
- Triassic
- European crust
- Eclogitized crust
- European mantle

Penninic nappe system

Upper Penninic nappes/Piemont Ocean

- Ophiolites (Viso)
- Mélange of ophiolites & sediments (Schistes lustrés)

Middle Penninic nappes/Briançon Rise

- Mesozoic
- Carboniferous – Permian
- Eclogite
- Crystalline rocks / upper crust

Adriatic continental margin

- Pliocene – Quaternary
- Mesozoic
- Adriatic crust
- Adriatic mantle

It is now important to consider the fact that, during active subduction, this channel was not vertical, as it is today, but plunged underneath the Adriatic plate towards the east – it was only the later folding of the Penninic nappe pile

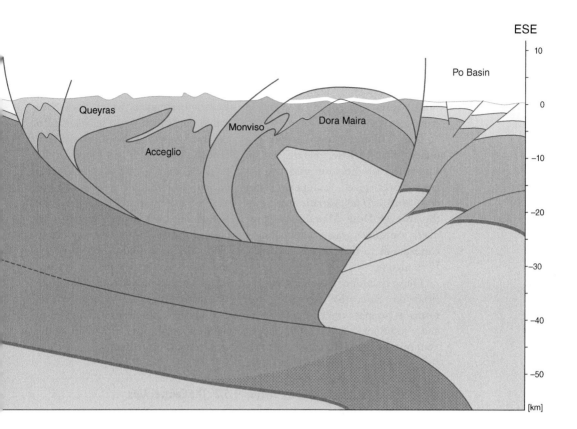

ESE

Po Basin

Queyras

Monviso   Dora Maira

Acceglio

▲ **Figure 5.1-10**
Geological cross-section
through the southern
Western Alps. Source:
Adapted from Lardeaux et al.
(2006) and Morag et al.
(2008).

that rendered it vertical. Two important shear zones are associated with the nappe contacts within the subduction channel and both possess shear-sense indicators that emphasize the uplift of the Penninic nappes relative to the Adriatic continental margin: the Gressonney shear zone is folded round the Gran Paradiso massif, such that a normal fault appears to be present at the surface; at the Insubric Fault, the Canavese Zone has been steeply thrust onto the Adriatic mantle and the lower crust.

Figure 5.1-10 shows a cross-section in the southern portion of the Western Alps. This is based on work carried out by Lardeaux et al. (2006) and Morag et al. (2008), but was also extended in the west based on data from Debelmas (1974) and Philippe et al. (1998). Of note is the fact that, according to Lardeaux et al. (2006), the core of the Dora Maira massif is made up of mantle rocks. This portion of the mantle is part of the Adriatic plate that was pushed into the crust of the Briançon Rise. The mantle was shortened on the Adriatic side by east-southeast-vergent thrusts. The Mesozoic sediments that lie above and the Cenozoic sediments of the Po Basin were possibly also affected by similar thrusts, as is indicated in the cross-section.

The European mantle plunges under the Adriatic mantle in an east-southeast direction and takes a layer of eclogized European lower crust with it. The Adriatic mantle in the core of Dora Maira above produces a positive gravity anomaly and corresponds to the Ivrea Body or the Ivrea mantle.

The crystalline basement of the Briançon Rise, the Acceglio unit, forms a complex imbricate structure and interlocks with the Upper Penninic nappes. The latter includes the Monviso Ophiolite and the Queyras nappe, which is composed of high-pressure metamorphic Mesozoic slates (schistes lustrés) and metabasalts. In the Dora Maira massif, ultra-high-pressure rocks form a shell around the core of the continental crustal rocks. There are also continental crustal rocks below the Queyras nappe. However, the contact is tectonic in nature and was later intensively folded. Permo-Carboniferous rocks are to be expected below the Mesozoic sediments in the westernmost Briançon-derived units and, below this, once again, continental crust. This corresponds to the Zone Houillère, as is discussed in the context of Fig. 5.1-9. A new situation arises at the contact to the Dauphinois in the west, as the Lower Penninic nappes are absent here. These nappes peter out towards the south because the Valais Trough gives way, in a

southwesterly direction, to the Vocontian Trough in the Dauphinois. In the cross-section in Fig. 5.1-10, the western portion of the Dauphinois is composed of the Pelvoux and Belledonne massifs which both have a thin cover of Jurassic sediments. In the case of the Pelvoux massif, the Jurassic sediments are overlain by thick Cenozoic flysch sediments (Champsaur Sandstone). Based on the situation in Fig. 5.1-6, the Belledonne massif is depicted as being thrust onto the foreland. The Mesozoic sediments between the Belledonne massif and the Valence Basin correspond to the fill in the Vocontian Trough. These sediments have been symmetrically shortened by west-northwest- and east-southeast-vergent thrust faults.

5.2 The Central Alps

The Central Alps essentially form the connection between the Western Alps that run north–south and the Eastern Alps that run east–west. Even though there is a gradual transition to the Western and Eastern Alps at both ends, the Central Alps are characterized by the peculiarity that the deepest tectonic units are exposed at the Earth's surface and the structure of all, the Helvetic, Penninic, Austroalpine and Southalpine nappe systems, is clearly visible due to the axial plunges. In addition, a good insight into the structure of the Jura Mountains and of the Molasse Basin is also provided here.

The tectonic map in Fig. 5.2-1 shows the Jura Mountains in the northwest, that are arc-shaped and become narrower towards their southern and eastern ends. The axes of the folds in the internal portion, the Folded Jura, emphasize this shape through their trend. They are cross-cut by strike-slip faults that were already active during the folding process. The Molasse Basin runs adjacent to this in the south, along the entire length of the Central Alps. This basin then continues in an easterly direction in the Eastern Alps. However, in the west, it becomes narrower at the transition to the Western Alps. The internal portion of the Molasse Basin was affected by Alpine deformation. This belt displays thrust faults and folds in the Molasse sediments and was once fully covered by Alpine nappes and is therefore called the 'Subalpine Molasse'.

In the Alps, the Helvetic nappe system is visible at the external edge. This is composed of an allochthonous nappe complex that has been displaced by up to 50 kilometres, the Helvetic nappes, and below this, a parautochthonous complex with folds and thrust faults with only moderate displacements, the Infrahelvetic complex. The crystalline uplifts forming the Aiguilles Rouges, Mont Blanc, Aar and Gotthard massifs have been partially thrust onto the autochthonous foreland over several kilometres and represent the crystalline basement of the Infrahelvetic complex. The origin of the Helvetic nappes lies in a region to the south or southeast of these crystalline uplifts. The Helvetic nappes exhibit erosional remnants of higher-level nappes in several locations. These klippen provide evidence for the fact that the Penninic and Austroalpine nappes once covered large parts of the Helvetic nappe system. Larger klippen composed of Penninic nappes are present in the Chablais (Haute Savoie) and in the Prealps of western Switzerland. There are even erosional remnants of the Austroalpine nappes in the, sometimes very small, klippen in central Switzerland. The Penninic nappe system is exposed in the central portion of the Central Alps. Nappes composed of crystalline rock predominate in the southern portion and sedimentary nappes are more prevalent in the north. The structurally deepest and most highly metamorphosed units are exposed in the Leventina and north of Domodossola in the so-called Lepontian Dome. The dipping nappe contacts around this area indicate that a dome-shaped bulge is actually present, in the core of which the structurally deepest and topmost metamorphosed rocks are exposed at the Earth's surface. Normal faults mark the western edge (Simplon Fault) and the eastern edge (Forcola and Turba normal faults) and contributed towards the uplift of the Lepontian Dome.

The outcrop pattern of the contact between the Penninic and Austroalpine nappe systems on the geological map is of interest because it exhibits a zigzag pattern in the Central Alps. In the extreme east, only Austroalpine nappes are exposed at the Earth's surface, crystalline rocks in the south and sediments in the north. In the Engadin, the Penninic nappes underlying the Austroalpine nappe system are exposed in a tectonic window, the Engadin Window. In the southeast, this window is delimited by a fault that uplifts the Penninic nappes relative to the Austroalpine nappes in the southeast.

▶ **Figure 5.2-1**
Simplified geological-tectonic map of the Central Alps with traces of cross-sections (green lines – numbers are figure numbers). Double arrows denote major fold axes.

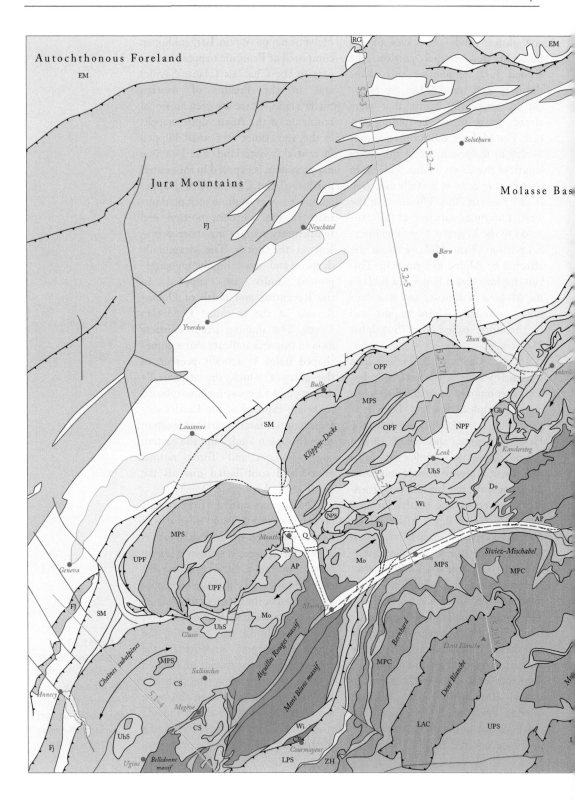

Legend to Figure 5.2-1

| Q | Quaternary fill of main valleys | | CP | Cenozoic pluton |

European (Northalpine) Foreland

RG	Rhine Graben
	Autochthonous Molasse		SM	Subalpine Molasse
EM	Mesozoic sediments
FJ	Folded Jura Mountains

Nappe systems of the European continental margin

Helvetic nappe system

UhS	Ultrahelvetic cover nappes
UhC	Ultrahelvetic crystalline nappes (Lucomagno, Verampio)
ShS	Southhelvetic cover nappes

CS	Chaînes subalpines		Wi	Wildhorn nappe		Db	Drusberg nappe		Sä	Säntis nappe
						Ax	Axen nappe			
			Di	Diablerets nappe		Gh	Gellihorn nappe		GSz	Gonzen schuppen zone
									GM	Glarus & Mürtschen nappe
						IV	Ilanz Verrucano		VGM	Verrucano of Glarus and Mürtschen nappe

Infrahelvetic complex

						SF	Subalpine Flysch
Mo	Morcles nappe		Do	Doldenhorn nappe		All	Allochthonous units
AP	Autochthonous-parautochthonous cover

	Chétif massif (Ch)
	Mont Blanc massif			Gotthard massif	Tavetsch massif (Ta)
	Belledonne massif			Aiguilles Rouges massif			Aar massif

Penninic nappe system

Lower Penninic nappes/Valais Trough

LPF	Flysch sediments
LPS	Mesozoic sediments
LPC	Crystalline basement

Middle Penninic nappes/Briançon Rise

MPS	Mesozoic sediments
ZH	Palaeozoic sediments (Zone Houillère)
MPC	Crystalline basement

Upper Penninic nappes/Piemont Ocean

| UPF | Flysch sediments |
| UPS | Mesozoic sediments & ophiolites |

Nappe systems of the Adriatic continental margin

Southalpine nappe system

SS	Mesozoic sediments
SK	Crystalline basement
IZ	Lower Crust (Ivrea Zone)

Austroalpine nappe system

Upper Austroalpine nappes

| UAM | Mesozoic sediments |
| UAC | Crystalline basement |

Lower Austroalpine nappes

| LAM | Mesozoic sediments |
| LAC | Crystalline basement |

Adriatic Foreland

| | Cenozoic of Po Basin |

The basal thrust of the Austroalpine nappe system has been uplifted to a domed shape throughout the remainder of the margin of the window, such that the deeper units were exposed by erosion. The Penninic–Austroalpine interface zigzags transversely across the Alps to the south of Lake Constance. The contact exhibits a tendency to dip towards the east, such that mainly Austroalpine nappes are exposed at the surface in the Eastern Alps. The encroachment of the Austroalpine nappe system in Vorarlberg and its retreat in the Prättigau (this is called the Prättigau half-window) is particularly conspicuous. A larger klippe of Austroalpine nappes on top of the Penninic nappe system is noted further west in the region of the Dent Blanche in western Switzerland/Italy. At the northern edge of the Austroalpine realm, to the east of Lake Constance, a thin band of Penninic and, in locations, Helvetic nappes run in an easterly direction to the south of the Subalpine Molasse. Ophiolitic rocks mark the boundary between the Penninic and Austroalpine realms in many areas. These are the Upper Penninic nappes that are composed of rocks from the former Piemont Ocean. The presence of windows and klippen and the transverse zigzag course of the Penninic–Austroalpine interface across the orogen can be explained as an erosional front. The thin band of Penninic nappes at the northern edge of the Alps and the Engadine Window are evidence for the fact that the Penninic (and Helvetic) nappe systems continue underneath the Austroalpine nappe system in the east.

A marked fault can be made out in the south of the Central Alps. This is part of the Peri-Adriatic fault system that can be followed from the Western Alps, through the Central Alps and to the eastern end of the Eastern Alps. In the region of the Central Alps, this fault system separates the Southalpine nappe system from the Penninic and Austroalpine nappe systems that lie to the north. This segment of the Peri-Adriatic fault system is referred to as the Insubric Fault. Two larger granodioritic intrusions (Bregaglia and Adamello) are located in the immediate vicinity of this fault system. Several south-vergent thrust faults with a southwest–northeast to east–west course are visible in the Southalpine nappe system. Those in the north thrust crystalline basement onto crystalline basement or Mesozoic sediments.

The cross-section in Fig. 5.2-2 summarizes the structure of the Central Alps along the NRP 20 Eastern Traverse (see Pfiffner et al. 1997a) and the European GeoTraverse (EGT) (see Blundell et al. 1992). The cross-section is located to the east of the Jura Mountains. The shallow dip of the Mesozoic and Cenozoic sediments in the direction of the Alps is visible in the Molasse Basin. The thickness of the Cenozoic sediments in the Molasse Basin increases in this direction and the imbricate structure in the Subalpine Molasse is visible at the boundary to the Helvetic nappe system. The boundary between the Subalpine and Plateau Molasse is demarcated by a nearly symmetrical anticlinal structure, a so-called triangle zone. The Helvetic nappes are visible in the form of an allochthonous nappe complex and its prolongation to the south was removed by erosion, and below them, in the Infrahelvetic complex, the uplift of the crystalline basement forming the Aar massif has created a dome that is 30 kilometres wide. To the south of this, the basement block of the Gotthard massif has been thrust in a northerly direction over a distance of more than

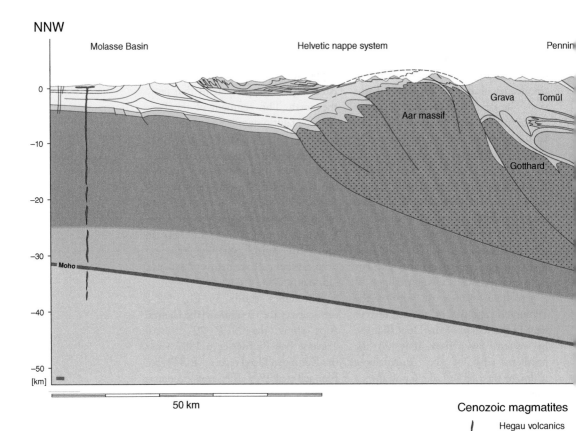

NNW

Molasse Basin   Helvetic nappe system   Pennin

Grava   Tomül

Aar massif

Gotthard

Moho

50 km

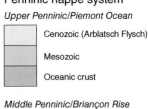

Cenozoic magmatites

/ Hegau volcanics

Bregaglia Pluton
Granodiorite/tonal

European continental margin

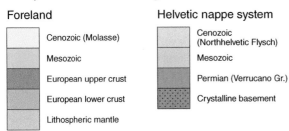

Foreland

☐ Cenozoic (Molasse)

☐ Mesozoic

☐ European upper crust

☐ European lower crust

☐ Lithospheric mantle

Helvetic nappe system

☐ Cenozoic
(Northhelvetic Flysch)

☐ Mesozoic

☐ Permian (Verrucano Gr.)

☐ Crystalline basement

Penninic nappe system

Upper Penninic/Piemont Ocean

☐ Cenozoic (Arblatsch Flysch)

☐ Mesozoic

☐ Oceanic crust

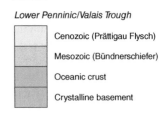

Middle Penninic/Briançon Rise

☐ Mesozoic

☐ Crystalline basement/Permo-Carboniferous

Lower Penninic/Valais Trough

☐ Cenozoic (Prättigau Flysch)

☐ Mesozoic (Bündnerschiefer)

☐ Oceanic crust

☐ Crystalline basement

Figure 5.2-2 Geological
cross-section through the
eastern Central Alps along
the seismic lines of the
Eastern Traverse of NRP20
(after Pfiffner & Hitz, 1997).
Source: Pfiffner et al. (1997).

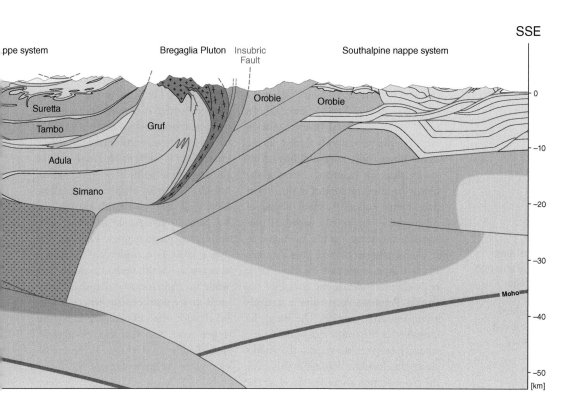

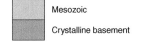

Adriatic continental margin

Austroalpine nappe system

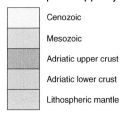

| | Mesozoic |
| | Crystalline basement |

Southalpine nappe system

	Cenozoic
	Mesozoic
	Adriatic upper crust
	Adriatic lower crust
	Lithospheric mantle

10 kilometres. In the Penninic nappes those composed of Mesozoic (and Cenozoic) sediments (Grava, Tomül, Schams) are clearly separated from those composed of crystalline basement (Simano, Adula, Tambo, Suretta). In many cases, the basal thrust faults in these nappes were tightly folded during a later phase (post-nappe folding). This is very clear, for example, in case of the Grava thrust or the contact in the roof of the Suretta nappe. A small klippe of Austroalpine nappe (Ela nappe) is preserved in the summit of Piz Toissa and an erosional remnant of the Upper Penninic Platta nappe is present on Piz Platta: both klippen indicate the course of the eroded Penninic-Austroalpine boundary in this cross-section. The Bregaglia intrusion in the south cross-cuts nappe contacts, but runs down into the depths along the Insubric Fault. In this case, the cross-section should make clear the inherent link between the emplacement of the intrusion and movement along the Insubric Fault. A thin sheet of Austroalpine nappes (Err-Bernina and Tonale nappes) lies sandwiched in a vertical orientation, immediately to the north of the Peri-Adriatic fault system. This steep zone developed due to steep, south-directed thrusting of the Penninic nappes pile relative to the Southalpine nappe system. This also led to the formation of so-called 'back-folds' (south-vergent folds with a steeply inclined axial surface), which are visible, for example, at the contact between the Simano and Adula nappes.

In the Southalpine realm, Fig. 5.2-2 shows the south-vergent nappe structure, which includes both the crystalline basement and the sediments. The lower crust of the Southalpine nappe system, the Adriatic continental margin, is indented into the European crust in the shape of a wedge towards the north.

Central Alps: Bivergent nappe structure, Asymmetric structure of crust-mantle boundary

The structure of the lower crust in Fig. 5.2-2 is based on teleseismic studies carried out by Waldhauser et al. (2002), Diel et al. (2009) and Wagner et al. (2012). The European lower crust plunges under the Adriatic crust, the European upper crust has been peeled off its lower crust and internally shortened. This is recognizable not only in the crystalline uplift of the Aar massif, but also in the allochthonous cystalline blocks of the Gotthard massif and the Simano and Adula Penninic crystalline basement nappes. The Adriatic upper crust is considerably thickened beneath the Southalpine nappe system, while the Adriatic lower crust is thickened at its northern tip where it collided with the off-scraped European upper crust, and furthermore it more than doubles in thickness at the southern end of the cross-section.

The Jura Mountains

In the Central Alpine section, the Jura Mountains are essentially large-scale folds of Mesozoic sediments that were sheared off their crystalline substratum in the Triassic evaporites and transported to the northwest. The tectonic map in Fig. 5.2-1 illustrates the trends of the anticlines, which are gathered closely together in the southeastern part, in the Folded Jura, whereas they are separated by broad synclines in the Plateau Jura in the northwestern part. Large strike-slip faults ('décrochements') cut through the folds. We can see that the strike-slip faults were active at the same time as folding occurred if we compare the two sides of these faults. The folds on either side do not match across these faults indicating that the folds formed independently.

The cross-section through the Central Jura Mountains in Fig. 5.2-3

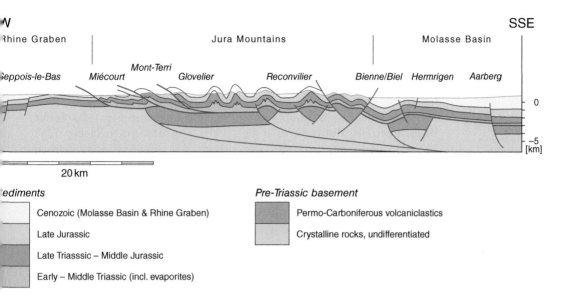

NW   SSE
Rhine Graben   Jura Mountains   Molasse Basin

        Mont-Terri
Seppois-le-Bas   Miécourt   Glovelier   Reconvilier   Bienne/Biel Hermrigen   Aarberg

                                                                                                    0

                                                                                                    −5
                                                                                                    [km]

        20 km

Sediments   Pre-Triassic basement

    Cenozoic (Molasse Basin & Rhine Graben)   Permo-Carboniferous volcaniclastics

    Late Jurassic   Crystalline rocks, undifferentiated

    Late Triasssic – Middle Jurassic

    Early – Middle Triassic (incl. evaporites)

▲ **Figure 5.2-3**
Geological cross-section through the central Jura Mountains.

shows the internal deformation of the Mesozoic–Cenozoic strata. In the north-northwest, the frontal portion, the strata have been thrust onto the autochthonous foreland. Two further thrusts are visible, immediately next to this in the south-southeast. The internal portion of the Jura Mountains is aptly termed the 'Folded Jura'. Those folds in which the centres are filled with evaporites can be interpreted as detachment folds, and n some cases their amplitudes are greater than 1 kilometre. Laubscher (1965) conducted a detailed analysis on the amount of shortening in the Jura Mountains. In his model, which assumes a rotation of the Jura Mountain chain around its eastern end, the internal shortening increases progressively from the eastern end of the Jura Mountains towards the west. Laubscher (1965) estimated the shortening at approximately 15 kilometres for the cross-section in Fig. 5.2-3, whereas retro-deformation of the cross-section shown in Fig. 5.2-3 results in a little less (10.5 kilometres). The amount of shortening that is determined is dependent on the interpretation of the types of folds: some anticlines contain thrusts in their cores, which results in a high value for shortening, whereas others are almost developed in the form of mushroom folds, which result in a lower value for shortening. In addition, the cross-sectional shape of the anticlines often changes along strike, which further complicates the determination of shortening.

A question that had preoccupied early researchers pertained to the mechanism of folding of the Jura Mountains and the incorporation of the pre-Triassic basement. During reconstructions of cross-sections of the fold structures, it soon became clear that it is impossible to fold the pre-Triassic basement in the same way as the layers above. Based on this observation, it was concluded that some form of decoupling must have taken place in the Triassic evaporites. However, as the shortening that is present in the Mesozoic layers must also be compensated for somewhere in the pre-Triassic basement, the 'Fernschub or 'distant-push' hypothesis

▶ **Figure 5.2-4**

▶ **Figure 5.2-4**

Geological cross-section through the Folded Jura along the Grenchenberg railroad tunnel (redrawn after Buxtorf, 1916). The photograph displays a view of the folds in the gorge of Moutier (Canton Jura, Switzerland) looking west-southwest. The syncline is a kink fold with straight limbs whereas the anticline is more open and contains several gentle kinks. The cliffs are made of Late Jurassic limestones. Source: Redrawn after Buxtorf (1916).

Folded Jura: Detached Mesozoic cover

was put forward, which was clearly formulated for the first time by Buxtorf (1907). According to this hypothesis, the impetus for the folding of the Jura Mountains came from the Alps and was transferred by the Mesozoic and Cenozoic sediments of the Molasse Basin. Shortening that occurred in the Mesozoic sediments in the Jura Mountains was compensated for in the crystalline basement in the Alps, namely, in the frontal nappes or under the basement uplifts (Aar and Aiguilles Rouges massifs) in the Helvetic nappe system. Laubscher (1961) discusses the historical development of the ideas and presented an in-depth analysis and starting points for solving the problem of the 'distant-push' hypothesis.

We now know that the decollement horizon lies in the Late Triassic evaporites in the western Jura Mountains and in those of the Middle Triassic in the eastern Jura Mountains. It then runs beneath the Molasse Basin and is rooted in the basement uplifts. The entire Molasse Basin was therefore also shifted as a 'piggy-back basin' during the folding of the Jura Mountains. This is revealed, for example, by the fact that the Upper Marine Molasse (UMM) in the western Molasse Basin now lies at an altitude of 1000 metres above sea level, while it is below sea level in the east. The UMM basically represents a 'fossil sea level' and the western Molasse Basin has therefore been uplifted substantially. The uplift can be at least partially explained by the thrusting of the Mesozoic–Cenozoic sediments of the Molasse Basin up a ramp inclined towards the Alps (see Laubscher 1961 and Pfiffner et al. 1997b). The geometry of this decollement horizon is only known approximately and indirectly through cross-sectional reconstructions. For example, Laubscher (1961) documented an irregular surface that lay

higher and almost horizontally in the central part of the Jura Mountains and dipped relatively steeply towards the southeast on the southern slopes of the Jura Mountains. A far simpler picture then emerged based on new cross-sectional reconstructions (Laubscher 1965). The cross-section in Fig. 5.2-3 shows that the decollement horizon exhibits small-scale irregularities, especially over partially squeezed out fills from Permo-Carboniferous grabens. The graben in the subsurface of Hermrigen is known from investigations linked to drilling for hydrocarbons and a reprocessing of the reflection seismic data (see Pfiffner et al. 1997b). Information on other Permo-Carboniferous basins is more speculative, as their geology has been extrapolated from known grabens in neighbouring areas, however, the higher position of the decollement horizon to the north-northwest of Bienne in Fig. 5.2-3 is inevitable based on the construction of the cross-section. The irregularities in the decollement horizon that are thus obtained indicate that the pre-Triassic basement beneath the Jura Mountains is also slightly shortened and warped. This shortening is minimal and occurred during a later phase of the formation of the Jura Mountains. There are also discontinuities in the course of the decollement horizon under the Molasse Basin, thus the basement was apparently slightly deformed during the final phase of formation of the Jura Mountains. This may be linked to the increased load that resulted from thickening of the Jura Mountains by folding and in the Molasse Basin due to emplacement of a thicker basin fill.

A classic example of the internal structure of the folds is provided by the cross-section along the Grenchenberg railway tunnel. The cross-section in Fig. 5.2-4 has been adapted from Buxtorf (1916), who documents the

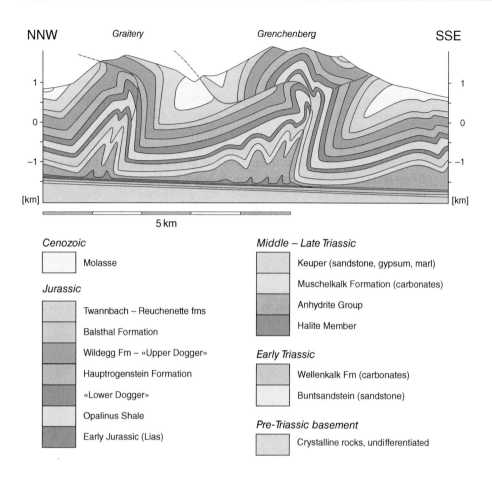

NNW   *Graitery*   *Grenchenberg*   SSE

5 km

Cenozoic

☐ Molasse

Jurassic

☐ Twannbach – Reuchenette fms

☐ Balsthal Formation

☐ Wildegg Fm – «Upper Dogger»

☐ Hauptrogenstein Formation

☐ «Lower Dogger»

☐ Opalinus Shale

☐ Early Jurassic (Lias)

Middle – Late Triassic

☐ Keuper (sandstone, gypsum, marl)

☐ Muschelkalk Formation (carbonates)

☐ Anhydrite Group

☐ Halite Member

Early Triassic

☐ Wellenkalk Fm (carbonates)

☐ Buntsandstein (sandstone)

Pre-Triassic basement

☐ Crystalline rocks, undifferentiated

advances in the knowledge of the geological structure of the Jura Mountains made by building the tunnel. The predictions and findings in the Hauenstein and Grenchenberg railway tunnels are compared in great detail and commented on. In the case of the Grenchenberg tunnel, a folded thrust was discovered in the core of the Grenchenberg anticline. Buxtorf (1916) outlines how this started as a thrust fault that was parallel to bedding in the decollement horizon (forming a flat), traversed the Mesozoic–Cenozoic sedimentary strata as a ramp and was then progressively folded into its present-day shape by the formation of the Grenchenberg anticline. This example shows that both thrust tectonics and fold tectonics shaped the structure of the Jura Mountains. Figure 5.2-3 shows that thrust faults predominate in the external part of the Jura Mountains and folds in the internal part. At the very eastern end of the Jura Mountains there are only thrust faults, while in the western Jura Mountains thrust faults are also found in the internal part. It has not yet been clarified which factors are decisive in the formation of folds versus thrusts. It is conceivable that these factors could be primary differences in the 'mechanical stratigraphy' of the Mesozoic rock suites, as has, for example, been demonstrated in the case of the Penninic and Helvetic nappe systems. The fold style is exhibited in deep gorges, called 'cluses', crossing the fold axes at right angle, which were cut by ancestral rivers that kept pace with folding (so-called antecedent rivers), for example the gorge of Moutier shown in Fig. 5.2-4.

The Molasse Basin

The Molasse Basin can be subdivided into three zones based on its tectonic structure: the Subalpine Molasse in the south, the flat Plateau Molasse or Foreland Molasse and the Subjurassic Zone in the north. **The Subalpine Molasse** is composed of a thrust belt that was overthrust by Alpine nappes. These are the Helvetic nappes to the east of the Aar Valley and the Penninic nappes to its west. The internal structure of the Subalpine Molasse and its contact in the north with the adjacent Plateau Molasse also changes along strike (see Fig. 5.2-1). In eastern Switzerland, the Subalpine Molasse has been thrust under the tilted Plateau Molasse forming a 'triangle zone'; a fold zone with nearly vertical fold limbs dominates this contact in central Switzerland, extending as far as the Aar Valley. The Plateau Molasse has been overthrust by the Subalpine Molasse to the west of the Aar Valley. The abrupt change in structures across the Aar Valley indicates the presence of a strike-slip fault that was active at the same time as the thrust faults.

The strata are mainly flat in the adjacent northern **Plateau Molasse or Foreland Molasse**, but are successively more tilted toward the Subalpine Molasse. Isolated broad and very open anticlines are discernible in the central part of the Plateau Molasse. These are more common in the western Molasse Basin, that is, in the rear part of the Jura Mountains.

The **Subjurassic Zone** extends along the southern edge of the Jura Mountains and is characterized by several anticlines that are slightly narrower than in the Plateau Molasse, and which also affect the underlying Mesozoic strata.

In spite of the apparent transition from the Plateau Molasse through the Subjurassic Zone into the Jura Mountains, the southernmost anticlines in the Jura Mountains exhibit a comparatively far greater amplitude. The apices of these anticlines are located at 1000–1700 metres above sea level and are thus 500–1000 metres higher than those in the Subjurassic Zone. Figure 5.2-1 shows that the anticlines in the southern foot of the Jura Mountains and in the Subjurassic Zone are characterized by extraordinary axial plunges. With the exception of the anticlines in the region of Lakes Neuchatel and Bienne (St Blaise, Cressier and Petersinsel), the anticlines are arranged en échelon and 'left-stepping' and plunge towards the east (Engelberg to the southeast of Olten, Chestenberg and Lägern). It is as yet unclear to what extent these plunges are connected to differences in the thicknesses of the associated sedimentary strata. It is also conceivable that the Molasse Basin fill was forced, as an indentor, into the southernmost folds, leading to shearing and rotation during the folding of the Jura Mountains. In this case, the core of the indentor would be located around the area of Grenchen.

Burkhard (1990) attempted to create a coherent kinematic model to describe the shortening in the Jura Mountains and in the Molasse Basin. According to this model, the total shortening of the Jura Mountains and the Molasse ought to be constant along the strike and have occurred simultaneously. His interpretation stated that the lesser amount of shortening observed in the Eastern Jura Mountains was compensated for by the greater amount of shortening in the eastern Subalpine Molasse. Burkhard (1990) reached the conclusion that the Plateau Molasse, in its guise as a 'deformable indentor', provided the best explanation for the different amounts of shortening in the Jura Mountains. A consequence of this model is that the Molasse Basin would have needed to be rotated clock-wise by 7°. Kempf et al. (1999) demonstrated using geomagnetic

data that a concordant rotation of between 7 and 14° can, indeed, be assumed. With reference to the internal deformation, Burkhard (1990) assumed that fault zones and strike-slip faults were present, and although there are no clear indications of this at the surface, the gentle folds in the Plateau and Subjurassic Molasse clearly point to internal deformation.

The two cross-sections of Figs 5.2-5 and 5.2-6 show the general structure of the Molasse Basin in the west and east, respectively, and in both a general increase in thickness of the Molasse sediments in the direction of the Alps is clearly visible. This applies, in particular, to the Lower Freshwater Molasse (LFM). In the east and west, the Subalpine Molasse thrusts must reach as far as the crystalline basement in a southerly direction. The cumulative amount of displacement is too great for it to simply lead into and disappear in the basal thrust of the Alpine nappes. It should also be noted that the Upper Marine Molasse (UMM) lies far deeper in the eastern cross-section than in the western cross-section. The elevated position of the strata of the Plateau Molasse in the west was attributed to the so-called 'residual uplift' (Resthebung) in older literature. We now know that this elevated position can be explained by the passive transport of the Molasse Basin on the inclined Triassic decollement horizon during the Pliocene folding phase of the Jura Mountains, as has already been discussed in the context of the 'distant-push' hypothesis in the Jura Mountains.

In the west and in the east, the Lower Marine Molasse strata (LMM) are limited to the Subalpine Molasse and those of the Upper Freshwater Molasse (UFM) to the Plateau Molasse. This shows how the axis of the foreland basin shifted from an internal to a more external position

Molasse Basin: Highly deformed inner part, Piggy-back basin in the West

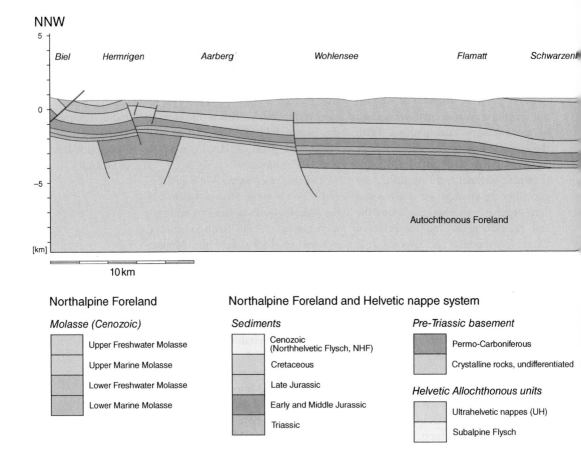

NNW

Biel   Hermrigen   Aarberg   Wohlensee   Flamatt   Schwarzen

Autochthonous Foreland

[km]

10 km

Northalpine Foreland

Molasse (Cenozoic)

- Upper Freshwater Molasse
- Upper Marine Molasse
- Lower Freshwater Molasse
- Lower Marine Molasse

Northalpine Foreland and Helvetic nappe system

Sediments

- Cenozoic (Northhelvetic Flysch, NHF)
- Cretaceous
- Late Jurassic
- Early and Middle Jurassic
- Triassic

Pre-Triassic basement

- Permo-Carboniferous
- Crystalline rocks, undifferentiated

Helvetic Allochthonous units

- Ultrahelvetic nappes (UH)
- Subalpine Flysch

▲ **Figure 5.2-5**
Geological cross-section through the western Molasse Basin along the seismic lines of the Western Traverse of NRP20.
Source: Pfiffner et al. (1997).

over time. The absence of younger sediments in the Subalpine Molasse can be explained by the fact that thrust faults at the front of the orogen at the time resulted in the internal part of the foreland basin being overthrust and covered and, therefore, it was not possible for any younger strata to be deposited. These processes are described in more detail in Pfiffner (1986) for eastern Switzerland and in Burkhard & Sommaruga (1998) for western Switzerland.

The cross-section in Fig. 5.2-5 is based on the reflection seismic investigations conducted by NRP 20 (Pfiffner et al. 1997b). An anticlinal structure is visible above a partially squeezed out Permo-Carboniferous graben fill near Hermrigen in the north-northwest. Two further Permo-Carboniferous basins can be seen below the Wohlensee and under the front of the Penninic nappes. In both cases, the Mesozoic sequence has been distorted by partial squeezing out (or partial inversion) of the graben fills. The decollement horizon in the Triassic evaporites that links to the

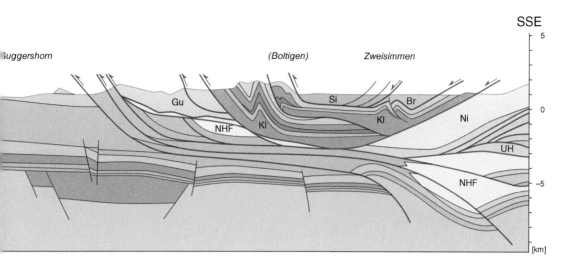

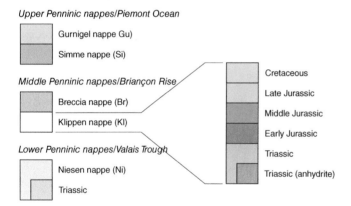

Penninic nappe system

Upper Penninic nappes/Piemont Ocean

Gurnigel nappe Gu)

Simme nappe (Si)

Middle Penninic nappes/Briançon Rise

Breccia nappe (Br)

Klippen nappe (Kl)

Lower Penninic nappes/Valais Trough

Niesen nappe (Ni)

Triassic

Cretaceous

Late Jurassic

Middle Jurassic

Early Jurassic

Triassic

Triassic (anhydrite)

Jura Mountain has also been affected by this deformation. This means that the shortening of the basement during the squeezing out of the Permo-Carboniferous graben fills occurred later than the deformation affecting the decollement horizon. The seismic data indicate the presence of Molasse thrust sheets in the footwall of the Penninic nappes in the Subalpine Molasse. These thrust sheets were apparently overthrust by the Penninic nappes after they had formed. A transition from the Northhelvetic Flysch sequence to the LMM must be present in the extreme south, but its precise geometry is a matter for speculation. However, the folds and thrust faults affecting the top of the crystalline basement do indicate more intensive shortening of the basement. The situation in the Penninic nappe pile is quite unusual. The northernmost unit, the Gurnigel nappe, originated in the Southpenninic realm and therefore should be assigned to the Upper Penninic nappes, but this cross-section shows it to be the lowest unit in the

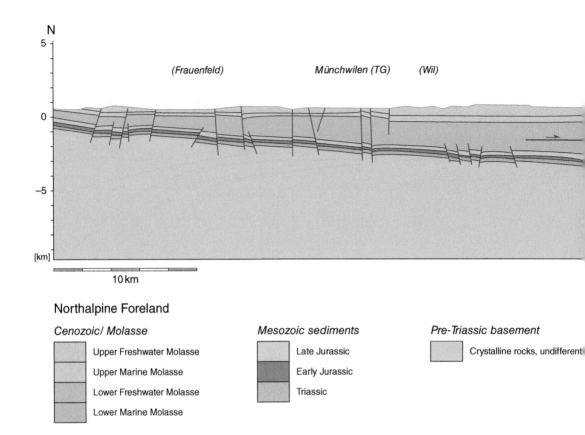

Northalpine Foreland

Cenozoic/ Molasse

	Upper Freshwater Molasse
	Upper Marine Molasse
	Lower Freshwater Molasse
	Lower Marine Molasse

Mesozoic sediments

	Late Jurassic
	Early Jurassic
	Triassic

Pre-Triassic basement

	Crystalline rocks, undifferent

▲ **Figure5.2-6**
Geological cross-section through the eastern Molasse Basin along the seismic lines of the Eastern Traverse of NRP20. Source: Pfiffner et al. (1997).

Penninic nappe system. Nappe stacking obviously did not occur in the usual manner here, whereby stacking starts in the internal portion and then shifts towards the external portions. The rule governing such 'in-sequence thrusting' dictates that units higher up stem from more internal palaeogeographical realms. This rule applies to the Northpenninic Niesen nappe (Valais Trough) and the mid-Penninic klippen nappe (Briançon Rise) in Fig. 5.2-5, but 'out-of-sequence thrusting' must be assumed for the Gurnigel nappe – Wissing & Pfiffner (2002) discuss this in greater depth.

The cross-section in Fig. 5.2-6 is also based on reflection seismic data (Pfiffner et al. 1997b). The transition zone between the Subalpine and Plateau Molasse is of interest here, which is characterized by an antiform (also called a 'triangle zone'). Two thrusts, each with different vergence, were rotated into a vertical position in the core of this antiform. In its core, the LMM has been thrust upwards in a northerly direction and pressed into the LFM; in the upper portion, the strata belonging to the lower LFM have been rotated into the vertical and isoclinally folded. The north-vergent thrusts further to

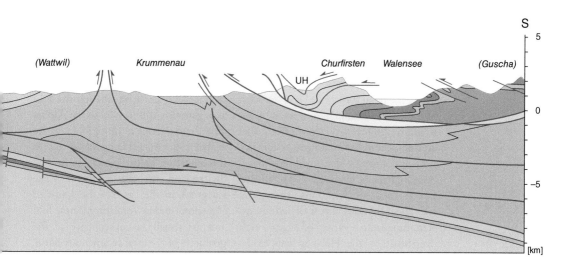

Helvetic nappe system

Mesozoic sediments

	Cretaceous
	Earliest Cretaceous (Palfris Shale)
	Late Jurassic
	Middle Jurassic
	Early Jurassic
	Triassic

Pre-Triassic basement

	Permian (Verrucano Group)

Allochthonous units

	Ultrahelvetic nappes (UH)
	Subalpine Flysch

the south are vertical at the Earth surface, but flatten off further down. These Molasse thrust sheets were sheared off along the argillaceous marls of the LMM. The thickness of the Subalpine Molasse is far greater in this cross-section than in that shown in Fig. 5.2-5, which is due to the greater thickness of the Molasse strata and not to the number of thrust sheets stacked on top of each other. The cross-section shows that the Helvetic nappes overthrust the uppermost thrust sheet almost parallel to the strata, but the map shows that the basal thrust of the Helvetic nappes cross-cuts internal thrusts in the Subalpine Molasse (see discussion on this in Pfiffner 1986).

The Helvetic Nappe System

The Helvetic nappe system can be divided into two very different tectonic zones: the Helvetic nappes at the top and the Infrahelvetic complex below. The **Helvetic nappes** include all allochthonous sediments that were overthrust in a northerly direction along the basal thrust fault and over many tens of kilometres. This basal thrust changes in character and changes its name from west to east: the Diablerets thrust in the west, the

Gellihorn thrust on either side of the Kander valley, the Axen thrust between Lauterbrunnen valley and Linth valley, and the Glarus thrust in the east.

The **Infrahelvetic complex** lies below the basal thrusts and is composed of crystalline basement and its autochthonous sedimentary cover. Both the crystalline and the sedimentary formations are strongly deformed, and have been folded and dismembered by thrust faults. In some locations, there are also allochthonous slivers of sediments stemming from a variety of facies zones overthrust by the Helvetic nappes. It should be noted that the Ultrahelvetic and Southhelvetic units are the most important elements.

The different Helvetic nappes are given different colours and marked with indices on the tectonic map in Fig. 5.2-1. A continuous band that runs along the entire length of the Central Alps and into the Eastern Alps is visible at the northern edge of the Helvetic realm, which is composed of the Wildhorn, Drusberg and Säntis nappes. These upper Helvetic nappes are essentially composed of Cretaceous sediments. In the case of the Wildhorn nappe, these are still connected to their Jurassic sedimentary substratum in the internal portion. In contrast, the Cretaceous layers in the Drusberg and Säntis nappes have been completely sheared off their Jurassic substratum. The lower Helvetic nappes sometimes contain Cretaceous sediments (Diablerets nappe, parts of the Axen nappe, Mürtschen nappe). Otherwise, only Triassic and Jurassic sediments are present. In the case of the Glarus nappe, Permian sediments (Verrucano) are also involved in its structure. In the extreme southwest, in the Western Alps, the Morcles nappe is overlain by the nappes of the Subalpine Chains (chaînes subalpines) of the Dauphinois. Even though their internal structure is

different, the structural position of the Dauphinois nappes equates precisely to that of the Helvetic nappes.

Figure 5.2-1 also shows the course of the fold axes of the most important large-scale folds. Even though the fold axes generally run parallel to the Alpine strike, curved axes forming fold arcs are visible in the west (Haute Savoie and to the south of Lake Brienz), in central Switzerland (to the south of Lake Lucerne), in the east (to the south of Walensee) and in Vorarlberg. These fold arcs are usually a consequence of the geometry of the local sedimentary basin. The original basin margins were reactivated during nappe formation, when the thicker basin fills were squeezed out and came to lie on a smoothed out thrust fault. The axial plunge of the folds also reveals this squeezing-out process (or basin inversion), as the fold axes plunge in both directions from the centre of the squeezed out basin.

The trend of the fold axes in the Morcles nappe is interesting: they run oblique to the axis (or crest) of the Aiguilles Rouges massif, culminate over the massif and then plunge down on both sides in opposite directions. Clearly, either the dome-like uplift of the Aiguilles Rouges massif occurred after the formation and emplacement of the Morcles nappe, or the Morcles nappe overthrust a pre-existing ridge while it was a ductile deformable body: the recent age for the uplift of the massifs, as revealed by fission track data, favours the first explanation.

The map in Fig. 5.1-1 shows that the Morcles nappe in the west, the Doldenhorn nappe in the Bernese Oberland and the Tschep nappe in the east have been attributed to the Infrahelvetic complex. These nappes were thrust over several kilometres and are thus more allochthonous than the other, parautochthonous units in the

Helvetic nappe system:
A master thrust fault, and
folds and thrusts in its hanging wall and footwall

Infrahelvetic complex. However, their thrust displacements (namely, for the Morcles and Doldenhorn nappes) are far less than those for the overlying Helvetic nappes.

The lithological compositions of the Late Palaeozoic, Mesozoic and Cenozoic formations and the regional differences between these successions played an important role in the nappe structure in the Helvetic nappe system (Pfiffner 1993 provides a more in-depth discussion). Figures 5.2-7 to 5.2-11 show a series of cross-sections through the Helvetic nappe system that illustrate these differences and indicate the extent to which lateral changes in structure are effective along strike.

Figure 5.2-7 shows a cross-section in western Switzerland, including the Morcles–Diablerets–Wildhorn 'trilogy'. The Wildhorn nappe exhibits an isoclinal fold structure in its southern portion. The amplitude of the Prabé synform is particularly high as shown by the Ultrahlevetic units that have been folded into its core. The Cretaceous strata are doubled by a thrust fault that subdivides the Wildhorn nappe into two digitations, but the thrust is lost in the isoclinal folds in the Jurassic strata. The photograph in Fig. 5.2-7 is from an anticline at the front of the upper digitation. The Jurassic strata are isoclinally folded and a thrust fault separates two digitations as well, the Sublage and Mont Gond digitations. Isoclinal folds also characterize the internal structure of the Diablerets nappe, but the extreme thinning of the limbs of the folds means that the connection to the folds in the Wildhorn nappe is almost obliterated. In the northern portion of the Wildhorn nappe the structure is simpler. The Diablerets nappe follows a decollement horizon in the shales of the

Middle Jurassic over a large distance and then climbs through the Late Jurassic and Cretaceous limestones and into the Cenozoic sediments in the form of a ramp – even the basal thrust of the Ultrahelvetic nappes is cross-cut by this fault. The layers in the hanging wall of this ramp form a broad dome that contains parasitic folds that form a 'ramp anticlinorium'. The Morcles nappe exhibits a large-scale recumbent anticline, the core of which is filled with thick Early and Middle Jurassic sediments. The inverted limb is shorter in Fig. 5.2-7 than is observed on the surface because the space between the Diablerets nappe and the crystalline basement beneath is too narrow to illustrate the entire structure that is visible at outcrop farther west. Information on the top of the crystalline basement is derived from the NFP 20 reflection seismic data (Pfiffner et al. 1997c), and the cross-section runs through the depression between the Aiguilles Rouges and Aar massifs.

It should be noted that even the basal thrust of the Penninic nappes was affected by the internal deformation in the Helvetic nappes. Internal folds in the Diablerets nappe appear to fold the Wildhorn thrust, and the Morcles nappe, as a whole, caused a dome-shaped bulge in the Diablerets thrust. These facts favour a sequence of nappe formation from 'top to bottom' or from the internal to the external domains – a process called 'in-sequence' thrusting.

The structure of the Helvetic nappes in the cross-section of the Jungfrau reveals a very different picture (Fig. 5.2-8). The basal thrust of the Helvetic nappes (the Axen nappe) climbs southwards over the dome formed by the Aar massif and plunges beneath the Penninic nappes on its south side, to the south of the Rhone Valley. The highest portion of the Axen

▶ **Figure 5.2-7**
Geological cross-section through the Helvetic nappe system along the transect of Sanetschpass. The internal structure is characterized by an interplay of isoclinal folds and thrust faults. The photograph gives a view of the frontal folds in the Cretaceous limestones of the Wildhorn nappe in the western flank of the Schluchhorn. Light coloured cliff is Schrattenkalk Formation. Drusberg Marl form the core of the fold.

Mechanical stratigraphy controls internal fold-and-thrust structure

Morcles, Wildhorn, Diablerets

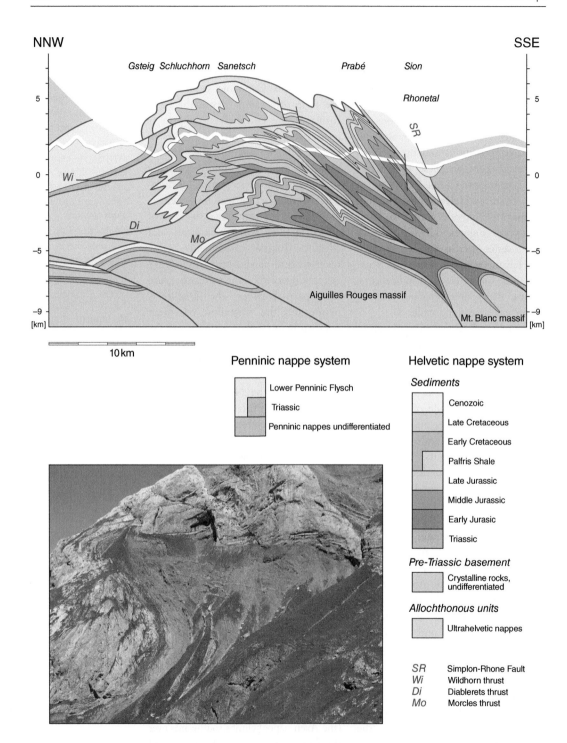

Penninic nappe system

Lower Penninic Flysch

Triassic

Penninic nappes undifferentiated

Helvetic nappe system

Sediments

Cenozoic

Late Cretaceous

Early Cretaceous

Palfris Shale

Late Jurassic

Middle Jurassic

Early Jurasic

Triassic

Pre-Triassic basement

Crystalline rocks, undifferentiated

Allochthonous units

Ultrahelvetic nappes

SR	Simplon-Rhone Fault
Wi	Wildhorn thrust
Di	Diablerets thrust
Mo	Morcles thrust

thrust has been eroded, but would be located above the peak of the Jungfrau, at above 5000 metres above sea level. The Axen thrust is vertical and even overturned to the north of the Jungfrau. From this, we can deduce that the Aar massif dome was uplifted to its present shape after the activity of the Axen thrust. The decollement horizon of the Axen nappe followed the shales in the Middle Jurassic and formed a ramp through the limestones of the Late Jurassic in the north. The internal structure of the Axen nappe is characterized by narrow, occasionally recumbent folds, and the highest of these folds has been cross-cut by normal faults that are younger than the fold structure. Three normal faults are visible in the lowest portion (Aabeberg, Bürgle and Sylere normal faults), which are interpreted to be reactivated synsedimentary normal faults (see the in-depth discussion in Hänni & Pfiffner 2001). The Cretaceous strata remained independent of the Jurassic strata, and this is easiest to recognize in the encroachment of the Cretaceous sediments onto the Subalpine Molasse in the north. The photograph in Fig. 5.2-8 gives a view of the situation across Lake Thun (Thunersee). These Cretaceous sediments were apparently sheared off in the Palfris Shale of the earliest Cretaceous and transported to the north as an independent unit. This Cretaceous unit extends further to the east, where it forms the Drusberg nappe. The most frontal portion of the Drusberg nappe forms the conspicuous Border Chain (Randkette), in which rugged limestone mountains lie on top of Molasse sediments that form a more gentle topography as depicted in the photograph of Fig. 5.2-8. The decollement horizon of this unit, that is, the Drusberg thrust, is slightly folded by the folds in the Jurassic strata in the

southernmost portion, which indicates that the Drusberg nappe was sheared off its Jurassic substratum at an early stage.

The north part of the cross-section reveals an imbricate structure in the Subalpine Molasse, which concludes with a type of antiform structure (or the 'triangle zone') in the extreme north, at the transition to the Plateau Molasse. This transition zone exhibits some similarities to that in Fig. 5.2-6.

In the south part of the cross-section, we can see the Goms massif and, finally, the Gotthard massif that are separated by Permian or Triassic sediments, which can be followed on the map over many kilometres, parallel to the strike. The southern Aar massif, the Goms massif and the Gotthard massif have lost most of their sedimentary cover, but it is unclear where exactly to search for which portions of the sheared off cover in the Helvetic nappes. This is why the Axen thrust is drawn such that it branches and runs into each of the three massifs in Fig. 5.2-8. It is worthy of mention here that the remainder of the sedimentary cover on the Goms massif is mainly in an inverted position, while the Gotthard massif exhibits more of an anticlinal structure with normal limbs dipping towards the north and south. The inverted limb in the Goms massif indicates a large-scale fold structure, where the younger sedimentary strata had apparently been sheared off at earler stage.

The geometry of the contact between the crystalline basement and the sedimentary cover in the northern declivity of the Aar massif is impressive. Figure 5.2-9 (adapted from Kammer 1985) shows an example of this for the Bernese Oberland. The contact between the crystalline basement and the sedimentary cover forms large-scale folds. The folds are recumbent or even plunging. Normal faults were also

Axen, Drusberg, Aar

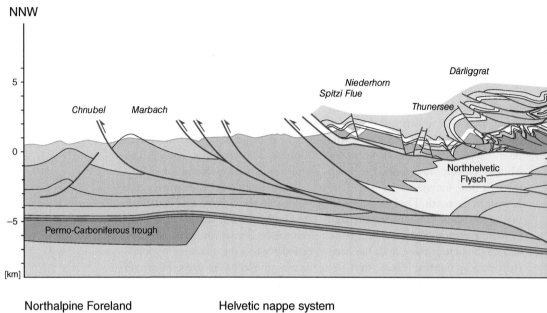

Northalpine Foreland

Cenozoic/ Molasse

- Upper Marine Molassse
- Lower Freshwater Molasse
- Lower Marine Molasse

Helvetic nappe system

Sediments

- Cenozoic
- Early Cretaceous
- Palfris Shale
- Late Jurassic
- Erzegg Formation
- Middle Jurassic
- Mols Member
- Early Jurassic
- Triassic

Allochthonous units

- Ultrahelvetic nappes
- Southhelvetic Flysch

Pre-Triassic basement

- Carboniferous - Permian (V: Verrucano Gr)
- Quartz-porphyry
- Late-/post-Variscan granite
- Gneisses and schist

Figure 5.2-8
Geological cross-section through the Helvetic nappe system along the transect of Jungfrau. The photograph shows the situation at the front of the Helvetic nappes where Cretaceous limestones of the border chain of the Drusberg nappe overly Subalpine Flysch and Subalpine Molasse units. Source: Adapted from Hänni & Pfiffner (2001) and amended.

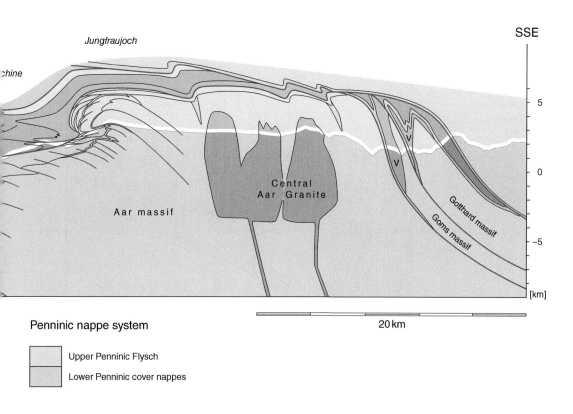

SSE

Jungfraujoch

chine

5

Central
Aar Granite

0

Aar massif

Gotthard massif

Goms massif

−5

[km]

20 km

Penninic nappe system

Upper Penninic Flysch

Lower Penninic cover nappes

Spitzi Flue

Niederhorn

Subalpine
Molasse

Southhelv.
Flysch

Figure 5.2-9 Geological cross-section through the northern margin of the Aar massif in the region of Grindelwald-Wetterhörner (ct. Bern, Switzerland). Source: Adapted from Kammer (1985).

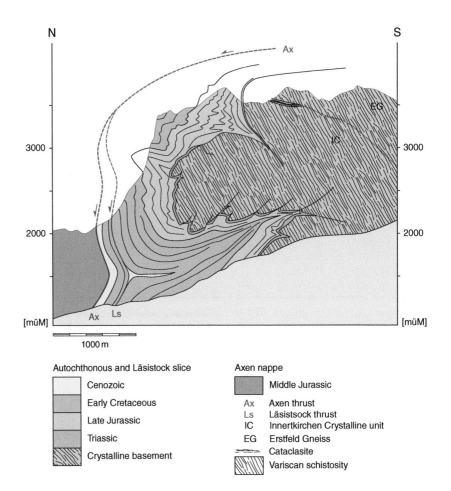

Autochthonous and Läsistock slice

	Cenozoic
	Early Cretaceous
	Late Jurassic
	Triassic
	Crystalline basement

Axen nappe

	Middle Jurassic
Ax	Axen thrust
Ls	Läsistsock thrust
IC	Innertkirchen Crystalline unit
EG	Erstfeld Gneiss
	Cataclasite
	Variscan schistosity

associated with their formation, which can be seen by the fact that the Late Jurassic limestones lie directly on top of the crystalline rocks in some locations. The Axen thrust has, indeed, been rendered vertical by the overall dome-like uplift of the Aar massif, but was not affected by the folds mentioned above. Conversely, these folds are also not cut by the Axen thrust from which it can be deduced that folding in the massif and motion along the Axen thrust were for a while active simultaneously and resulted in the formation of an 'antiformal stack' (in an antiformal stack, folding and overthrusting occurs in the core of an antiform that is being passively uplifted into

a dome). During a later phase, the Axen thrust was passively rotated into its present-day vertical position by deformations deep underground. The Läsistock thrust sheet is a sliver of the original southern autochthonous cover of the Aar massif that was shaved off and transported northwards by the Axen nappe.

Figure 5.2-10 shows a cross-section of central Switzerland from Pilatus into the central Gotthard massif (adapted from Menkveld 1995 and amended). The sheared off Cretaceous strata (Drusberg nappe) also need to be distinguished from the tightly folded Jurassic strata (Axen nappe) in these Helvetic nappes. In contrast to the simple structure of the gently dipping strata in the Border Chain, shown in Fig. 5.2-8, the frontal portion of the Drusberg nappe forms a fold-and-thrust complex in the Pilatus. The rear portions of the sheared off Cretaceous strata have also been compressed into folds. The photograph in Fig. 5.2-10 shows the internal structure of the Drusberg nappe as seen along the shore of the Uri branch of Lake Lucerne. A relatively tight syncline within the Cretaceous strata with an overturned limb is cut by a thrust fault. The thrust fault follows the Palfris Shale as the detachment horizon and places a normal sequence of Cretacous strata onto the underlying syncline. The normal sequence forms a frontal anticline farther to the north that has been eroded.

The structure of the Axen nappe in the Jurassic strata below is characterized by multiple isoclinal, recumbent to plunging folds. Shales and sandstones from the Middle Jurassic are found in their cores and rocks from the Early Jurassic are also present in the lowest folds. The Late Jurassic Quinten Limestone forms the mechanically strong, unyielding layer. Multiple normal faults are associated with these folds, some of which dip northwards and cross-cut the Quinten Limestone at an acute angle to the layering, others dip southwards and cross-cut the Quinten Limestone at an obtuse angle. Both types cause horizontal stretching of the Axen nappe, which is not immediately comprehensible in the core of an orogen that was formed by horizontal shortening. It is conceivable that this stretching represents the collapse of an orogen that became too high, such that it fell apart under its own weight, a process referred to as gravity spreading. Such events are observed in numerical

models of orogenic processes (Pfiffner et al. 2000, Selzer et al. 2008).

The Axen thrust first climbs in a southerly direction and then plunges underground to the south of the Aar massif. The autochthonous sedimentary cover is absent in the southern Aar massif, as is indicated in Fig. 5.2-10. In the case of the Gotthard massif, only Permo-Carboniferous rocks are present on the north side and Triassic and Early Jurassic sediments are present on the south side. Similar to the cross-section in Fig. 5.2-8, the Goms massif is present in the form of a thin, overturned rock unit. The southern Aar massif, the Goms and the Gotthard massif could therefore all potentially be the crystalline substrate for the Helvetic nappes.

We can gain a better insight into the structure of the Aar massif in the cross-section of central Switzerland, from the Mythen into the eastern Gotthard massif. The northern portion of the cross-section in Fig. 5.2-11 has been adapted from Schmid in Funk et al. (1983) and the deep structure is based on unpublished reflection seismic data. In this cross-section, the Aar massif is thrust on top of the autochthonous foreland. The surface of this massif is folded in the southern portion and cross-cut by thrust faults. A larger, recumbent fold is observed on the south-southeast flank of Windgälle, which is shown in the photograph of Fig. 5.2-11. It has a core made up of (Windgällen) quartz-porphyry, which is first overlain by Middle Jurassic sediments and was therefore raised above its flat surroundings as a high in the Triassic: this high may have caused the formation of a fold in exactly this location. The thrusts in the Late Jurassic Quinten Limestone above provide a slight indication of the recumbent fold, which can be interpreted as two originally shallow-dipping

▶ **Figure 5.2-10**
Geological cross-section through the Helvetic nappe system along the transect of Pilatus-Goms. The photograph is taken a few kilometres east of the cross-section and shows the view across the Uri branch of Lake Lucerne (view is to the west). An internal thrust fault (highlighted in red) in the Drusberg nappe places a normal sequence of Early Cretaceous onto a syncline formed by Cretaceous strata. Bedding is indicated by dotted lines. Pa, Palfris Shale (detachment horizon); Kk, Kieselkalk Formation; Sr, Schrattenkalk Formation.

Axen,
Drusberg,
Aar,
Gotthard

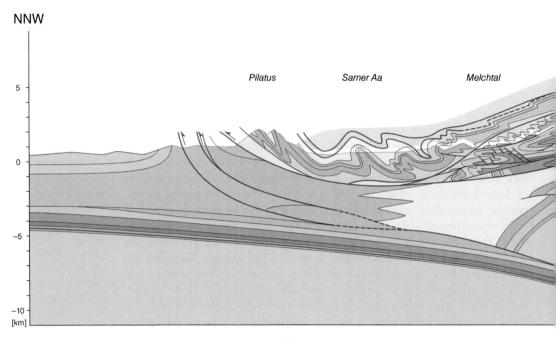

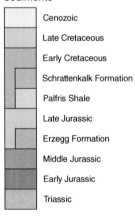

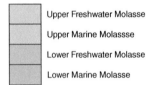

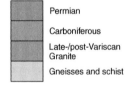

Northalpine Foreland

Cenozoic/ Molasse

- Upper Freshwater Molasse
- Upper Marine Molassse
- Lower Freshwater Molasse
- Lower Marine Molasse

Helvetic nappe system

Sediments

- Cenozoic
- Late Cretaceous
- Early Cretaceous
- Schrattenkalk Formation
- Palfris Shale
- Late Jurassic
- Erzegg Formation
- Middle Jurassic
- Early Jurassic
- Triassic

Pre-Triassic basement

- Permian
- Carboniferous
- Late-/post-Variscan Granite
- Gneisses and schist

Allochthonous units

- Ultrahelvetic nappes
- Subalpine Flysch

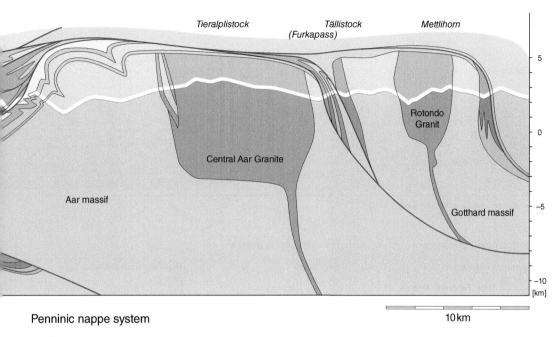

SSE

Tieralplistock Tällistock (Furkapass) Mettlihorn

Rotondo Granit

Central Aar Granite

Aar massif

Gotthard massif

10 km

Penninic nappe system

Upper Penninic Flysch

Lower Penninic cover nappes

S N

Niederbauenstock

Sr

Kk

Pa

Kk

Sr

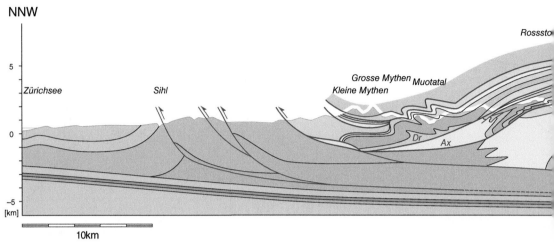

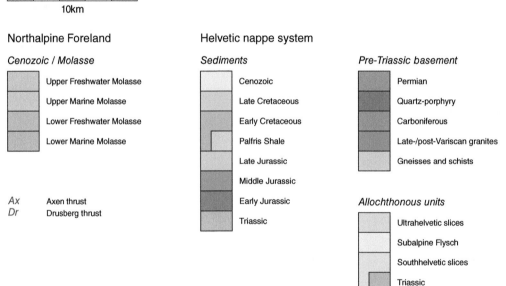

Northalpine Foreland

Cenozoic / Molasse

- Upper Freshwater Molasse
- Upper Marine Molasse
- Lower Freshwater Molasse
- Lower Marine Molasse

Ax Axen thrust
Dr Drusberg thrust

Helvetic nappe system

Sediments

- Cenozoic
- Late Cretaceous
- Early Cretaceous
- Palfris Shale
- Late Jurassic
- Middle Jurassic
- Early Jurassic
- Triassic

Pre-Triassic basement

- Permian
- Quartz-porphyry
- Carboniferous
- Late-/post-Variscan granites
- Gneisses and schists

Allochthonous units

- Ultrahelvetic slices
- Subalpine Flysch
- Southhelvetic slices
- Triassic

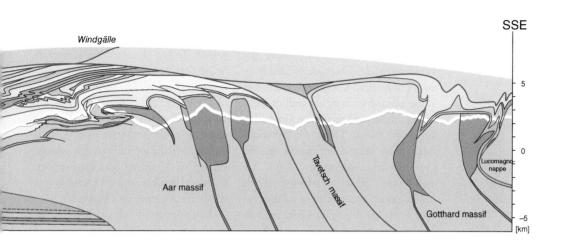

Penninic nappe system

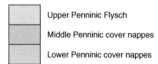

Upper Penninic Flysch

Middle Penninic cover nappes

Lower Penninic cover nappes

◀ **Figure 5.2-11**

◀ Figure 5.2-11
Geological cross-section
through the Helvetic nappe
system along the transect of
Mythen-Lukmanierpass. The
photograph shows the
Windgällen fold located on
the south-southeast flank of
Windgälle (view is to the
west-southwest). The
recumbent anticline has
Permian Windgällen
Porphyry in its core (Pe);
Middle Jurassic strata (MJ;
brown) directly overly the
Permian; the Late Jurassic
Quinten Limestone (Qi; grey)
highlights the fold hinge.

Glarus,
Säntis,
Aar

thrust faults stacked the autochthonous
cover sediments on top of each other and
were then later folded to produce the
Windgällen fold. There are isolated fragments of Jurassic–Cretaceous sediments
in the immediate hanging wall of these
folded thrusts, and also in the footwall of
the Axen thrust. These allochthonous
slices were shaved off south of the
Cenozoic cover sediments of the Aar
massif and thrust northward. Their
emplacement occurred prior to the formation of the Windgällen fold as indicated by the fact that their basal thrust
fault is folded.

The internal structure of the Axen
nappe is characterized by a normal
sequence that is many kilometres long
and also includes Cretaceous and
Cenozoic sediments in the north. The
Drusberg nappe above is also composed
of a normal sequence, but only made up
of Cretaceous strata. Folds in the frontal
portion also affect the thrusts of the
Ultrahelvetic nappes and the Penninic
nappes above it. Once again, the higher
nappes of internal origin were first
thrust onto the future Helvetic nappes
and then later passively folded.

As the rear parts of the Axen and
Drusberg nappes have disappeared due
to erosion, the location of where their
basal thrusts merge in the cross-section
is a matter for speculation, as is also the
case for the branching of the overthrust
above the Aar, Tavetsch and Gotthard
massifs. A plausible original location
for the Axen nappe could be on top of
the Tavetsch massif, as the Early Jurassic
in the northern corner of the Gotthard
massif is very similar to that of
the southern end of the Axen nappe.
The Gotthard massif has also lost most
of its sedimentary cover, apart from
a few remnants of Triassic sediments.
Based on a palinspastic reconstruction (Kempf & Pfiffner 2004), we can

assume that the original location for the
Drusberg nappe is on the Gotthard
massif. Today, Ultrahelvetic sediments
lie on the Gotthard massif, some in an
inverted position. The thrust at the base
of these Ultrahelvetic nappes and the
overlying Penninic nappes has also been
affected by the intensive folding, as is
indicated by the top surface of the massif.
Figure 5.2-11 shows how the southern
Gotthard massif was overturned in a
southerly direction by a back fold.

In contrast to the previous cross-
section, Fig. 5.2-12 shows a layer of
Permian sediments (Verrucano Group)
at the base of the Helvetic nappes. The
Permian sediments are not very thick
in this cross-section, but are almost
2 kilometres thick just a few kilometres
to the west. The basal thrust of the
Helvetic nappes is the Glarus thrust.
Overall, a normal sequence overlies the
Glarus thrust in the Glarus nappe. In
the Jurassic strata, multiple repetition
of the Late Jurassic Quinten Limestone
is visible in the north (the Walenstadt–
Gonzen schuppen zone) and a fold cas-
cade is visible in the south. The
photograph in Fig. 5.2-12 shows the
Gonzen, which marks the transition
from imbricate thrusting to folding.
The Cretaceous strata, the Säntis
nappe, has been sheared off the Jurassic
substratum and transported about
12 kilometres to the north. The Säntis
thrust used the Early Cretaceous Palfris
Shale as a decollement horizon (see
photograph in Fig. 5.2-12). The frontal
part of the Säntis nappe is compressed,
with imbricatue thrusts in the extreme
north and folds in the south. The rear
parts of the Säntis nappe are present as
a simple normal sequence. Figure 5.2-12
shows very clearly that the Jurassic and
Cretaceous strata were completely
independent. The fairly extensive
internal shortening in the Jurassic

strata resulted in the displacement along the Säntis thrust decreasing from about 12 kilometres in the north to about 1–2 kilometres in the south. Partial decoupling of the Permian sediments and the Late Jurassic Quinten Limestone occurred in the thick shaly sediments of the Early and Middle Jurassic. This is one reason why the Permian strata extend further back than the southern end of the Jurassic strata in the south. The fact that the contact between the Helvetic and the Penninic units is hidden beneath the Quaternary of the Rhine Valley at the southern end of the cross-section must also be taken into consideration when interpreting the nappe geometries.

We can see the dome formed by the Aar massif in the footwall of the Glarus thrust. Compared with the situation further west (Figs 5.2-8 and 5.2-11), the dome reaches an altitude of only about 1000 metres above sea level, as is revealed by the exposed rocks in the Vättis Window. The geometry of the dome is fairly well defined by the NFP 20 reflection seismic data (Pfiffner et al. 1997c). Either way, the Aar massif is not thrust onto the foreland in this cross-section. Folds and small thrusts deform the crystalline basement, which is particularly strong on the apex of the dome, as indicated by the multiple repetition of the Mesozoic sediments. Above the thick Paleogene sediments of the Northhelvetic Flysch, there are also two thick allochthonous units of Southhelvetic and Ultrahelvetic sediments in the footwall of the Glarus thrust. Their basal thrusts are cut by the thrust faults in the parautochthonous cover of the Aar massif, so must be older. These allochthonous slices apparently slid onto the Northhelvetic Flysch at an early stage and were later overthrust by the Helvetic nappes, a process

that in Alpine literature is called 'out-of-sequence thrusting'.

Based on the NFP 20 seismic data, the structure of the Subalpine Molasse in the north of the cross-section is now also better understood. Similar to the situation in Fig. 5.2-5, there is an anticlinal structure at the transition from the imbricate structure in the Subalpine Molasse to the tilted Plateau Molasse.

As mentioned at the start of this chapter, the Glarus thrust is of great importance in the history of geology, namely, to nappe theory, and it continues to be the subject of structural investigations even today. In July 2008, the area in the southern portion of the cross-section in Fig. 5.2-12 was included in the UNESCO list of World Heritage Sites (the official name is the 'Swiss Tectonic Arena Sardona'): a retrospective view of research and the current status can be found in Herwegh et al. (2008). Figure 5.2-13 is designed to give an impression of the situation. The Quinten Limestone in the footwall of the Glarus thrust has been transformed into calc-mylonite (also called Lochsitenkalk), about one metre thick, only in the upper-most portion. The entire limestone unit is a piece that was dragged along at the base of the Helvetic nappes and was ripped out of the footwall further to the south. This is why there is also a tectonic contact at the base of this unit, at the contact with the ultrahelvetic Sardona Flysch. The Quinten Limestone even has been shortened internally by thrusting, as is indicated by the thin band of flysch that runs up to the Martin's Hole. This ripping out of the unit and the internal thrusting are probably linked to the fact that the thrust surface of the Glarus thrust was increasingly smoothed out over the course of its activity: bumpy, elevated areas were ground down and

▶ **Figure 5.2-12**
Geological cross-section through the Helvetic nappe system along the transect of Säntis-Flims. The photograph is a view across Seez valley at Sargans (canton St. Gall, Switzerland) looking northeast, showing the transition from imbricate thrusting to folding. The Gonzen anticline has a core of Middle Jurassic strata. Northwest of Gonzen, the Quinten Limestone is repeated by thrust faults following a detachment horizon in the Middle Jurassic. The Säntis thrust follows the Palfris Shale, shown here at the type locality Palfris. MJ, Middle Jurassic; Qi, Quinten Limestone; Pa, Palfris Shale; Kk; Kieselkalk Formation.

Glarus Thrust

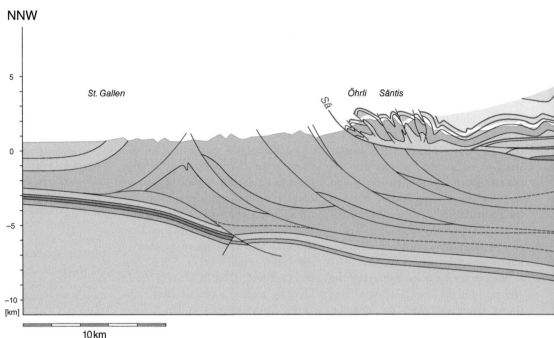

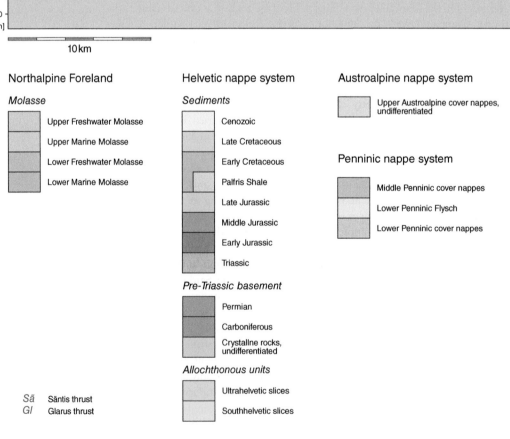

Northalpine Foreland

Molasse

Upper Freshwater Molasse

Upper Marine Molasse

Lower Freshwater Molasse

Lower Marine Molasse

Helvetic nappe system

Sediments

Cenozoic

Late Cretaceous

Early Cretaceous

Palfris Shale

Late Jurassic

Middle Jurassic

Early Jurassic

Triassic

Pre-Triassic basement

Permian

Carboniferous

Crystalline rocks,
undifferentiated

Allochthonous units

Ultrahelvetic slices

Southhelvetic slices

Austroalpine nappe system

Upper Austroalpine cover nappes,
undifferentiated

Penninic nappe system

Middle Penninic cover nappes

Lower Penninic Flysch

Lower Penninic cover nappes

Sä   Säntis thrust
Gl   Glarus thrust

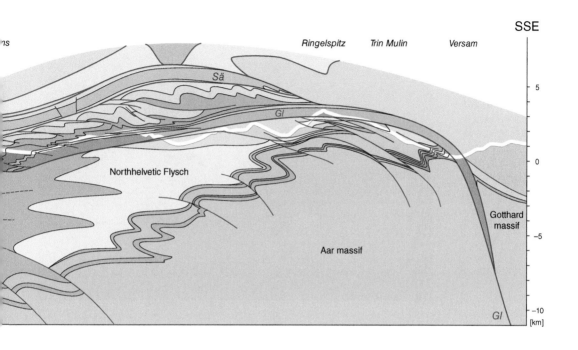

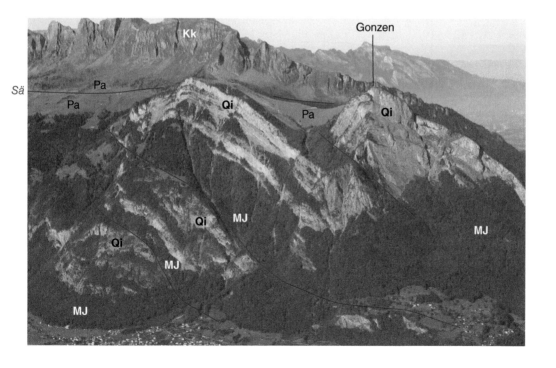

Figure 5.2-13 The Glarus Thrust in the Tschingelhoren (looking South), a center piece of the UNESCO World Natural Heritage site 'Tectonic Arena Sardona'. The Quinten Limestone beneath the Glarus Thrust is a sliver dragged along the base of the Helvetic nappes, which came to lie on top of the Sardona Flysch. This sliver consists itself of smaller slivers as indicated by the tongue of Sardona Flysch near the Martin's Hole.

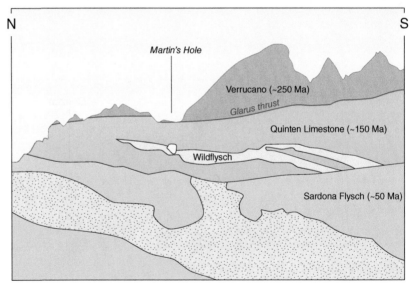

depressions filled in. This process explains many of the phenomena that are observed along the length of the Glarus thrust.

In the following, some of the basic deformation types in the Helvetic realm will be discussed in more detail. A com-

parison between the cross-section of western Switzerland (Fig. 5.2-7) and that of eastern Switzerland (Fig. 5.2-12) reveals fundamental differences: in the west, the rear part is dominated by isoclinal folds, whereas a normal sequence is present in the east, which is

(A)

(B)

Figure 5.2-14 Fold-and-thrust tectonics in the Helvetic nappe system.

(A) Inverted limb of Morcles fold in the Dents de Morcles (western Switzerland). Source: Photograph courtesy of Bernhard Edmaier. Note extreme thinning indicated by Garschella Formation on the lowermost inverted limb. Kk, Kieselkalk Formation; Sr, Schrattenkalk Formation; Ga, Garschella Formation; Eo, Eocene (Roc champion and Vivipares Mb). View is to the ENE.

(B) Recumbent folds at the front of the Doldenhorn nappe (Gasterntal, Canton Bern, Switzerland). Late Jurassic and Early Cretaceous are folded more or less harmonically. Qi, Quinten Limestone; Ze, Cementstone Formation; Öh, Öhrli Limestone; Kk, Kieselkalk Formation. View is to the east-northeast.

shortened by imbricate thrusting. On the other hand, large-scale inverted sequences are present in the Morcles nappe and in the Doldenhorn nappe, and in both cases, the cores of the folds are filled with thick shaly sediments of the Early and Middle Jurassic. Figure 5.2-14A and B displays the inverted limb of the Morcles nappe and the frontal recumbent folds of the Doldenhorn nappe. The cores of the smaller-scale isoclinal folds in the Axen nappe in central Switzerland (Fig. 5.2-10 and 5.2-14C) are also made up of the same rock. These differences in structure reflect the stratigraphic succession of the Mesozoic sediments. In the Jurassic strata, the Quinten Limestone

Fold-and-thrust tectonics

(C) Plunging fold in the flank of Graustock (west of Engelberg, central Switzerland). The recumbent fold with a core of (Early and) Middle Jurassic sediments was tilted in the course of the latest updoming of the Aar massif. EJ, Early Jurassic; MJ, Middle Jurassic; Qi, Quinten Limestone (Late Jurassic); Ax, Axen thrust. View is to the West.

(C)

(D) Imbricate thrusting in the area of Vättis (Canton St Gall, Switzerland). Thrust fault (marked red) puts two normal sequences on top of each other. Qi, Quinten Limestone; Öh, Öhrli Limestone; Kk, Kieselkalk Formation; Sr, Schrattenkalk Formation; Se, Seewen Limestone; Bü, Bürgen Formation; Gl, Gobigerina Marl. The Garschella Formation forms the thin dark band between Schrattenkalk Formation and Seewen Limestone. View is to the north. Asterisk, location of Drachenberg Cave with bones of extinct cave bears assembled some 45 000–40 000 years ago (see Fig. 7.4).

(D)

forms the mechanically strong, unyielding unit that generally controls the folding process. The shaly sediments of the Early and Middle Jurassic are more easily deformed and behave in a ductile manner. If they are thick, then they can fill the cores of anticlines, if not, then they can serve as a decollement horizon and lead to imbricate thrusting. Examples of imbricate thrusting are given in Fig. 5.2-14D and E.

Analysis has demonstrated that the critical ratio of thicknesses of incompetent, soft strata to competent, strong strata is about 0.5 (Pfiffner 1993b): imbricate thrusting develops if the ratio is smaller than this and larger folds that are easily sheared off their substrate develop if the ratio is higher. As the Early and Middle Jurassic sediments exhibit abrupt changes in thickness, this also results in

(E)

Crap Mats (2946 m)

(E) Imbricate thrusting in the eastern flank of Crap Mats (north-northwest of Reichenau, canton Graubünden, Switzerland). Tschep thrust (marked red) places Late Jurassic Quinten Limestone onto a manifold stack of Cretaceous limestones. The imbrication is outlined by the repetition of the Schrattenkalk–Garschella pair (light Schrattenkalk Formation and dark Garschella Formation above). The pairs are marked by a black filled circles. View is to the west.

(F)

Piz d'Artgas (2786 m)

X
Rö
VN
Bü
Gl
Se
Gl
Ga
Sr

(F) Thrusting and folding in the western flank of Piz d'Artgas (north of Breil/ Brigels, canton Graubünden, Switzerland). The inclined north-vergent anticline beneath the summit of Piz d'Artgas is well outlined by the brown Bürgen Formation (Eocene). The summit area above the thrust fault marked in red consists of an inverted sequence with crystalline basement (diorite) forming the peak and overlying Triassic Röti Dolostone. The Val Nauscha mélange (VN) in the footwall of the thrust fault consists of multiple repetitions and blocks of Cretaceous and Eocene strata. View is to the east. X, crystalline basement (Punteglias submassif of Aar massif); Rö, Röti Dolostone; Sr, Schrattenkalk Formation; Ga, Garschella Formation; Se, Seewen Limestone; Bü, Bürgen Formation; Gl, Globigerina Marl; VN, Val Nauscha mélange.

rapid changes in the structure of the nappes. The abrupt changes in thickness are a result of the Jurassic extensional tectonics on the European continental margin as discussed in Chapter 3. There is a tendency towards thicker sediments in the southern portion. This demonstrates clearly why the type of structure in the Infrahelvetic complex (see the examples in Fig. 5.2-14D and E) is so different from that in the Helvetic nappes.

In a similar fashion, we can explain why the Upper Helvetic nappes, namely, the Drusberg and Säntis nappes, are present as independent units. In this case, active subsidence that lasted into the Cretaceous plays a role: the higher rates of subsidence and greater water depth in the south led to greater thickness of the

Palfris Shale. However, the boundary for changes in thickness was not in exactly the same location as in the Jurassic sediments (see Fig. 3.29), which is why the Cretaceous in the Helvetic nappes remained attached to its Jurassic substratum in a few isolated cases (Mürtschen nappe and parts of the Axen nappe).

The past edges of basins can also be revealed on maps, and two examples of this are presented. In western Switzerland, the Morcles nappe and the Doldenhorn nappe exhibit great similarities in their internal structure and their positions in the nappe stack are also equivalent, and they are therefore often directly correlated. However, the tectonic map in Fig. 5.2-1 reveals that the fold axes in both nappes do not point towards each other at all, but instead the two nappes lie next to each other, in a left-stepping or 'en échelon' fashion, which is easily explained by the original basin geometry. The second example pertains to the Verrucano Group in the Glarus nappe. The contact between the Verucano Group and the Triassic sediments is folded in several locations. The folds include overturned or steep limbs and are west-vergent in the western edge of the Glarus nappe and east-vergent in its eastern edge. The fold axes plunge to the north, but they bend round in an east–west direction towards the north, and form a full arc. The folds in the Jurassic sediments to the east and west of the Verrucano Group of the Glarus nappe plunge towards east-northeast and west-southwest, respectively. This geometry is best explained by a complete squeezing out (or inversion) of the Verrucano graben fill. The lateral edges of the trough ran from north to south and acted as lateral ramps when the fill was squeezed out. During emplacement of the graben fill on the smooth thrust surface, the edges of the

Which cover on which basement

trough were tilted to form an anticline. The axis of this anticline followed the graben edge and thus acquired an arcuate shape (see Fig. 2.21B). The squeezed out trough fill caused a local high. Accordingly, the folds in the rocks above, that run transversely to the fill, plunge away in both directions from this high.

Deformation of the Aar massif and its autochthonous cover sediments is a notable feature, where folding and thrusting go hand in hand. The basement-cover contact displayed in Fig. 5.2-9 shows folds that are sustained by faults on the northern flank of the Aar massif, but in the southern part of the Aar massif where temperatures were higher, more ductile folds developed: an example is given in the photograph in Fig. 5.2-11. A more intriguing example is illustrated in Fig. 5.2-14F Here, a relic of an inverted limb consisting of crystalline basement (forming the summit of Piz d'Artgas) and Triassic sediments is thrust onto a mélange of Cretaceous and Cenozoic sediments, which in turn overlies an anticline–syncline pair outlined by Cretaceous and Centozoic strata. The inverted limb with crystalline basement can be traced westward into an anticline cored by crystalline basement, and the Val Nauscha mélange in Fig. 5.2-14F is an expression of high shear strains asscociated with the thrust fault that developed beneath this anticline.

A question that has previously arisen is that posed by the crystalline substratum of the Helvetic nappes. In the case of the Glarus nappe, the Tavetsch massif can be followed directly into the Glarus Verrucano to the east of Pigniu/Panix. Only very few remnants of the Mesozoic cover can be found on the south side of the Aar massif (Surselva, Urseren, Goms), but on the northern

Gotthard massif the Verrucano Group is overlain by Triassic (with evaporites) and Early Jurassic (facies similar to the southern end of the Axen nappe on the Klausen Pass), although only by remnants of the Triassic in the south. In order to clarify the question of the crystalline substratum, a palaeogeographical reconstruction of the entire Helvetic realm was constructed, taking into consideration the internal deformation in the nappes (Kempf & Pfiffner 2004). In a first step, the crystalline uplifts were pushed back relative the foreland (the Black Forest was used as the reference point) and the Permo-Carboniferous troughs were then fitted into this (Fig. 5.2-15, top). In a second step, the Helvetic nappes were retro-deformed (unfolded) based on a competent limestone layer (Schrattenkalk Formation and Quinten Limestone) and, finally, the nappes were pushed back towards the south (eastern Switzerland), south-southeast (western Switzerland) and southeast (Haute Savoie). Figure 5.2-15 (bottom) shows the palaeogeographical position, as well as the present-day position of the nappe front. The transport distance to the east is greatest (almost 100 kilometres) and then drops off progressively towards the west, to 50 kilometres in the Haute Savoie, as is shown by the arrows labelled with kilometres. The reconstruction in Fig. 5.2-15 also shows that the southernmost portion of the Helvetic nappes (southern edge of Drusberg and Säntis nappes) originated on the Gotthard massif whereas the origin for the South- and Ultrahelvetic units is to be considered as being the more internally located crystalline nappes (Lucomagno, Leventina, Verampio).

Figure 5.2-16 illustrates the disharmonic structure of the Jurassic and Cretaceous strata based on the example of the Helvetic nappes in eastern Switzerland: the large fold visible in the Sichelkamm is formed by Cretaceous limestones of the Säntis nappe. The Early Cretaceous Schrattenkalk Formation clearly traces the geometry of the syncline in the form of a massive grey band. The Palfris Shale in the lower portion of the picture forms the substratum for the flat parts that are sunlit. Two grey cliffs are visible at the very bottom, which are composed of two units of Late Jurassic Quinten Limestone that have been stacked on top of each other, with Middle Jurassic shales (Mols Mb) forming the decollement horizon. Apparently, the imbricate structure found in the Quinten Limestone does not continue up into the overlying Cretaceous limestones and, conversely, the syncline in the Sichelkamm does not extend down into the Late Jurassic limestones. The thick Palfris Shale filled the cores of the anticlines in the Cretaceous limestones above and allowed independent internal shortening in the Jurassic strata below and in the Cretaceous strata above.

The Penninic Nappe System

The Penninic nappes in the tectonic map in Fig. 5.2-1 form an irregular band that lies adjacent to the Helvetic nappe system in the south. The subdivision of the Penninic nappes into Lower, Middle and Upper Penninic nappes relates somewhat to the palaeogeographical origin of the Mesozoic rocks within the nappes. The Lower Penninic nappes comprise Mesozoic–Cenozoic sediments on one hand, and nappes consisting essentially of crystalline rocks on the other hand. The former are composed of sediments

▶ **Figure 5.2-15**
Palaeogeographical map of the pre-Triassic basement (upper diagram) and the Mesozoic sediments of the Helvetic nappe system (lower diagram) in the Central Alps (redrawn and supplemented after Kempf & Pfiffner 2004). Ax, Di, Do, Dr, Ge, Gl, Mo, Wi: Axen, Diablerets, Doldenhorn, Drusberg, Gellihorn, Glarus, Morcles, Wildhorn nappes; Chs, Chaînes subalpines (allochthonous); Bc, Border chain (of the Drusberg nappe). Source: Kempf & Pfiffner (2004). Reproduced with permission of John Wiley & Sons.

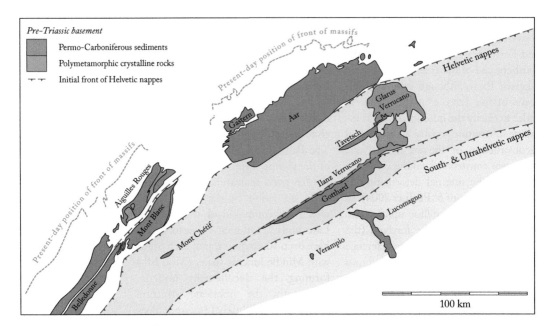

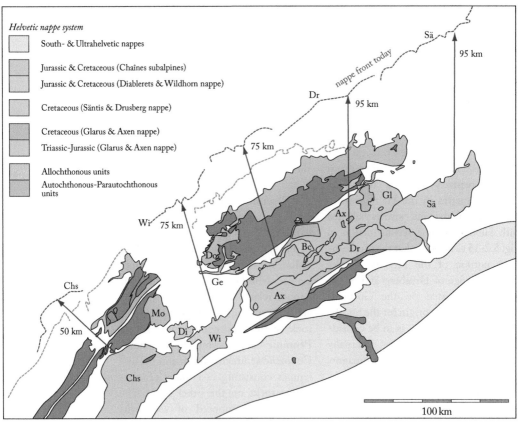

Figure 5.2-16 Stockwork tectonics in the area Walensee–Sichelchamm (canton St Gall, Switzerland). The Cretaceous limestones of the Säntis nappe are detached from their Jurassic substratum and folded as detachment folds above the Säntis thrust. The Jurassic limestones in the footwall of the Säntis thrust display a completely different style characterized by imbricate thrusting. The individual imbricates are detached along shales at the base of the Middle Jurassic.

from the Valais Trough that were sheared off their crystalline basement and the latter represent slabs of the uppermost portion of thinned continental crust that were incorporated into the nappe stack: examples of the latter include the Simano and Adula nappes. The Middle Penninic nappes originate from the region of the Briançon Rise, and in this case the Mesozoic sediments also form an independent nappe pile (e.g. the Klippen nappe) that was sheared off its original crystalline basement: the latter was also incorporated into the nappe structure, as revealed, for example, in the Bernhard nappe complex in western Switzerland and in the Tambo and Suretta nappes in eastern Switzerland. The situation is slightly different in the Upper Penninic nappes, which are composed of rocks from the former Piemont Ocean, an ocean that had previously been fragmented by transform faults and synsedimentary normal faults in the Mesozoic, and during Alpine orogeny it was fragmented even further. For this reason, the Upper

Penninic nappes exhibit a more complex internal structure: in some locations large-scale complexes of ophiolitic rocks containing serpentinites, gabbros and basalts are visible, which represent exhumed mantle and newly formed oceanic crust (e.g. the Platta nappe or the area around Saas–Zermatt); at other locations, large-scale complexes of Mesozoic sediments are visible, which usually are linked to mafic or ultramafic rocks (Avers Bündnerschiefer, Gurnigel nappe or Tsaté nappes); at yet other locations a mixture of very different rocks is present, including fragments of continental crustal rock and associated sediments – the term 'mélange' is used in such cases and examples of this are the Arosa Zone and the Simmen nappe.

Figure 5.2-1 shows that the Upper Penninic nappes run along the boundary to the Austroalpine nappe system, which is also revealed in the footwall of the klippe formed by the Dent Blanche nappe in western Switzerland and the frame around the Engadine Window in eastern Switzerland. There are, however,

Penninic nappe system:
Cover nappes,
Basement nappes,
Oceanic fragments
and mélanges

exceptions: for example in the north-west of the Klippen nappe there are Southpenninic units (Gurnigel nappe) lying on top of the Subalpine Molasse and below the Middle Penninic Klippen nappe. The term 'Upper Penninic' seems obviously inappropriate here, but special tectonic position can be explained by an out-of-sequence thrusting of the nappes. Normally, propagation of the thrust sheets occurred from internal to external, or the initiation of the thrusts faults from top to bottom. In the case of the Gurnigel nappe, however, it can be assumed that it was initially transported a long way in the external direction into a position north of the Briançon Rise and the Valais Trough, but during a later phase, the Gurnigel nappe was then overthrust by the Middle Penninic nappes and thus ended up in its present-day position: this point will be discussed in more detail in Chapter 6 (see Fig. 6.12).

The structure of the Penninic nappes changes far more rapidly along the strike than is the case, for example, in the Helvetic nappe system, and in the following, this structure will be described based on a selection of typical case studies.

Klippen nappe

The Klippen nappe in the Prealps of western Switzerland and in the Chablais region in Haute Savoie is an example of a Penninic sedimentary nappe that was completely detached from its crystalline basement and transported a long distance in a northerly direction. Large-scale folds characterize the internal structure in the north (Fig. 5.2-17, adapted from Wissing & Pfiffner 2002), whereas imbricate thrusting predominates in the south: the terms 'Médianes plastiques' and 'Médianes rigides' for the northern and southern portions, respectively, emphasize this difference in structure. The reason for the difference can be found in the stratigraphic

succession: the evaporites in the north were thick enough to fill the cores of the large anticlines and the Jurassic sequence with marly and calcareous strata was well suited to folding. Interestingly, the thrusts between the folds and the adjacent southern Heiti and Gastlosen thrusts are associated with primary discontinuities in the stratigraphic succession, which were caused by synsedimentary faults (see Wissing & Pfiffner 2002). The formation of the imbricate structure in the south was favoured by the thick, mechanically strong, unyielding Triassic carbonates and the absence of the marly Early and Middle Jurassic sediments. The basal thrust in the Southpenninic Simme nappe and the Middle Penninic Breccia nappe was affected by the internal deformation of the Klippen nappe. As mentioned above, the Southpenninic Gurnigel nappe is in an unusual position due to being overthrust by the Middle Penninic Klippen nappe at a later stage. The evaporites in the decollement horizon of the Klippen nappe are composed of intercalated layers of anhydrite and dolostones, and during nappe formation the anhydrite was able to flow at fairly low temperatures and the thin layers of dolostone broke up (see Wissing & Pfiffner 2002). Due to the effects of percolating groundwater, the anhydrite was dissolved or transformed into gypsum and there was local formation of calcite. Finally, so-called cargneules were produced from the anhydrite–dolostone–gypsum–calcite mixtures (Schaad 1995). The effects of weathering continued in these rocks: dissolution caused the formation of hollows, several metres in dimension, which are visible at the surface in the form of sinkholes; fragments of dolostone collected in hollows lying deeper down; later these fragments were bound together

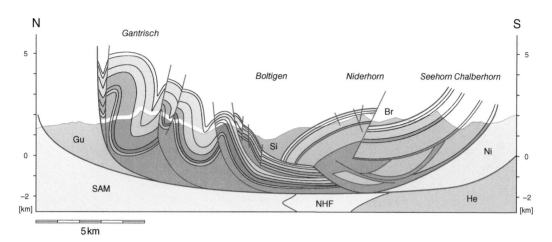

N S

Gantrisch

Boltigen   Niderhorn   Seehorn Chalberhorn

Br

Gu

Si

Ni

SAM

NHF   He

5 km

Penninic nappe system

Upper Penninic nappes/Piemont Ocean

Gurnigel nappe (Gu)

Simme nappe (Si)

Middle Penninic nappes/Briançon Rise

Breccia nappe (Br)

Klippen nappe

Lower Penninic nappes/Valais Trough

Niesen nappe (Ni)

Cenozoic

Cretaceous

Late Jurassic

Middle Jurassic

Early Jurassic

Triassic (cabonates)

Triassic (evaporites)

Molasse

Subalpine Molasse undifferentiated (SAM)

Helvetic nappe system

Cenozoic Northhelvetic Flysch (NHF)

Helvetic nappes undifferentiated (He)

to form breccia, and create the impression of thick layers of dolostone – rocks that would not be assumed to be present in a decollement horizon. Mosar et al. (1996) and Plancherel (1979) have provided detailed discussions on the structure of the Klippen nappe, and Masson (1976) provided an overview of the history of research conducted on the nappe.

Figure 5.2-18 shows a cross-section through the main body of the Penninic nappes in western Switzerland, adapted from Escher et al. (1997) and Scheiber et al. (2013), and based on the NFP 20 reflection seismic data. The Helvetic nappes (Mont Blanc massif) plunge

down at depth in the northwest, and is shown to be affected by a back-fold. The Penninic nappes are cross-cut by the Rhone–Simplon Fault: at the Simplon Pass, it is a normal fault (see Fig. 5.2-1), becoming a strike-slip fault in the Rhone Valley, and then makes a transition to a thrust fault in the Haute Savoie (see Fig. 5.1-4). The varying nature of the fault can be explained by the orientation of the fault suface, on which the southeastern block was moved towards the southwest. In the cross-section in Fig. 5.2-18, the Rhone–Simplon Fault caps the base of the Middle and Lower Penninic nappes and brings them into direct

▲ **Figure 5.2-17**
Geological cross-section through the Penninic Klippen nappe in the transect of Gantrisch (canton Bern, Switzerland). Imbricate thrusting prevails in the rear (south), whereas thrust faults at the front are associated with folds. Source: Data from Wissing & Pfiffner (2002).

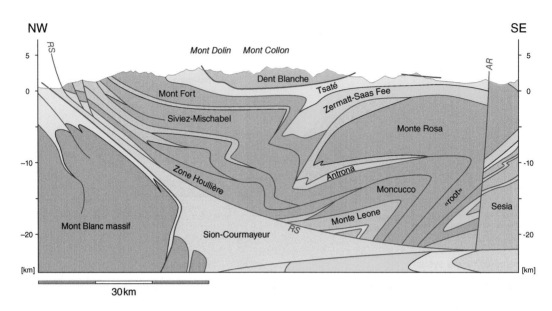

NW   SE

Mont Dolin   Mont Collon

Dent Blanche
Mont Fort   Tsaté
Siviez-Mischabel   Zermatt-Saas Fee

Monte Rosa

Zone Houllière   Antrona

Moncucco

Monte Leone   «root»   Sesia

Mont Blanc massif
Sion-Courmayeur   RS

30 km

European continental margin

Helvetic nappe system

Mesozoic

Crystalline rocks,
undifferentiated

Penninic nappe system

Upper Penninic nappes/Piemont Ocean

Mesozoic (Tsaté)

Oceanic crust

Middle Penninic nappes/Briançon Rise

Mesozoic

Crystalline rocks/
Permo-Carboniferous

Lower Penninic nappes/Valais Trough

Mesozoic

Crystalline rocks,
undifferentiated

Adriatic continental margin

Austroalpine nappe system

Crystalline rocks,
undifferentiated

Southalpine nappe system

Adriatic lower crust

RS   Rhone-Simplon Fault
AR   Aosta-Ranzola Fault

contact with the Bündnerschiefer Group (or schistes lustrés) in the Sion-Courmayeur Zone – a sequence that originated in the Valais Trough and for which the thickness and internal structure are largely unknown. The Bernhard nappe complex lies in the hanging wall of the Rhone–Simplon Fault and can be divided into the Siviez–Mischabel and Mont Fort digitations, which are composed mainly of pre-Triassic crystalline basement, overlain by Permo-Triassic quartzites. Escher et al. (1997) considered the internal structure to represent enormous isoclinal folds, the rear portions of which have been folded by large-scale south-vergent back-folds. Research by Scheiber et al. (submitted) revealed that the 'inverted limbs' of these 'folds' consist of a stack of small thrust sheets made up of normal sequences. The Zone Houillère is the bottom unit in the Middle Penninic nappe pile and in the northern part its Permo-Carboniferous sediments and the overlying Mesozoic sediments (formerly identified as the Pontis nappe) have been cross-cut by several normal faults. These normal faults can be interpreted as branches of the Rhone–Simplon Fault. In the south, the Mont Fort and Siviez–Mischabel digitations interlock with the ophiolite-bearing Antrona and Zermatt–Saas Fee zones, which, themselves, encase the Monte Rosa nappe, which is mainly composed of granite of the same name. This Variscan intrusive body was subjected to such weak Alpine overprinting that it was regarded as a Cenozoic intrusion by early researchers. As is clear from Fig. 5.2-18, the southern portion of the Monte Rosa nappe and the nappes that lie below it form an enormous, south-vergent back-fold. The thinned limb that dips towards the northwest is regarded as a classic 'nappe root zone'. This thinned fold limb is visible at the surface near Villadossola. The Monte Rosa granite has been transformed to an orthogneiss displaying a pronounced stretching lineation ('Stängelgneis'), the magmatic origin of which is unmistakable due to the plagioclases that measure up to several centimetres in size (see photograph in Fig. 5.2-18). The back-fold in the Monte Rosa nappe that was mentioned above is cross-cut by the Aosta–Ranzola Fault. This fault is part of an extensional fault system that was active in Late Oligocene times (Bistacchi et al. 2001).

In the portion of the cross-section in Fig. 5.2-18 that is close to the surface, the Tsaté nappe overlies both the Mont Fort digitation of the Bernhard nappe complex and the Zermatt–Saas Fee Zone. According to Escher et al. (1997), the Tsaté nappe contains basaltic pillow lavas at its base that probably date from the Jurassic, but that are complexly folded internally. The Dent Blanche nappe completes the sequence at the top, a klippe composed of Lower Austroalpine nappes. The equivalent of the Dent Blanche nappe in the inverted limb in the Monte Rosa back-fold may be found in the Lower Austroalpine Sesia Zone.

A unit composed of Penninic nappes and Southalpine lower crust is indicated right at the very base of the cross-section, in the southeast. These units are part of a larger complex of rocks with an abnormally high seismic velocity. The complex is wedge-shaped and is interpreted as an element that was sheared off the subducted European plate. The Penninic nappes in the hanging wall of this wedge have certainly been thrust in a southeasterly direction, which could be linked to a phase of back-folding. The syncline at a depth of 20 kilometres adjacent to the Aosta–Ranzola Fault has been visualized by downward projection of surface structures (Escher et al. 1997).

The cross-section in Fig. 5.2-19 runs alongside the NFP 20 East Traverse and its deep structure is fairly well supported by the reflection seismic data (see Pfiffner & Hitz 1997, Schmid et al. 1996, 1997a). In the north, the crystalline basement of the Helvetic nappes and the Southhelvetic to Ultrahelvetic units (Aar massif, Gotthard massif and Lucomagno–Leventina basement block) plunge down at depth in a southerly direction.

The Simano nappe and the Adula nappe complex above it belong to the **Lower Penninic nappes,** and based on their lithological composition are allocated to the crystalline nappes that once belonged to the thinned European

◀ **Figure 5.2-18**
Geological cross-section through the Penninic nappes along the seismic lines of the Western Transect of NRP20. Thin slivers of sediments mark contacts between crystalline basement nappes. The photograph shows a highly deformed augengneiss from the 'root' of the Monte Rosa nappe taken near Villadossola (Italy). Pencil gives scale. Source: Data from Escher et al. (1997).

Folded thrusts, Backfolds

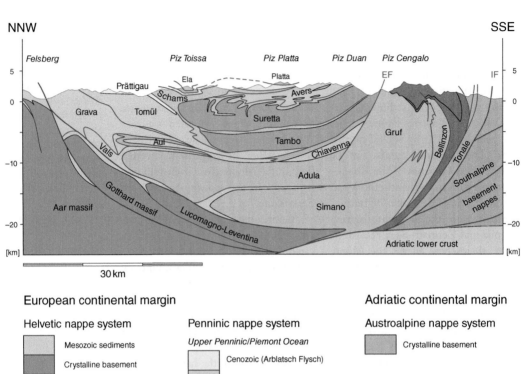

European continental margin

Helvetic nappe system

	Mesozoic sediments
	Crystalline basement

Penninic nappe system

Upper Penninic/Piemont Ocean

	Cenozoic (Arblatsch Flysch)
	Mesozoic sediments
	Oceanic crust

Middle Penninic/Briançon Rise

	Mesozoic sediments
	Permocarboniferous & Crystalline basement

Lower Penninic/Valais Trough

	Cenozoic (Prättigau Flysch)
	Mesozoic sediments
	Oceanic crust
	Crystalline basement

Adriatic continental margin

Austroalpine nappe system

	Crystalline basement

Southalpine nappe system

	Crystalline basement
	Adriatic lower crust

Bregaglia pluton

	Granodiorite, tonalite

EF	Engadin Fault
IF	Insubric Fault

continental margin and formed the substratum of the Valais Trough. The sedimentary fill of the Valais Trough was scraped off and transported to the north, such that these Mesozoic sediments, mainly from the Bündnerschiefer Group, form a separate nappe pile, which includes the Grava and Tomül nappes in the cross-section, as well as the smaller units underneath (Vals, Aul and Chiavenna units). It should be noted that some of the thrusts (e.g. those at the base of the Grava nappe) are isoclinally folded and provide evidence for a multiphase tectonic history. Ophiolitic rocks are also found at the base of the Tomül nappe and in the Aul and Chiavenna nappes, which indicate an association with small pull-apart basins in the Valais Trough (see Fig. 3.31). The Prättigau Flysch lies above, in a normal stratigraphic succession, in the roof of the Bündnerschiefer Group in the Tomül nappe. The Adula nappe complex contains internal thrusts (not indicated in Fig. 5.2-19), indicated by thin layers of marble in the crystalline rocks, which point to a complex imbricate structure produced during nappe formation, probably at great depth in the subduction channel (called 'tectonic accretion channel' by Engi et al. 2001). A precise assignment of the sedimentary basins composed of the Bündnerschiefer Group from the Valais Trough to their crystalline basement is not possible in any detail, but retro-deformation shows that the Bündnerschiefer Group indicates a width of approximately 200 kilometres for the Valais Trough (Steinmann 1994, Schaad 1995), which is at least of the same magnitude with regard to the entire length of the crystalline nappes (see Fig. 3.31).

In the south, the Lower Penninic crystalline nappes were affected by a large-scale backfold, and its overturned limb forms a broad shear zone that also includes Austroalpine units that have been rendered vertical (Tonale in Fig. 5.2-19), in addition to the Lower Penninic nappes (Adula–Gruf and Bellinzona zones). The steep, vertical zone is part of the Peri-Adriatic fault system, which demarcates the northern boundary of the Southalpine nappe system. The intensively sheared rocks in this steep zone point to both back-thrusting and dextral strike-slip motion. The Insubric Fault (IF in Fig. 5.2-19) corresponds to a later brittle faulting phase in the Peri-Adriatic fault system and separates the Austroalpine from the Southalpine nappe systems here. The tonalites and granodiorites of the Bregaglia Intrusion also welled up along the steep zone (see Fig. 4.8).

The Lower Penninic nappes are overlain by the **Middle Penninic nappes**. In this case, the Mesozoic sediments of the former Briançon Rise have also been separated from their crystalline substratum. In the cross-section in Fig. 5.2-19, these are the Schams nappes, an allochthonous unit that is delimited on all sides by folded tectonic contacts: a detailed discussion of the relationships can be found in Schmid et al. (1996, 1997b). Two fold limbs characterize the internal structure of the Schams nappes, each with an entirely different composition: one contains thick successions of Jurassic breccias that were deposited at synsedimentary normal faults at the edge of the Briançon Rise; the other, predominately sediments deposited on the rise itself. The abrupt lateral changes in the sedimentary sequences took place during Alpine deformation and resulted in the formation of an inverted limb containing breccias. Equivalent formations to the Schams nappes can also be found further to the north (Sulzfluh and Falknis nappes), but are not present in the current cross-section due to erosion. The Tambo and Suretta nappes (see Fig. 5.2-19) are mainly composed of pre-Triassic crystalline basement rocks. Both nappes exhibit remnants of an autochthonous cover of Triassic quartzites, evaporites, dolostone and calcareous marbles, and there is a thrust contact at the base of the crystalline rocks. Similar sedimentary remnants made up of quartzites and dolostone marbles are also found between the Simano und Adula nappes and on top of the Adula nappe complex. These thin sedimentary zones that can be followed over many kilometres are also called 'nappe separators' ('Deckenscheider'). It is often unclear whether these sedimentary remnants are normal or inverted strata, especially in the Lower

◀ **Figure 5.2-19**
Geological cross-section through the Penninic nappes along the seismic lines of the Eastern Transect of NRP20. Cover rocks were detached from their crystalline substratum and form a nappestack of their own (Vals, Grava, Tomül, Schams). The crystalline basement was piled up to four major nappes (Simano, Adula, Tambo, Suretta). Source: Data from Pfiffner & Hitz (1997).

Penninic nappes. The roof of the Suretta nappe is also of interest because the contact between crystalline and sedimentary rocks, as well as the thrust where it is in contact with the overlying Avers nappe, is overprinted by tight folds. The axial surfaces of these folds dip towards the north, but become steep towards the base of the Suretta nappe and, finally, bend round to dip towards the south, and so these folds clearly were overprinted by back-folding. The ductile nature of these folds obscures the fact that the base of the nappe is a thrust fault that formed under more brittle conditions, at temperatures that were still relatively low.

The Schams dilemma

The origin of the Schams nappes, with tectonic contacts on all sides, was for a long time a controversial topic: Do they stem from the roof of the Suretta nappe? From underneath the Tambo nappe? Or even from the area in between (the Splügen Zone)? The investigations accompanying NFP 20 revealed that the sedimentary remnants on the Tambo and Suretta nappes exhibit similarities to the sediments in the Schams nappes, but that the Schams nappes run deep underneath the Tambo nappe. A palinspastic retro-deformation (in which folds and thrusts are reversed) indicates that the original extent of the Schams nappes, which corresponds to the breadth of the Briançon Rise, is greater than that of the Tambo and Suretta nappes combined, and apparently part of the crystalline substrate of the Briançon Rise was subducted during nappe formation (see Schmid et al. 1996, 1997a).

The Upper Penninic nappes include the Avers nappe (composed of sediments resembling the 'Bündnerschiefer Group'), the Arblatsch Flysch and the Platta nappe, which is mainly composed of ophiolites. The Avers nappe also contains lenses of gabbros and basalts (or prasinites), and lenses of evaporites are to be found in locations at its base, but the age of these metamorphic 'Avers Bündnerschiefer' remains unknown. The thrust at the base has been folded in conjunction with the sedimentary remnants on the Suretta nappe, which is evidence for an early emplacement of this unit on the Suretta nappe. A small erosional remnant of the Lower Austroalpine Ela nappe remains on top of the ophiolites of the Platta nappe at Piz Toissa, and its basal thrust is also folded. If we extended the cross-section a little further to the east, then there would be an almost continuous cap of Austroalpine nappes on top. The cross-section in Fig. 5.2-19 shows that the Middle and Upper Penninic nappes climb towards the south and the exposed rocks further east provide evidence for the fact that they were also involved in the large backfold that affects the Lower Penninic nappes, but in the cross-section, this picture is impaired by the fact that the nappes are offset by the Engadin Fault (EF in Fig. 5.2-19). Overall, the Penninic nappes exhibit a synformal structure, due to the southerly dip of the massifs in the north and the backfold in the south. The backfolding north of the Insubric Fault carries on in the south-vergent nappe structure in the Southalpine nappe system to the south.

The Austroalpine Nappe System

Figure 5.2-1 provides an overview of the Central Alps, with the Austroalpine realm covering large areas in the extreme east. To the north of this, the Northern Calcareous Alps (a huge complex of Mesozoic sediments) continue into the

Eastern Alps and will be discussed in more detail in this context. Further south, this is followed by areas in which crystalline basement dominates the picture in the map, although interrupted by sedimentary deposits. The continuous outcrop of the Austroalpine nappe system is also interrupted, by the Engadin Window (located farther east, see Fig. 5.3-1), in which the Penninic nappes are exposed in the footwall of the Austroalpine nappe system, indicating that the total thickness of the Austroalpine nappe pile amounts to only a few kilometres: emphasized by the irregular contact between the Austroalpine and Penninic nappe systems, which crosses the Central Alps from north to south (see Fig. 5.2-1).

The cross-section in Fig. 5.2-20 provides a good impression of the orogenic lid formed by the Austroalpine nappes. This has been adapted from Schmid et al. (1997b) and amended, and shows the slightly irregular nappe boundary between the Penninic and Austroalpine nappe systems. The photograph in Fig. 5.2-21 demonstrates the irregular contact between the Arosa Zone (derived from the Piemont Ocean) and the Lower Austroalpine nappes (derived from the Adriatic continental margin). As is evident from the tectonic map of Fig. 5.2-1, it is mainly Upper Penninic nappes assigned to the Piemont Ocean that straddle this contact plane all the way across the Alps. This contact should be regarded as a former plate boundary because the Austroalpine nappes represent the past Adriatic continental margin, as previously stated by Trümpy (1975). In the Central Alps, the Austroalpine nappe system is divided into two, into Lower and Upper Austroalpine nappes (Fig. 5.2-20), and both the pre-Triassic basement and its Mesozoic sedimentary cover are involved

in the nappe structure of both subdivisions. The internal structure of these nappes is strongly determined by Jurassic synsedimentary normal faults, which in many places dip down very flatly and were long interpreted as thrusts. Their importance as Jurassic normal faults has already been discussed in detail in connection with Figs 3.23–3.25: the most important of these are the Ducan and Trupchun normal faults, both with offsets of several kilometres.

Jurassic faults influence nappe structure

In the extreme west of the cross-section, the Austroalpine nappe system overlies the Arosa Zone, which, in turn, overlies the Prättigau Flysch. The slightly chaotic internal structure of the Arosa Zone cannot be distinguished any further at this scale, but as mentioned in Chapter 3, its internal structure is shaped by Jurassic extensional tectonics, its position in an accretionary wedge in a subduction zone in the Cretaceous, and the Alpine thrust tectonics in the Cenozoic. The Lower Austroalpine nappes include the Rothorn slice in the west of the cross-section. The Silvretta nappe above this is composed of Permian and Mesozoic sediments overlying a thick crystalline body: the Mesozoic sediments had previously been rotated into vertical fault blocks associated with listric Jurassic normal faults, which dip towards the east and their offset runs counter to the westerly thrust direction. The Engadin Fault, a sinistral strike-slip with a normal fault component, cross-cuts the Austroalpine nappe pile. To the east of this, the Ortler and Quattervals nappes of the Engadin Dolomites are mainly composed of sediments. Although their crystalline substratum may be in the Campo and Languard nappes, the contact between the basement and cover nappes is, once again, associated with normal faults.

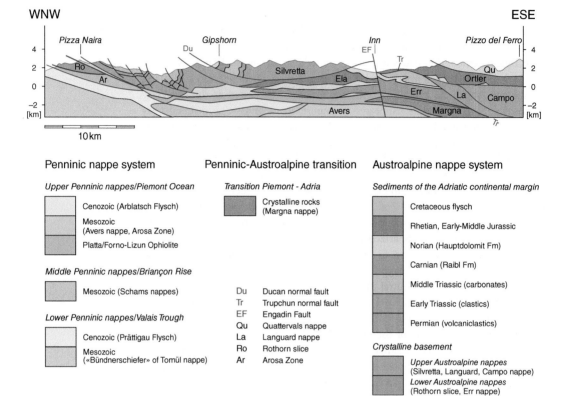

Penninic nappe system

Upper Penninic nappes/Piemont Ocean

Cenozoic (Arblatsch Flysch)

Mesozoic
(Avers nappe, Arosa Zone)

Platta/Forno-Lizun Ophiolite

Middle Penninic nappes/Briançon Rise

Mesozoic (Schams nappes)

Lower Penninic nappes/Valais Trough

Cenozoic (Prättigau Flysch)

Mesozoic
(«Bündnerschiefer» of Tomül nappe)

Penninic-Austroalpine transition

Transition Piemont - Adria

Crystalline rocks
(Margna nappe)

Du	Ducan normal fault
Tr	Trupchun normal fault
EF	Engadin Fault
Qu	Quattervals nappe
La	Languard nappe
Ro	Rothorn slice
Ar	Arosa Zone

Austroalpine nappe system

Sediments of the Adriatic continental margin

Cretaceous flysch

Rhetian, Early-Middle Jurassic

Norian (Hauptdolomit Fm)

Carnian (Raibl Fm)

Middle Triassic (carbonates)

Early Triassic (clastics)

Permian (volcaniclastics)

Crystalline basement

Upper Austroalpine nappes
(Silvretta, Languard, Campo nappe)
Lower Austroalpine nappes
(Rothorn slice, Err nappe)

▲ **Figure 5.2-20**
Geological cross-section
through the Austroalpine
nappes of Graubünden
(eastern Switzerland).
Source: Data from Schmid
et al. (1997).

*South-vergent
thrusting,
basement involved*

The Southalpine Nappe System

The Southalpine nappe system is delimited in the north by the Peri-Adriatic
fault system and plunges under the
Cenozoic Po Basin fill in the south (see
Fig. 5.2-1). The Ivrea Zone dominates
the western part of the Southalpine
nappe system: the rocks of the lower
crust in the Ivrea Zone have already been
discussed in greater detail in connection
with Figs 2-11–2-13, as has the adjacent
upper crust rock unit in the Strona-
Ceneri Zone to the east. In the eastern
part of the Southalpine nappe system in
the Central Alps, crystalline basement
predominates in the north, whereas sediments predominate in the south, and
Fig. 5.2-1 shows that the crystalline
rocks have been thrust on top of the
sediments in a southerly direction. The

thrusts in the Southalpine nappe system
are arranged en échelon, which can be
explained by the simultaneous dextral
strike-slip at the Peri-Adriatic fault
system. The internal structure of the
Southalpine nappe system is illustrated
based on two cross-sections.

The first cross-section, based on NFP
20 reflection seismic data, Fig. 5.2-22
has been adapted according to Schumacher (1997) and amended. The geological interpretation shows multiple
repitition of the Triassic sediments, and
a lack of inclusion of fragments of the
crystalline basement. In contrast to the
Penninic nappe system to the north of
the Peri-Adriatic fault system, the individual nappes were overthrust in a
southerly direction. The structure indicated in the cross-section is directly visible at outcrop in the case of the Lower

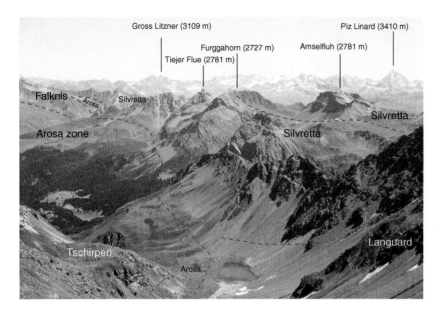

Figure 5.2-21 The contact between the Austroalpine (Adriatic margin) and Penninic nappes (Piemont Ocean and Briançon Rise) in the area around Arosa (canton Graubünden, Switzerland). Amselfluh and Furggahorn are made of Permian and Mesozoic sediments of the upper Austroalpine Silvretta nappe. The sediments are repeated by a nappe-internal thrust fault. The upper Austroalpine Languard nappe consists of crystalline basement and pinches out towards the lake at the bottom of the photograph (Älplisee). The Lower Austroalpine Tschirpen nappe in the lower left conists of Mesozoic sediments and pinches out in the distance. The Arosa Zone is a mélange of ophiolitic rocks and oceanic sediments derived from the Piemont Ocean that underlies the Silvretta nappe, but is visible in the flank of the Furggahorn, and in turn overlies the Falknis nappe visible at the left margin of the photograph. The Mesozoic sediments making up the Falknis nappe were deposited on the Briançon Rise. View is from Parpaner Rothorn toward the east-northeast.

and Upper Orobic nappes. Based on the reflection patterns yielded by the reflection data, an extrapolation of this structure to deeper levels is plausible. The amount of transport along the thrusts in the lowest units, the parautochthonous unit, the Lombardian and the Upper Orobic nappes, is relatively low, at 5–10 kilometres for each unit, but is far greater for the Lower Orobic nappe, at just under 30 kilometres. The crystalline rocks of the Strona–Ceneri Zone have been pushed into the rear, northern part of the Upper Orobic nappe in the form of a wedge. Two further faults with normal fault components fragment the crystalline rocks in the Upper Orobic nappe: the Val Colla Fault, which is dissected by the Tesserte Fault. In the south, the Mesozoic sediments of the Upper Orobic nappe plunge beneath the Po Basin, but this general plunge is accompanied by north-vergent thrusts, one of which actually lies within the plunging Mesozoic sediments. In the case of a second thrust, the Cenozoic Po Basin fill (the Gonfolite–Lombarda Group) has been thrust on

top of the plunging Upper Orobic nappe. The basal thrusts in the Orobic nappes combine into one thrust at depth, which is dissected by the Peri-Adriatic fault system in the extreme north. With regard to age, this can be determined from the fact that the Orobic thrusts in the east are cut by the Adamello Intrusion, so must be older then 30 million years, and from the fact that the Peri-Adriatic fault system was still active in more recent times.

The second cross-section, Fig. 5.2-23, adapted and slightly changed according to Schönborn (1992), is based on data from wells in the Po Basin and runs further to the east in Lombardy. The cross-section essentially reveals a similar structure to that shown in Fig. 5.2-22, but provides far more information on the structure of the substratum of the Po Basin. Numerous, south-vergent thrusts are responsible for multiple stacking of the Mesozoic sediments, and the crystalline basement is present in this nappe structure only in the north. Within the sediments, these

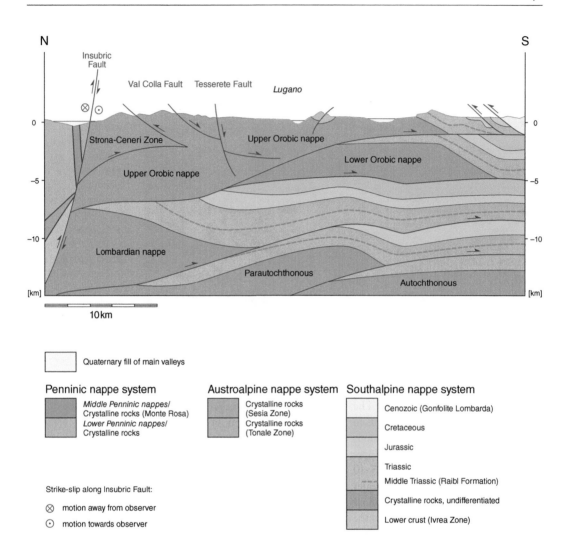

▲ **Figure 5.2-22**

Geological cross-section through the Southalpine nappe system along the seismic lines of the Southern Traverse of NRP20 (redrawn and modified after Schumacher, 1997). Source: Data from Pfiffner et al. (1997).

thrusts run parallel to the strata over large distances in the evaporitic Triassic layers, in the form of so-called flats. The thrust faults then climb in a southerly direction through the Triassic carbonates in the form of ramps and also affect the older portion of the Cenozoic Po Basin fill in the south. Schönborn (1992) demonstrated that the activation of the thrusts occurred in the fashion of in-sequence thrusting from top to bottom, or from north to south. The Orobic thrust (thrust system 1 in Schönborn 1992) is the oldest and was probably

activated as early as the Late Cretaceous. The Coltignone thrust (thrust system 2) branches: the upper branch is probably also pre-Adamello (Late Cretaceous) in age; the later, lower, branch was crosscut by a younger thrust from thrust system 3. The Milan thrust (system 4 in Schönborn 1992) is known only from seismic data and boreholes, but is associated with an actual fold-and-thrust belt, which runs in an east–west direction in the deep substratum of the Po Basin, where it affects the older preMessian sediments: the Milan thrust

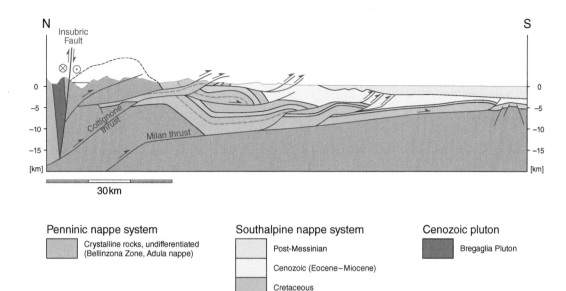

Penninic nappe system
- Crystalline rocks, undifferentiated (Bellinzona Zone, Adula nappe)

Austroalpine nappe system
- Crystalline rocks, undifferentiated (Tonale Zone)

Southalpine nappe system
- Post-Messinian
- Cenozoic (Eocene–Miocene)
- Cretaceous
- Jurassic
- Triassic
- Middle Triassic (Raibl Formation)
- Crystalline rocks, undifferentiated

Cenozoic pluton
- Bregaglia Pluton

Strike-slip along Insubric Fault:
- ⊗ motion away from observer
- ⊙ motion towards observer

must therefore date from the Middle to Late Miocene. The sealing of the reverse faults that split off from the Milan thrust fault by the post-Messinian sediments is clearly visible in Fig. 5.2-23. At the northern end of the cross-section, the Insubric Fault has been cut off below by a younger thrust. This sequence of faulting can be derived from the fact that the Coltignone and the Milan thrust system post-date activity along the Insubric Fault.

5.3 The Eastern Alps

The mountain chain that forms the Eastern Alps extends from east to west, disappears into the Vienna Basin in the east, and then continues in the Carpathians. Of note is a band of Mesozoic sediments, about 50 kilometres wide, in the northern part of the

Eastern Alps (see Fig. 5.3-1). This forms the **Northern Calcareous Alps**, a nappe pile composed of Austroalpine nappes. Within this, three nappe complexes are distinguished. The lowest of these, the Bavarian nappe complex (or the Bajuvaricum), make up the western portion of the Northern Calcareous Alps, but run eastwards in the form of a narrow band at its northern margin. The overlying nappes form the Tyrolean nappe complex (or the Tirolicum), which is mainly exposed in the central portion of the Northern Calcareous Alps. Finally, the latest, the Julian nappe complex (or the Juvavicum), is found in the southeastern portion of the Northern Calcareous Alps. In the west, the Northern Calcareous Alps have generally been sheared off their pre-Triassic basement. This basement may be found in the adjacent Silvretta and Ötztal nappes to the south, which are

▲ **Figure 5.2-23**
Geological cross-section through the Southalpine nappe system of Lombardy (Italy). Source: Redrawn and adapted from Schönborn (1992).

▶ **Figure 5.3-1**
Simplified geological–tectonic map of the Eastern Alps and Dolomites with locations of cross-sections (green lines–numbers are figure numbers). Source: Drawn after geological interpretations by Auer & Eisbacher (2003), Lammerer et al. (2008), Castellarin et al. (2006b), as well as Lammerer & Weger (1998), Doglioni (2007), Brandner (1980), Ortner et al. (2006) and own interpretation.

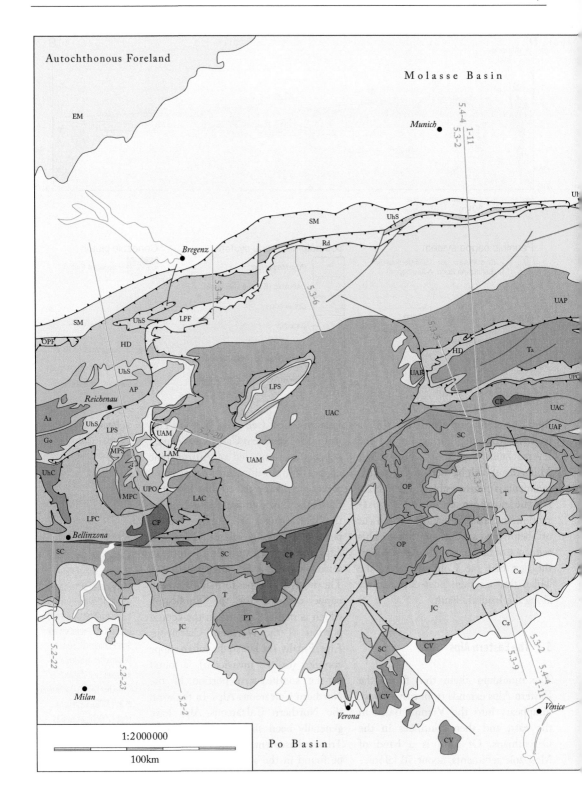

Autochthonous Foreland

Molasse Basin

5.4-4 1-11
5.3-2

Munich

EM

Bregenz

SM

UhS

Rd

5.3-6

5.3-5

UAP

UhS

LPF

Ta

HD

UAP

OPF

SM

UhS

HD

UPC

Reichenau

UhS

AP

LPS

UAC

CP

UAC

Aa

UhS

LPS

UAM

5.2-20

SC

UAP

Go

MPS

LAM

UAM

UhC

MPC

UPO

OP

T

LPC

LAC

Bellinzona

CP

OP

Cz

SC

SC

CP

Milan

T

JC

Cz

5.3-2
5.4-4
1-11

PT

JC

5.2-22

5.2-23

5.3-9

JC

SC

CV

5.2-2

Verona

CV

Po Basin

CV

Venice

1:2 000 000

100km

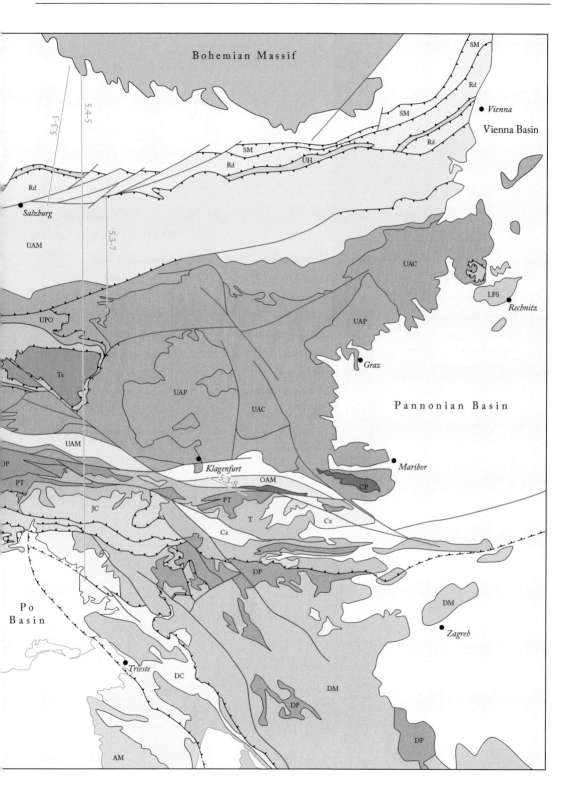

Legend to Figure 5.3-1

European (Northalpine) Foreland

Cenozoic sediments

SM	Subalpine Molasse
	Autochthonous Molasse
EM	Autochthonous Mesozoic
	Crystalline basement (Bohemian massif)

Nappe systems of the European continental margin

Helvetic nappe system

UhS	Ultrahelvetic cover nappes
UhC	Ultrahelvetic crystalline nappes
HS	Helvetic cover nappes
AP	Autochthonous-parautochthonous cover
	Crystalline basement blocks Gotthard (Go), Aar (Aa), Tauern (Ta)

Penninic nappe system

Upper Penninic nappes/Piemont Ocean

UPF	Flysch nappes
OUPO	Ophiolite nappes

Middle Penninic nappes/Briançon Rise

MPS	Cover nappes
MPC	Crystlline nappes

Lower Penninic nappes/Valais Trough

LPF	Flysch nappes Renodanubian Flysch (Rd)
LPS	Cover nappes
LPC	Crystalline nappes

Adriatic Foreland

Cenozoic sediments

	Cenozoic of Po Basin
AM	Autochthonous Mesozoic

Nappe systems of the Adriatic continental margin

Austroalpine nappe system

Upper Austroalpine nappes

UAM	Mesozoic
UAP	Palaeozoic
UAC	Crystalline basement

Lower Austroalpine nappes

LAM	Mesozoic
LAC	Crystalline basement

Southalpine nappe system/Dolomites

Cz	Cenozoic
JC	Jurassic – Cretaceous
T	Triassic
PT	Permian – Triassic
OP	Ordovician – Permian
SC	Crystalline basement

CV	Cenozoic volcanic rocks
CP	Cenozoic plutons

Extraalpine units

Intramontane basins in the East

Vienna Basin, Pannonian Basin

Dinarides

DC	Cenozoic
DM	Mesozoic
DP	Palaeozoic

two large crystalline blocks that in some locations are covered in Permo-Triassic sediments. In the east, However, the Palaeozoic substratum of the Northern Calcareous Alps is exposed at its southern margin: the Innsbruck Quartz Phyllites and the Greywacke Zone in the east (see Fig. 5.3-1). A narrow band of Helvetic and Penninic nappes can be followed along the entire northern margin of the Northern Calcareous Alps. In the west, in Vorarlberg, these are essentially Helvetic nappes, whereas further to the east, a band of Penninic nappes that varies in width or, more precisely, the Rhenodanubian Flysch, extends as far as Vienna. The Helvetic and the Penninic nappes have been thrust onto the Subalpine Molasse in the north. This latter unit is very narrow in the middle part of the Eastern Alps, such that the Alpine nappes rest almost directly on the autochthonous Foreland Molasse, but boreholes provide evidence that dislodged slices of allochthonous thrust sheets of the Subalpine Molasse are present beneath the Alpine nappes.

The substratum of the Austroalpine nappes inside the mountain chain is exposed in three large-scale tectonic windows: the Engadin and Rechnitz windows in the extreme west and east, respectively, in which the Penninic nappes are exposed, and the Tauern window in the central part, in which not only Penninic nappes are exposed, but also the underlying Helvetic nappe system. The deeper tectonic units in the window have been uplifted relative to the surrounding rock at a marked normal fault in the western margin of the Tauern window, the Brenner normal fault, and at the extensional Katschberg shear zone system in its eastern margin (see Ratschbacher & Frisch 1993, Scharf et al., 2013), which is a situation

that can be compared to the Lepontian Dome in the Central Alps.

Mainly units of crystalline basement or Palaeozoic sediments are encountered in the southern part of the Austroalpine realm, as is shown in Fig. 5.3-1. Only in the extreme south is an elongate complex of Mesozoic sediments to be found that runs along the Peri-Adriatic fault system, from the Carnian Alps into the Karawanks. Numerous strike-slip faults cross-cut the different Austroalpine units, in both the south and in the Northern Calcareous Alps, which developed due to simultaneous north–south-directed shortening and east–west-directed extension of the mountain chain (see Ratschbacher et al. 1991).

Towards the south, the Austroalpine nappe system is separated from the Southalpine nappe system, or the Dolomites, by the **Peri-Adriatic fault system**. This fault system contains a segment with a SSW–NNE strike, the Giudicaria Fault, which sinistrally offsets the general east–west course of the fault system. Numerous smaller and larger plutons have intruded along the Peri-Adriatic fault system, mostly around 30 million years ago. There is an inherent link between these intrusions and the position and activity of the Peri-Adriatic fault system, as has been discussed in the context of Fig. 4.7.

In the **Dolomites**, the pre-Triassic basement is exposed at the surface in the northern part, just to the south of the Peri-Adriatic fault system, and also appears in the central part of the Dolomites, in the hanging wall of the Valsugana thrust. Apart from this, Triassic carbonates predominate, which have given this mountain range its name. Numerous thrusts cross the Dolomites from east to west, accompanied by two larger strike-slip faults that run from north–south, and the transport direction of thrusting was usually towards the south.

Two strike-slip faults delimit the Giudicaria fault zone that lies in the west of the Dolomites: the western one of the two marks the contact with the Cenozoic Adamello intrusion and the boundary of the Southalpine nappe system. The entire fault zone was active over a long period and in many different ways and caused an offset of the Peri-Adriatic fault system in the region of the Dolomites by 70 kilometres (see Prosser (1998) and Castellarin & Cantelli (2006) for in-depth discussions). Multiple thrusts organized en échelon within the Giudicaria fault zone indicate sinistral transpressive shearing (see Fig. 5.3-1).

In the east of the Dolomites, the thrusts bend round into a northwest–southeast direction, and vertical faults with the same orientation are associated with this. This fault system marks the transition to the neighbouring Dinarides in the south.

The cross-section that runs from north–south in Fig. 5.3-2 summarizes the structure of the Eastern Alps in the cross-section through the Tauern window. The cross-section is based on the TRANSALP reflection seismic data (Lüschen et al. 2004) and interpretations of this data (Auer & Eisbacher 2003, Castellarin et al. 2006, Lammerer et al. 2008,), other publications (Brandner 1980, Lammerer & Weger 1998, Doglioni 2007, Ortner et al. 2006) and my ideas: the structure of the lower crust is again based on the teleseismic studies by Waldhauser et al. (2002) and Diel et al. (2009).

The pre-Triassic crystalline basement with its autochthonous cover of Mesozoic sediments in the substratum of the Molasse Basin dips down in a shallow manner towards the south and beneath the Alps. Two larger normal

faults, indicated by the reflection seismic data, offset the Mesozoic strata beneath the Northern Calcareous Alps. A thin band of Subalpine Molasse, Helvetic slices and a unit of Penninic Rhenodanubian Flysch dips down under the Northern Calcareous Alps and peters out in the south. The internal structure of the Northern Calcareous Alps (adapted from Auer & Eisbacher 2003) reveals a typical nappe pile. Two thin slices of Austroalpine sediments (also called 'Border imbricates') mark the basal thrust of the Northern Calcareous Alps, and are overlain in this cross-section by the Allgäu nappe, which is 2 kilometres thick. The Northern Calcareous Alps are mainly composed of the Lechtal nappe, which lies at the top. At its base, the Lechtal nappe also exhibits a stack of small imbricates made up of Middle Triassic sediments, while the upper portion of the nappe exhibits large-scale folds and steep, north- and south-vergent thrusts faults. This structure is strongly controlled by the thick, unyielding dolostone successions of the Late Triassic 'Hauptdolomit' Formation. In the south of the nappe, the Palaeozoic rocks of the Greywacke Zone or the Innsbruck Quartz Phyllite nappe underlie the Triassic until, finally, the steep south-dipping Inntal Fault makes a clean cut through the entire nappe pile.

The middle portion of the cross-section in Fig. 5.3-2 is dominated by a large-scale antiform. In its core, the lowest unit, the pre-Triassic crystalline basement, of the Tauern massif is exposed (the composition of its rocks is discussed in more detail in the context of Fig. 2.8). Isoclinal folds and plunging folds indicate ductile deformation under high temperatures. Lammerer & Weger (1998) estimate a horizontal, ductile shortening of the original dome-like

Eastern Alps:
Thick Australpine
nappe pile,
Nearly symmetric
crust-mantle
boundary,
Tauern massif as
pop-up

bulge by about half its original width. The autochthonous sedimentary cover of the crystalline basement has similarities to the Helvetic facies zone. There is a thrust contact in its roof, where the Penninic nappes follow above, which are composed of a metamorphic shale–sandstone sequence that also contains prasinites. The only possible origin of these Penninic nappes is the Penninic Ocean, which constitutes the common continuation of the Valais Trough and the Piemont Ocean in the Central Alps. The relationship between these Bündnerschiefer-like rocks and the Penninic Rhenodanubian Flysch also remains unclarified. The basal thrust fault in the Penninic nappes was also affected by the ductile deformation in the core of the Tauern window and was folded. Emplacement of the Penninic nappes onto the Helvetic realm must therefore have occurred earlier. The Austroalpine nappes form the roof of the above-mentioned antiform: in the north, these are Palaeozoic formations (Innsbruck Quartz Phyllite nappe and Greywacke Zone), with crystalline formations in the south.

A large deeply buried thrust is responsible for the fact that the crystalline basement has been uplifted from a depth of approximately 7 kilometres below sea level in the foreland (below the Northern Calcareous Alps) to over 3 kilometres above sea level in the core of the Tauern window. A cluster of reflections in the TRANSALP seismic line indicates this south-dipping thrust (the so-called Tauern ramp).

In the southern portion of Fig. 5.3-2, in the Dolomites, multiple thrusts that dip flatly towards the north indicate a shortening of the upper crust. The crystalline basement is also affected by this shortening. Contrary to the situation in the Lombardian Southalpine nappe system, there was no substantial decoupling between the crystalline basement and the Mesozoic sediments in the Dolomites. The units that dip steeply towards the north immediately to the north of the Dolomites (the Pustertal Fault of the Peri-Adriatic fault system, itself, and the adjacent Austroalpine basement nappes) indicate that the Adriatic crust was pushed northwards underneath the nappe pile in the Tauern window.

The geophysical data show that the European crust–mantle boundary (Moho) plunges towards the south under the Alps and reaches a depth of 50 kilometres beneath the Tauern window. The Adriatic Moho plunges towards the north under the Alps, also reaches a depth of 50 kilometres and abuts the European Moho beneath the Tauern window. There is thus no determinable overlap between the two Mohos, as is the case for the Central Alps. Figure 5.3-2 shows the European lower crust as thickened in the south and even offset by the Tauern antiform thrust fault, while the Adriatic crust seems thickened at the contact zone to the European lower crust only. The geometry of the Adriatic crust will be encountered again in the discussion of the deep lithospheric structure of the Alps in section 5.4.

The Molasse Basin

The structure of the Molasse Basin in the foreland of the Eastern Alps is illustrated based on a cross-section (adapted from Steininger & Wessely 2000 and Hamilton et al. 2000) from the Bohemian massif to the Northern Calcareous Alps (Fig. 5.3-3). The thickness of the Molasse sediments increases in the direction of the Alps, similar to the situation in the Central Alps. In the north, the sediments of the

▶ **Figure 5.3-2**
Geological cross-section through the Eastern Alps and Dolomites along the seismic lines of TRANSALP (Lüschen et al. 2004). Source: Drawn after geological interpretations by Auer & Eisbacher (2003), Lammerer et al. (2008), Castellarin et al. (2006b), as well as Lammerer & Weger (1998), Doglioni (2007), Brandner (1980), Ortner et al. (2006) and own interpretation.

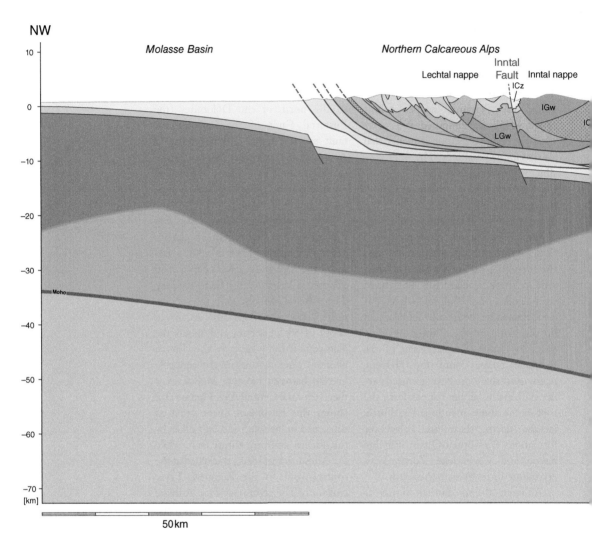

NW

Molasse Basin   Northern Calcareous Alps

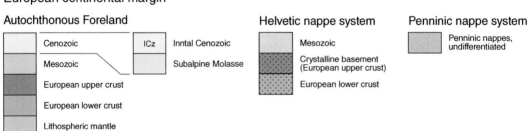

European continental margin

Autochthonous Foreland

	Cenozoic
	Mesozoic
	European upper crust
	European lower crust
	Lithospheric mantle

| ICz | Inntal Cenozoic |
| | Subalpine Molasse |

Helvetic nappe system

	Mesozoic
	Crystalline basement (European upper crust)
	European lower crust

Penninic nappe system

| | Penninic nappes, undifferentiated |

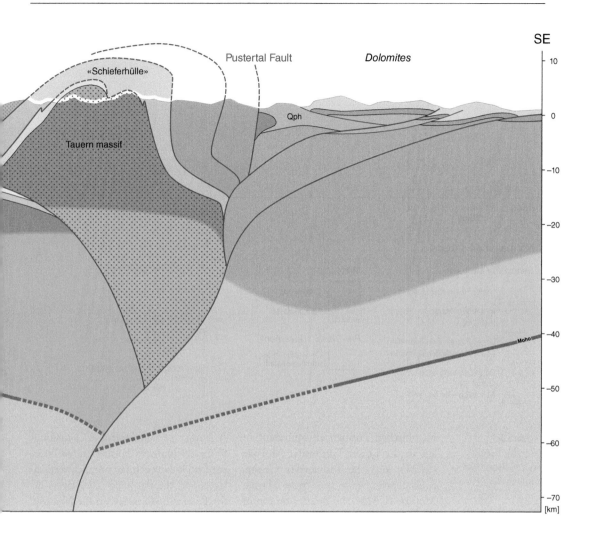

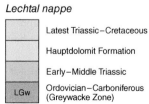

Adriatic continental margin

Austroalpine nappe system

Inntal nappe

| IGw | Ordovician–Carboniferous (Greywacke Zone) |

Lechtal nappe

	Latest Triassic–Cretaceous
	Hauptdolomit Formation
	Early–Middle Triassic
LGw	Ordovician–Carboniferous (Greywacke Zone)

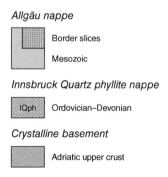

Allgäu nappe

| | Border slices |
| | Mesozoic |

Innsbruck Quartz phyllite nappe

| IQph | Ordovician–Devonian |

Crystalline basement

| | Adriatic upper crust |

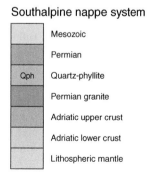

Southalpine nappe system

	Mesozoic
	Permian
Qph	Quartz-phyllite
	Permian granite
	Adriatic upper crust
	Adriatic lower crust
	Lithospheric mantle

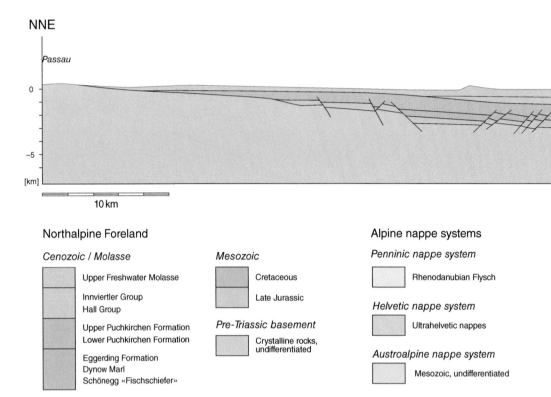

NNE

Passau

0

−5

[km]

|———— 10 km ————|

Northalpine Foreland

Cenozoic / Molasse

Upper Freshwater Molasse

Innviertler Group
Hall Group

Upper Puchkirchen Formation
Lower Puchkirchen Formation

Eggerding Formation
Dynow Marl
Schönegg «Fischschiefer»

Mesozoic

Cretaceous

Late Jurassic

Pre-Triassic basement

Crystalline rocks,
undifferentiated

Alpine nappe systems

Penninic nappe system

Rhenodanubian Flysch

Helvetic nappe system

Ultrahelvetic nappes

Austroalpine nappe system

Mesozoic, undifferentiated

▲ **Figure 5.3-3**
Geological cross-section
through the Molasse Basin
of Bavaria and Austria.
Source: Adapted from
Steininger & Wessely (2000)
and Hamilton et al. (2000).

Molasse Basin:
Strongly deformed
inner part,
onlap in outer part

Puchkirchen Formation, equivalent in age to the Lower Freshwater Molasse (LFM), and the Innviertler Group, equivalent in age to the Upper Marine Molasse (UMM), lie directly on top of the pre-Triassic crystalline basement. The (lower) Puchkirchen Formation also lies directly on top of crystalline rock to the north of the edge of the Alps. The Mesozoic sediments have been fragmented by numerous normal faults that form horst and graben structures. Cretaceous sediments are present only in a broad graben structure in the middle portion of the cross-section in Fig. 5.3-3. However, in the south, the Cretaceous sediments then appear under the Rhenodanubian Flysch and continue across a horst.

An antiform (a triangle zone) is outlined by the Molasse strata close to the contact with the Alpine nappes.

A wedge of conglomerates belonging to the Puchkirchen Formation has been jammed in between the younger, vertical layers and the flat-lying older layers of the Foreland Molasse, with the Molasse sediments continuing under the Alpine nappes. This Subalpine Molasse is composed of older strata (LMM and LFM). Two thrusts parallel to the strata have caused a three-fold stacking of the Molasse sediments and indicate substantial shortening. The Alpine nappes have been thrust as far as the tilted Foreland Molasse. The frontal portion is composed of a pile of Ultrahelvetic nappes and Penninic nappes (Rhenodanubian Flysch) that also has been subjected to internal imbrication. The thin layer of Ultrahelvetic nappes at the base of the Rhenodanubian Flysch reaches as far as the Earth's surface and is visible in the form of a striped pattern on

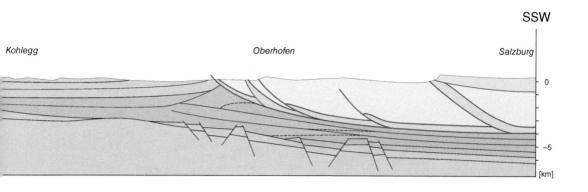

a map (we also call this structure 'stripe windows'). The internal schuppen structure must therefore have been created after the emplacement of the Penninic nappes on top of the Ultrahelvetic nappes. The basal thrust over the highest Alpine nappes, the Austroalpine Northern Calcareous Alps, cross-cuts both nappe contact types in the footwall, indicating more recent emplacement.

The Helvetic Nappe System

We can gain deeper insight into the structure of the Eastern Alpine Helvetic nappe system in the extreme west, in Vorarlberg. The cross-section in Fig. 5.3-4 shows the pronounced fold structure in the Cretaceous strata. Towards the west, across the Rhine Valley, these folds can be more or less correlated with the folds in the eastern

Swiss Säntis nappe (see Pfiffner 1993, 2011, Pfiffner et al. 2010. In contrast to the situation in eastern Switzerland (see Fig. 5.2-12), the Jurassic strata in Vorarlberg extends much further to the north. In the Canisfluh (Fig. 5.3-4), the folded Quinten Limestone is exposed at the surface, but at depth, these limestones extend even further to the north and into the core of the frontal Cretaceous folds of the Niedere. Although the thick marly successions of the earliest Cretaceous allowed somewhat independent folding in the Jurassic and Cretaceous strata, actual shearing-off did not occur, as is the case in the eastern Swiss Säntis nappe. The basal decollement horizon of the Helvetic nappes is hidden deep down in the cross-section of Vorarlberg. The interpretation given in Fig. 5.3-4 is based on borehole data and an extrapolation of the structures

Vorarlberg:
Buried nappe pile

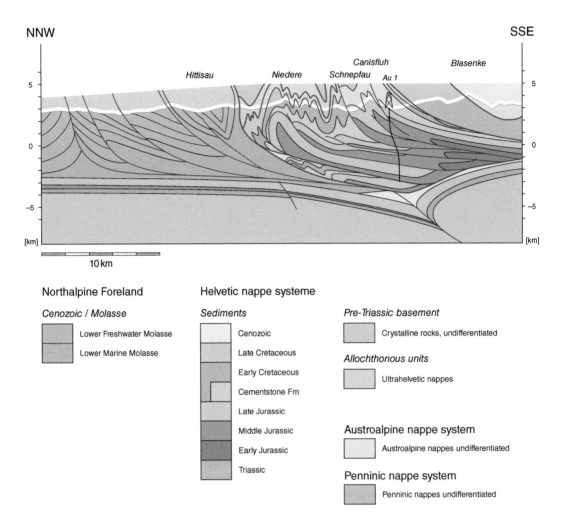

▲ **Figure 5.3-4**

Geological cross-section through the Helvetic nappe system of Vorarlberg (western Austria).

visible in the Central Alps. Both the argillaceous layers at the bottom in the Early or Middle Jurassic could form the decollement horizon and potential Triassic evaporites also cannot be completely excluded. The basal thrust faults in the Ultrahelvetic and Penninic nappes in the roof of the Cretaceous are included in all the folds, which proves that the emplacement of these nappes on top of the Helvetic nappe system occurred before this latter was dislocated and internally deformed. In addition to the imbricate structure, the Subalpine

Molasse in the northern part of the cross-section also exhibits a larger syncline immediately in front of the Helvetic nappes, indicated by the distribution of the Lower Marine Molasse (LMM). The formation of such a syncline, which is unique in the Subalpine Molasse, infers the presence of a very thick LMM sequence in the subsurface. The transition of the LMM to the Northhelvetic Flysch deep underground is similar to the situation that is probably present in the Central Alps. The illustration of the basement–cover contact at the base is derived from the information obtained in the NFP 20 East Traverse (Pfiffner & Hitz 1997) in the Central Alps.

The continuation of the Helvetic nappe system towards the east is composed of a narrow band of Cretaceous rocks that can be followed between the Subalpine Molasse and the Penninic or Austroalpine nappes. In addition, the crystalline basement of the Helvetic nappe system and remnants of its Triassic–Jurassic autochthonous sedimentary cover are exposed under the Austroalpine and Penninic nappes in the Tauern massif (see below).

The Penninic Nappe System

Similar to the Helvetic nappes, the Penninic nappes are exposed both in a narrow band at the northern edge of the Austroalpine nappes and in the Tauern and Rechnitz windows. The narrow band in the north is the Cretaceous Rhenodanubian Flysch. The sediments of the Glockner nappe in the Tauern window are similar to the Bündnerschiefer Group in the Central Alps. Together with the neighbouring ophiolites they can be allocated to the Penninic Ocean, which equates to the merged Valais Trough and Piemont Ocean. The Briançon Rise, which separates these domains in the Western and Central Alps does not extend eastwards beyond the Engadin window.

The cross-section in Fig. 5.3-5 summarizes the nappe structure in the western Tauern window and is based on Lammerer & Weger (1998), Ortner et al. (2006) and the geological and tectonic map of Tyrol (Brandner 1980). The crystalline core of the Tauern window, the Tauern massif, is composed of three submassifs (Ahorn, Tux and

Zillertal), of which the majority of the volume is made up of post-Variscan granites (see Fig. 2.8), and a thin strip of crystalline rocks is present at the southern edge. The Permo-Mesozoic autochthonous cover of this crystalline basement exhibits similarity to the Helvetic sediments. Taken together with the structural proximity to the autochthonous crystalline basement, which can be followed from the Bohemian massif, under the Molasse Basin, and as far as below the northern edge of the Tauern massif (see Fig. 5.3-2), this favours assignment of the crystalline and the Permo-Mesozoic rocks to the European margin, or more precisely the Helvetic realm. The three submassifs form relatively narrow, upright folds with vertically dipping southern limbs. The northern boundaries of the Tux and Zillertal submassifs and the southernmost strip of crystalline outcrop correspond to thrust faults that are folded in the roof of the submassifs. These thrust faults indicate that the internal deformation of the crystalline basement was brittle in nature during an early phase and was later overprinted by ductile folding at high temperatures. Lammerer & Weger (1998) determined a horizontal shortening to 50% of the original length in the late, ductile portion of the deformation. The thrust faults in the overlying Penninic and Austroalpine nappes have also been passively folded by the ductile deformation. The basal thrust fault of the Penninic nappes runs almost parallel to the upper edge of the Helvetic sediments below. The lower portion of the Penninic nappes consists of a thrust sheet of Permian–Triassic sediments. The overlying the Glockner nappe is composed of metamorphic sandstones, shales and limestones, a suite comparable to the Bündnerschiefer

Thin thrust sheets of cover rocks, folded above Tauern massif

Group of the Valais Trough in the Central Alps. In the upper portion of the Penninic nappes, the Matrei Zone, prasinites are also associated with these sediments in many locations and the entire suite is more reminiscent of that in the Piemont Ocean in the Central Alps. The Penninic nappes apparently also exhibit extensive internal imbrications. In case of the Matrei Zone, imbrication resulted in the formation of a mélange composed of ophiolites, Bündnerschiefer and fragments of Austroalpine units. Both the content and the structural position at the Penninic–Austroalpine boundary conform to that in the Arosa Zone in the Central Alps. This mélange marks the former plate boundary between the Adriatic continental margin and the Penninic Ocean, with the chaotic mixture being created by intensive shearing in the subduction zone. Blocks of Austroalpine crystalline basement (Deferegger–Schober Altkristallin intruded by the Rensen Pluton) that dip steeply towards the north are exposed at the southern edge of the Tauern Window. At the northern edge, the basal thrust in the Austroalpine nappe system dips more shallowly towards the north. The structure of the Austroalpine nappe system itself is made up of the Palaeozoic units of Innsbruck Quartz Phyllite. A tectonic lens at the very base of the Austroalpine nappe system exhibits Permo-Mesozoic sediments over the quartz phyllite.

Northern
Calcareous Alps

The Pustertal Fault, a segment of the Peri-Adriatic fault system, forms the boundary to the Southern Alps in the extreme south of Fig. 5.3-5. A continuation of the Rensen granite that is exposed at the surface along the Pustertal Fault is indicated along the fault plane, deep down and within the Austroalpine basement to the north.

The Austroalpine Nappe System

As mentioned at the start of this section, the Austroalpine nappes are essentially composed of a nappe complex of Mesozoic sediments, the Northern Calcareous Alps in the north, and a nappe complex made up of crystalline nappes in the south: the eastern and western Greywacke Zones and different units of quartz phyllites form a nappe complex comprising a type of transitional zone between. The Greywacke Zone, which has been discussed in the context of Fig. 2.15, is composed of Palaeozoic sediments and volcanics and represents the stratigraphic substratum of part of the Northern Calcareous Alps. The same applies to the quartz phyllites, which were discussed in detail in the context of Fig. 2.16. The cross-section in Fig. 5.3-2 shows the relative positions of these three nappe complexes.

As an example, the structure of the Northern Calcareous Alps is elucidated in more detail in two cross-sections. The first, Fig. 5.3-6, is adapted from Eisbacher et al. (1990) and shows the situation in the west. Three nappes composed of Mesozoic sediments overlie each other here, which are internally shortened by folds and bivergent (i.e. north- and south-directed) thrusts: from bottom to top these are the Allgäu, Lechtal and Inntal nappes. The Allgäu and Lechtal nappes are referred to together as the Bavarian nappe complex (Bajuvaricum) and the Inntal nappe is assigned to the Tyrolean nappe complex (Tirolicum). Two thick, mechanically strong, unyielding carbonate sequences and multiple decollement horizons substantially control the internal structure of the Northern Calcareous Alps. The former include the Middle Triassic Muschelkalk Group and the interlocking Wetterstein Formation, as well as the

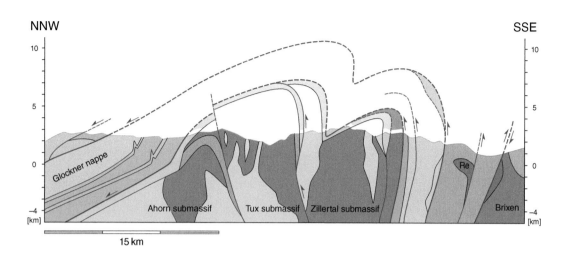

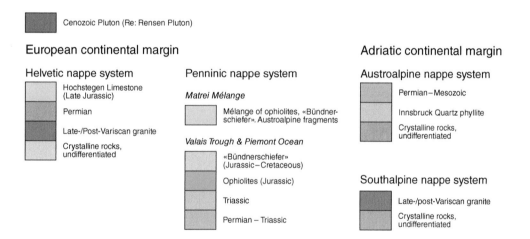

Late Triassic 'Hauptdolomit' Formation. Decollement horizons are present in the Permo-Triassic evaporitic successions (Haselgebirge, Buntsandstein and Reichenhall formations), in the shales of the Partnach Formation that interlock with the Wetterstein Formation, in the Late Triassic evaporitic argillaceous Raibl Formation and in the argillaceous and thinly bedded limestones of the Jurassic–Cretaceous successions (Eisbacher et al. 1990). Figure 5.3-6 shows how the Allgäu and Lechtal nappes have been sheared off in the Middle Triassic Partnach Formation. However, we can see that the pre-Triassic basement (crystalline rocks or Palaeozoic sediments) is also incorporated into the nappe structure at the southeastern end of the Allgäu nappe. In the case of the Inntal nappe, the basal thrust fault also climbs ramp-like towards the northwest from the crystalline basement, through the Triassic sediments and into the shales of the Partnach Formation. In the extreme southeast, the Inntal nappe has been overthrust by the crystalline rocks of the Ötztal nappe. According to Eisbacher et al. (1990), the Northern Calcareous Alps in this cross-section were shortened to 62 kilometres by internal deformation and nappe stacking

▲ **Figure 5.3-5**
Geological cross-section through the western Tauern window (Austria–Italy).

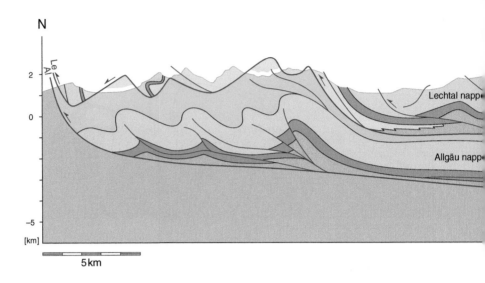

N

Lechtal nappe

Allgäu nappe

2

0

-5

[km]

5 km

Austroalpine nappe system

L. Cretaceous		Clastic sediments
E. Cretaceous–Late Triassic		Kössen Fm & younger limestones, shale, radiolarian chert
Late Triassic		Hauptdolomit Formation
		Raibl Formation
Middle Triassic		Wetterstein Formation, Muschelkalk Group
		Partnach Formation
Early Triassic		Reichenhall Formation Buntsandstein Formation
		Crystalline basement

Penninic nappe system

Penninic nappes, undifferentiated

Thrust faults:

Al Allgäu
Le Lechtal
In Inntal
Öz Ötztal

▲ **Figure 5.3-6**
Geological cross-section through the Northern Calcareous Alps in Tyrol–Bavaria (Austria–Germany). Source: Eisbacher et al. (1990). Reproduced with permission of Swiss Geological Society.

from an original length of 108 kilometres. The formation of the lowest nappe in the Northern Calcareous Alps, the Allgäu nappe, started at the end of the Early Cretaceous, approximately 97 million years ago. The internal deformation of the uppermost nappe, the Inntal nappe, started slightly later, around 90 million years ago, and lasted until the Late Cretaceous (65 million years ago).

The cross-section in Fig. 5.3-7 has been adapted from Mandl (2000) and amended. This shows the structure of the Northern Calcareous Alps in the east and its contact with the units in the

footwall. The Bavarian nappe complex is present only in the form of small lenses of Triassic sediments in the north and overlies a substratum composed of Rhenodanubian Flysch. The Tyrolean nappe complex has a substratum composed of Palaeozoic sediments in its southern portion, the so-called Greywacke Zone. The nappes are mainly composed of Triassic sediments that have been sheared off their Palaeozoic substratum and, accordingly, lie further north in the cross-section. They were sheared off in the Permian clastics and evaporites. Within the Jurassic succession, larger and

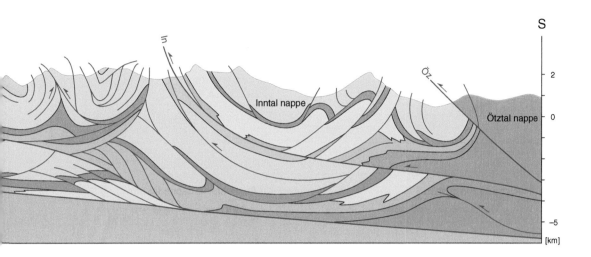

smaller bodies of rock are to be found in the form of exotic fragments: radiolarian cherts and carbonates from the basin region of the Juvavicum, that entered into the limestones and marls of the future Tyrolean nappe as olistoliths in the Oxfordian (approximately 155 million years ago). However, the entire Jurassic strata within the Tyrolean nappe complex was also later sheared off its former Triassic substratum, in the Early Cretaceous (Mandl 2000), and is now present on top of the Triassic carbonates or even on the Early Cretaceous clastic sediments in the form of a separate, internally deformed unit. This shearing process was probably triggered by the emplacement of the uppermost unit, the Juvavic nappe complex. The Juvavic nappe complex is visible in the cross-section in the form of a slightly folded normal sequence. The folds in the middle portion of the cross-section are due to stacking within the Tyrolean nappe complex. At the southern end of Fig. 5.3-7, the Greywacke Zone of the Tyrolean nappe complex overlies a larger block of crystalline basement, the so-called Middle Austroalpine Crystalline. At its northern front, this contains remnants

Greywacke Zone

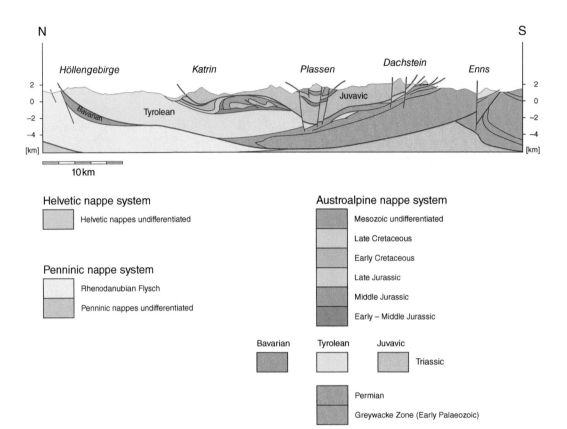

▲ **Figure 5.3-7**
Geological cross-section
through the Northern
Calcareous Alps of Styria,
Upper Austria. Source:
Mandl (2000). Reproduced
with permission of
Österreichischen
Geologischen Gesellschaft.

Karawanks

of a Permian and Mesozoic sedimentary cover and is part of the original substrate of the Northern Calcareous Alps. According to Mandl (2000), the Northern Calcareous Alps were sheared off this substrate in the Late Cretaceous (Turonian; approximately 90 million years ago) and the shearing process continued in the Late Eocene (approximately 35 million years ago).

In the extreme southeast of the Eastern Alps, the Karawank mountain range runs along the Peri-Adriatic fault system and the boundary to the Dolomites. With reference to their facies, the Northern Karawanks belong to the Austroalpine realm, the Southern Karawanks to the Southalpine realm. Based on large-scale facies comparisons, Polinski (1991) estimates the dextral offset

in the Karawanks at the Peri-Adriatic fault system as 150 kilometres. This strike-slip was transpressive and caused a bivergent system of thrusts and folds, which indicate a WNW–ESE-directed shortening within the Karawanks (Polinski 1991). Figure 5.3-8 shows a slightly offset cross-section through the Northern and Southern Karawanks, adapted from (Polinski 1991) and amended. The thrust faults on both sides of the central strike-slip become steeper towards it, whereby the crystalline basement ended up in a far higher location in the central region. This structure, also called a 'positive flower structure', is typical of transpressive deformation regimes.

In the Northern Karawanks, the Middle Miocene clastic sediments of the Klagenfurt Basin lie unconformably

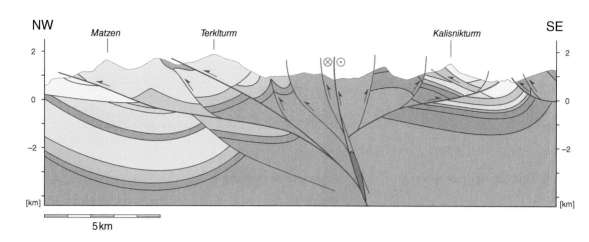

NW   SE

Matzen   Terklturm   Kalisnikturm

5 km

Karawanks Tonalite (Oligocene)

Austroalpine nappe system

Northern Karawanks

	Cenozoic (Miocene)
	Jurassic – Early Cretaceous
	Kössen Formation
	Hauptdolomit Formation
	Raibl Formation
	Wetterstein Formation
	Muschelkalk Group
	Permoskyth Sandstone
	Crystalline basement

Triassic

Permian

Southalpine nappe system

Southern Karawanks

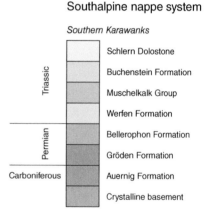

	Schlern Dolostone
	Buchenstein Formation
	Muschelkalk Group
	Werfen Formation
	Bellerophon Formation
	Gröden Formation
	Auernig Formation
	Crystalline basement

Triassic

Permian

Carboniferous

above the tilted and folded Triassic sediments, but in turn have also been overthrust by a unit of Triassic sediments.

The west-northwest-directed thrust faults and folding therefore started before the Middle Miocene (13 million years ago, according to Polinski 1991). The transpressive activity of the Peri-Adriatic fault system is emphasized by the intrusion of the Oligocene Karawank Tonalite (28–29 million years old). The intrusion, which is exposed at the surface, is located to the east of the cross-section, but the intrusive body marked in Fig. 5.3-8 is hypothetical. The Klagenfurt Basin was formed by extension in the Late Miocene, contemporaneous with the subsidence of the Pannonian Basin. The basin sediments must have been overthrust (as mentioned above) in the latest Miocene (11–8 million years ago) and this is linked to the formation of numerous northwest–southeast striking strike-slip faults, which cross-cut the Peri-Adriatic fault system. According

▲ **Figure 5.3-8**
Geological cross-section through the the Karawanks. Source: Redrawn and adapted from Polinski (1991).

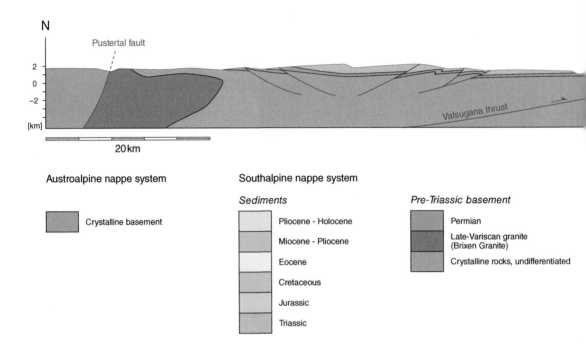

N

Pustertal fault

2
0
-2
[km]

Valsugana thrust

20 km

Austroalpine nappe system

☐ Crystalline basement

Southalpine nappe system

Sediments

☐ Pliocene - Holocene

☐ Miocene - Pliocene

☐ Eocene

☐ Cretaceous

☐ Jurassic

☐ Triassic

Pre-Triassic basement

☐ Permian

☐ Late-Variscan granite
(Brixen Granite)

☐ Crystalline rocks, undifferentiated

▲ **Figure 5.3-9**
Geological cross-section
through the Dolomites
(Italy). Source: Based on
Castellarin et al. (2006a)
and Doglioni (2007).

to Polinski (1991), the most recent
movement affected glacial sediments
and probably continues to this day.

Polinski (1991) proposed that the
Paleogene deformation (60–40 million
years old) in the Southern Karawanks
could be linked to the formation of
the Dinarides, but was overprinted
by more recent movement. As shown
in Fig. 5.3-8, these later movements
resulted in northwest- and southeast-
directed thrust faults, which are dated
to the Middle Miocene (approximately
16 million years ago).

The Southalpine Nappe System
and Dolomites

The cross-section through the Southalpine nappe system and the Dolomites (Fig. 5.3-9) shows two prominent
south-vergent thrust faults, the Valsugana and the Montello thrusts. In both
cases, the crystalline basement has also
been included in the nappe structure.

In the northern part, the actual
Dolomites, the Triassic sediments form
a broad synform that is flanked on either
side by north- and south-vergent thrust
faults. In the south, the Montello thrust
climbs towards the surface and overthrusts Plio-Pleistocene sediments.
However, its southernmost continuation
is cut off by a north-vergent thrust fault
that pushed the Miocene–Pliocene Po
Basin fill northwards onto the Mesozoic
sediments in the hanging wall of the
Montello thrust. The resultant bivergent
'triangle structure' is reminiscent of
the situation at the northern edge of the
Subalpine Molasse. Finally, in the
extreme south, the Plio-Pleistocene
Po Basin fill is cross-cut by a south-
vergent thrust.

According to Castellarin et al. (2006),
shortening within the Dolomites
amounts to about 40–55 kilometres. Of
this, approximately 20 kilometres are
accounted for by the Montello thrust
and its other associated thrusts, and

*Gentle basement
upwarp*

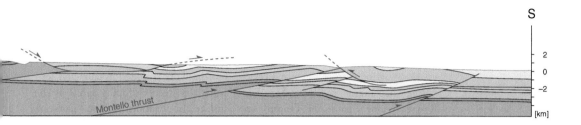

10–20 kilometres are accounted for by the Valsugana thrust. A further approximately 15 kilometres are to be expected in the crystalline basement (the 'Altkristallin') in the northern portion of the cross-section. With reference to age, progradation of the thrusting can be assumed to have propagated from north to south. The bivergent thrusts in the north cross-cut the Dinaridian structures that date from the Eocene to Oligocene. The Valsugana thrust was active in the Middle to Late Miocene (Serravallian to Tortonian, according to Castellarin et al. 2006), the more southern Montello thrust only became active in the past 6 million years (Late Miocene to Pleistocene; possibly still active).

5.4 The Deep Structure of the Alps

In combination with structural data, the controlled-source (reflection and refraction) seismology investigations in the Western Alps (ECORS-CROP), the Central Alps (NFP 20) and the Eastern Alps (TRANSALP) have enabled the construction of geological cross-sections at greater depths than had been possible previously. In many cases, the reflectors were the interfaces between Mesozoic dolostones or limestones, on the one hand, and the crystalline basement below or the argillaceous–marly sedimentary successions, on the other. Substantial differences in seismic velocity (approximately 6–7 kilometres per second in the carbonates compared with 5–6 kilometres per second in the gneisses and less than 5 kilometres per second in shales and marls) ensured the necessary impedance contrasts. Reflectors of this type are to be expected in those parts of the Alps where the crystalline basement was incorporated into the nappe structure (external massifs, Penninic crystalline nappes, Southalpine nappe system) or where flysch or molasse sequences are overlain by thick carbonates.

*Enormous
basement uplifts*

A particularly relevant interface was the contact between the crystalline basement and Mesozoic cover in the subsurface of the Molasse Basin and its continuation into the external crystalline basement uplifts (Aiguilles Rouges–Mont Blanc–Aar–Tauern). In this context, the network of lines produced by NFP 20 (Pfiffner et al. 1997a–c) was especially productive, as several lines traversed these structures and it was also possible to integrate the ECORS-CROP line into this network. Figure 5.4-1 shows the contours for the top of the crystalline basement in the Northalpine foreland and in the northern part of the Central Alps and the neighbouring Western and Eastern Alps. The information underlying this was based on single boreholes in the foreland, the reflection seismic lines, the geological and topographic maps of the external massifs and structural analyses. In the foreland, the isohypses (lines at the same level as the top basement surface) run about parallel to the Alpine strike. They indicate that the position of this contact gradually increases in depth towards the Alps and reaches a maximum depth of 7 kilometres below sea level just to the northwest of the external massifs. The external massifs are elongate, dome-like bulges, on the ridges of which the surface of the massif can be found at over 5 kilometres above sea level. The amplitude of the basement uplift thus measures up to 12 kilometres! The reflection seismic lines at the northeastern end of the Aiguilles Rouges massif (W5 and W6) and in the central part of the Aar massif (NEAT, C1 and other industrial lines in central Switzerland) provide evidence that the external massifs have been thrust several kilometres onto the foreland. These thrusts and the large amplitude of the basement uplifts prove that the external

massifs are the result of substantial shortening that involved a large portion of the upper crust. On the map, the Aiguilles Rouges–Mont Blanc and Aar–Gotthard are arranged in pairs and separated by steeply dipping faults at the surface. In addition, the two pairs are arranged en echelon in a left-stepping fashion: the structures at the northeastern end of the Aiguilles Rouges massif and those at the western end of the Aar massif cannot be directly connected to each other. At its eastern end, the Aar massif initially exhibits a depression and then a small dome, which is exposed in the Vättis Window along line E1. The subsequent plunging of the Aar massif and the Gotthard massif under the Eastern Alps is, again, characterized by an en échelon arrangement. Overall, the shortening of the massifs and their en échelon arrangement point to a dextral transpressive regime.

These controlled-source seismic studies enabled determination of the structure of the Moho – the crust–mantle boundary beneath the Alps. With the aid of local earthquake tomography studies more information could be gained on the structure of the boundary between lower and upper crust, the so-called Conrad discontinuity (Saldhauser et al., 2002; Diel et al., 2009; Wagner et al., 2012). Furthermore, teleseismic tomography was used to investigate the deep structure of the lithospheric plates involved in the formation of the Alps (Lippitsch 2002, Lippitsch et al. 2003). This method is based on the analysis of earthquake wave travel times, which are recorded using permanent and temporary installations. Analysis of the travel times allows the subducted lithospheric plates to be traced far into the asthenospheric mantle. The deep structure of the Alps is a three-dimensional problem, but in spite of this we will attempt to determine the essential characteristics based on four two-dimensional cross-sections. The deep structure of the crust has been adopted from the appropriate geological cross-sections in Figs 5.1-2, 5.2-2 and 5.3-2, but simplified. To make things clearer, the lower crust, upper crust and sediments have been given different colours in the two lithospheric plates that are involved.

The first cross-section (cross-section 1 in Fig. 5.4-2 follows the ECORS-CROP line in the Western Alps, and the structure in the upper crust reveals that the volume of the European Continental margin in the Alpine nappe structure is far greater than that of the Adriatic continental margin. On the European side, the different units have ended up lying on top of each other in an imbricate fashion towards the northwest. The lower crust plunges down towards the southeast and is thickened under the Alps. On the Adriatic side, the upper and lower crust have been stacked up such that these crusts now overlie each other vertically. The lithospheric mantle on the Adriatic side almost reaches the Earth's surface and has also been affected by the crustal thrusts. The high position of the mantle (the Ivrea Body) is attributed to Jurassic extensional tectonics. The Ivrea mantle body must have acted like a buffer when the European continental margin was compressed by plate convergence. However, it must be noted that the steep contact plane on the northwestern side of the Ivrea mantle body was also active as a strike-slip fault and we are not now seeing the geometry as it was during subduction. Of further interest is the fact that Lippitsch (2002) and Lippitsch et al. (2003) found a slab of a lithospheric plate that had broken off and can be followed (visible in Fig. 5.4-2, bottom right) from a depth of just under

150 kilometres to over 300 kilometres. If we consider the original total length of the European crust, then it stands to reason that a mantle substrate for it must be present somewhere, as the length of the lithospheric mantle that lies immediately below the European crust is far too small (Schmid & Kissling (2000) provide an in-depth discussion of this calculation). The tomography indicates that an additional piece of lithosphere lies at an even greater depth (over 400 kilometres), which can be interpreted as oceanic lithosphere from the subducted Piemont Ocean that has broken off. In summary, we can state that the Western Alps arose from a subduction zone that dipped towards the east and that the European continental margin, which has been subjected to Alpine deformation, was originally substantial in width.

In the Central Alps, shown in cross-section 2 in Fig. 5.4-2, the upper crust of both continental margins exhibits a far more symmetrical structure, with north- and south-vergent thrust faults and a comparatively similar volume for both deformed continental margins, but this picture looks different in the lower crust, where there is an asymmetric subduction geometry. Here, the Adriatic lower crust has been wedged into the European crust and thrust over the south-dipping European lower crust. A similar situation is also found at the lithospheric scale. The European lithospheric plate plunges beneath the Adriatic plate and can be followed to a depth of just over 200 kilometres using teleseismic tomography. The length of the mantle lithosphere is, again, too short to act as a substrate for the crust of the entire European margin that has been shortened by Alpine orogeny. The lower tip of the European lithosphere would correspond approximately to the

Detached slabs of lithosphere, European lithosphere sinks beneath Adriatic lithosphere

Figure 5.4-1
Structure contour map of
the top of the pre-Triassic
basement in the foreland
and the crystalline basement
uplifts of the Central Alps
and neighbouring areas.

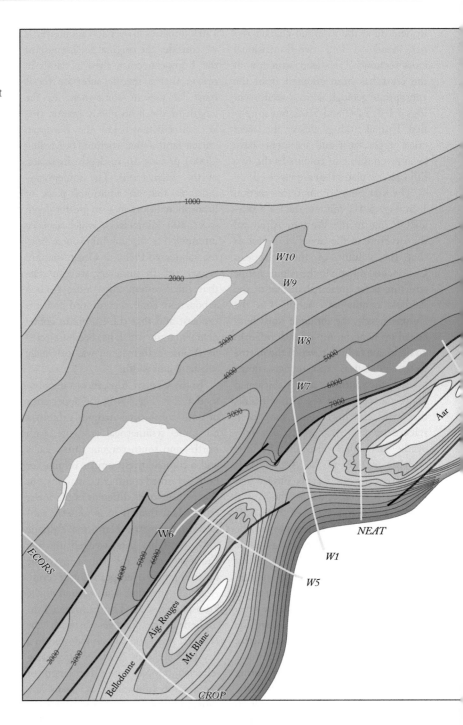

Figure 5.4-1
Structure contour map of the top of the pre-Triassic basement in the foreland and the crystalline basement uplifts of the Central Alps and neighbouring areas.

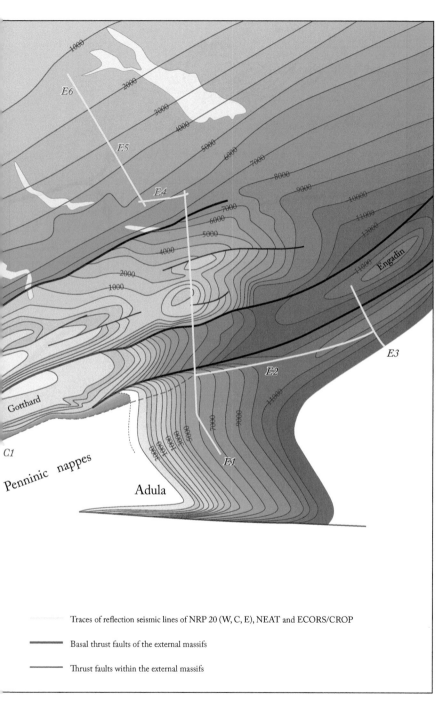

Traces of reflection seismic lines of NRP 20 (W, C, E), NEAT and ECORS/CROP

Basal thrust faults of the external massifs

Thrust faults within the external massifs

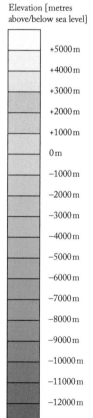

Elevation [metres above/below sea level]

+5000 m
+4000 m
+3000 m
+2000 m
+1000 m
0 m
−1000 m
−2000 m
−3000 m
−4000 m
−5000 m
−6000 m
−7000 m
−8000 m
−9000 m
−10000 m
−11000 m
−12000 m

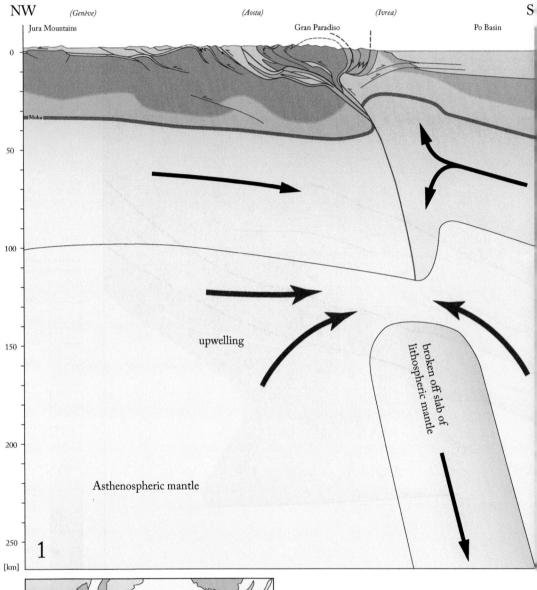

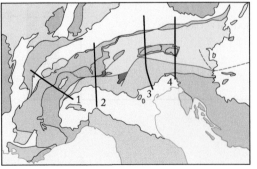

Figure 5.4-2 Lithosphere-scale geological cross-sections through the Western and Central Alps along the seismic lines of ECORS/CROP and the Eastern Traverse of NRP20. Crustal structure is based on Waldhauser et al. (2002), Diel et al. (2009) and Wagner et al. (2012), and mantle structure after Lippitsch et al. (2003) and Kissling (personal communication, 2013). The Conrad discontinuity between lower and upper crust is taken at a seismic velocity of 6.5 kilometres per second. Source: Adapted from Waldhauser et al. (2002), Diel et al. (2009), Wagner et al. (2012) and Lippitsch et al. (2003).

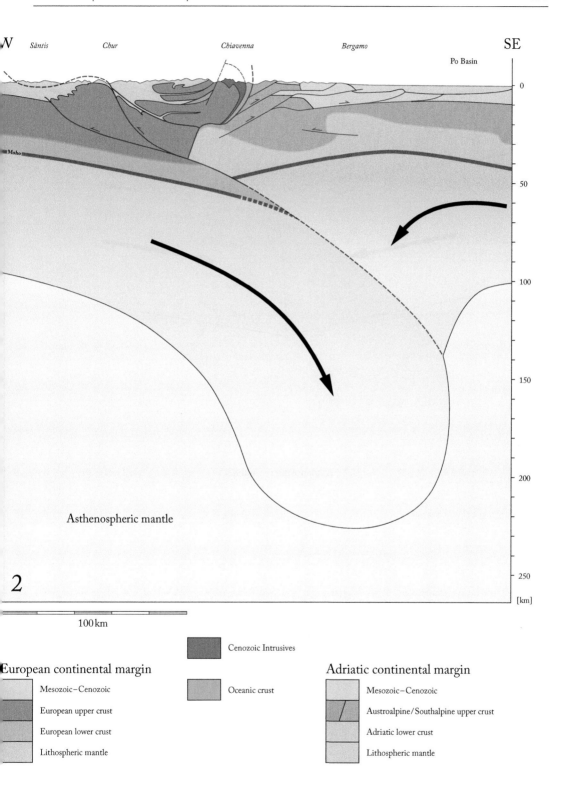

NW *Säntis*   *Chur*   *Chiavenna*   *Bergamo*   SE

Po Basin

Moho

Asthenospheric mantle

2

100 km

Cenozoic Intrusives

European continental margin

Mesozoic–Cenozoic

European upper crust

European lower crust

Lithospheric mantle

Oceanic crust

Adriatic continental margin

Mesozoic–Cenozoic

Austroalpine/Southalpine upper crust

Adriatic lower crust

Lithospheric mantle

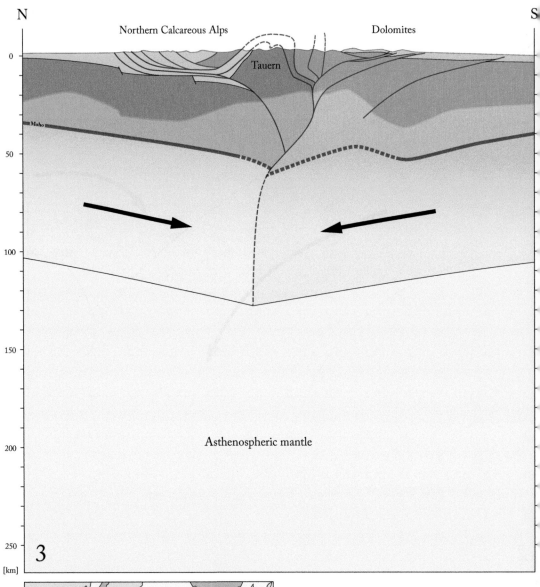

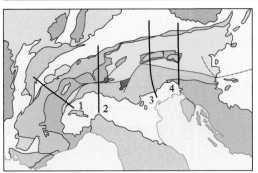

Figure 5.4-3 Two lithosphere-scale geological cross-section through the Eastern Alps, one along the seismic lines of TRANSALP, the other farther east. Crustal structure is based on Waldhauser et al. (2002), Diel et al. (2009) and Wagner et al. (2012), and mantle structure after Lippitsch et al. (2003) and Kissling (personal communication, 2013). The Conrad discontinuity between lower and upper crust is taken at a seismic velocity of 6.5 kilometres per second. Source: Adapted from Waldhauser et al. (2002), Diel et al. (2009), Wagner et al. (2012) and Lippitsch et al. (2003).

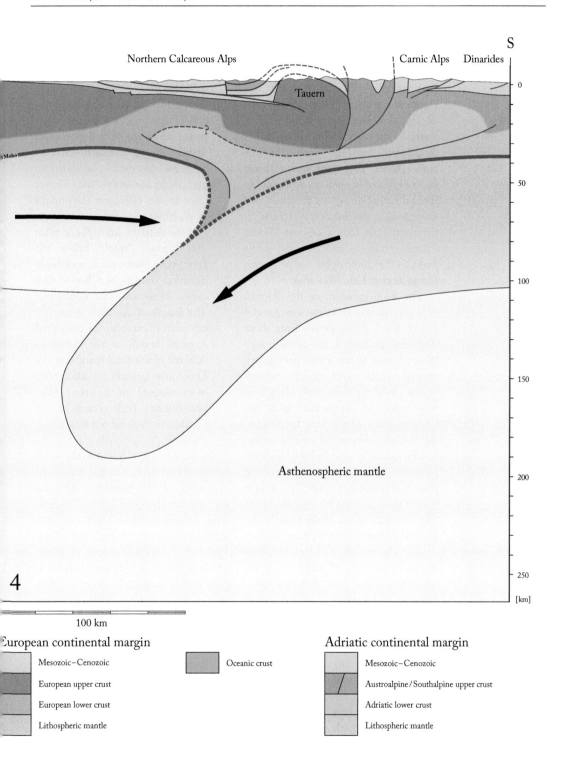

S

Northern Calcareous Alps   Carnic Alps   Dinarides

Tauern

Moho

Asthenospheric mantle

0

50

100

150

200

250

[km]

4

100 km

European continental margin   Adriatic continental margin

Mesozoic–Cenozoic   Oceanic crust   Mesozoic–Cenozoic

European upper crust   Austroalpine/Southalpine upper crust

European lower crust   Adriatic lower crust

Lithospheric mantle   Lithospheric mantle

lower crust of the Ultrahelvetic realm, but nothing remains for the Penninic realm (Valais Trough, Briançon microcontinent and Piemont Ocean). We therefore can assume that a slab of subducted lithosphere several hundred kilometres long also broke off and has disappeared into the depths in this cross-section. The most recent thrust faults in this cross-section correspond to the Aar massif thrust and the associated shortening in the Subalpine Molasse on the north side (dated as post-Middle Miocene) and the southernmost thrust faults in the Southalpine nappe system (dated as post-Late Miocene).

The deep structure of the Eastern Alps displayed in cross-sections 3 and 4 in Fig. 5.4-3 differs greatly from those that have just been discussed. Although there is, once again, a bivergent nappe structure in the upper crust, the volume of the deformed European margin is very small in comparison with the Western and Central Alps. In contrast, the Adriatic continental margin makes up a far greater volume and the Mesozoic sediments in its northern, frontal portion (the Northern Calcareous Alps) almost overlie the Cenozoic Molasse Basin fill. The structure of the lower crust in the central portion of cross-section 3 is characterized by a thickened lower crust in the contact zone between the two. As for the lithosphere, the asthenosphere–lithosphere boundary forms a trough and it is evident that the lengths of the two lithosphere segments is far shorter than the retrodeformed upper crust above, but the missing pieces have still to be found. Finally, in cross-section 4, which is located farther east, the lower crust and lithospheric structures reveal something completely different.

Adriatic lithosphere sinks beneath European lithosphere

(1) The lower crust forms two bulges located beneath the Northern Calcareous Alps and the Carnic Alps. These bulges are shown to be associated with thrust faults of opposite sense, which are interpreted to be caused by buoyancy effects that prevent the crust from being subducted. Such a sideways-upward escape is termed 'channel flow' and has been postulated to explain the upward escape of lower crust in the Himalaya (Beaumont et al., 2004).

(2) It is the Adriatic lithospheric plate that plunges steeply below the European plate in a northerly direction and can be followed to a depth of almost 250 kilometres. The length of the plate (over 150 kilometres) exceeds the calculated required length of the deformed Adriatic continental margin in the Dolomites (around 50 kilometres of shortening) and the offset of the peri-Adriatic fault system by the Giudicarie Fault (about 80 km).

In cross-sections 3 and 4, the most recent thrust fault on the European side ends in the Subalpine Molasse and is probably post-Middle Miocene. The youngest thrust fault on the Adriatic side cross-cuts the Plio-Pleistocene sediments of the Po Basin and may currently still be active. It is therefore substantially younger than that on the north side of the Alps. For this reason, the plunging Adriatic lithospheric plate in cross-section 4 is illustrated such that it cross-cuts the European lithospheric plate. However, it is unclear where the deepest portion of the plunging mantle rocks originated. Is it only Adriatic margin? Does it contain anything of the Piemont Ocean? Or are any sheared-off portions of the European mantle lithosphere present?

References

Affolter, T., Faure, J.-L., Gratier, J-P. & Colletta, B., 2008, Kinematic models of deformation at the front of the Alps: new data from map-view restauration. Swiss Journal of Geosciences, 101, 289–303.

Auer, M. & Eisbacher, G. H., 2003, Deep structure and kinematics of the Northern Calcareous Alps (TRANSALP Profile). International Journal of Earth Sciences, 92, 210–227.

Beaumont, C., Jamieson, R. A., Nguyen, M. H. & Medvedev, S., 2004, Crustal channel flows: 1. Numerical models with applications to the tectonics of the Himalayan–Tibetan orogeny. J. Geophysical Research, 109, B06406, doi: 10.1029/2003JB002809.

Bistacchi, A., Dal Piaz, G. V., Massironi, M., Zattin, M., Balestrieri, M.L., 2001, The Aosta–Ranzola extensional fault system and Oligocene – Present evolution of the Austroalpine-Penninic wedge in the northwestern Alps. International Journal of Earth Sciences (Geol. Rundsch.), 90, 654–667, DOI 10.1007/s005310000178.

Blundell, D., Freeman, R. & Mueller, S. (eds), 1992, A Continent Revealed: The European Geotraverse, Cambridge University Press, 275 pp.

Brandner, R., 1980, Tirolatlas: Geologische und Tektonische Übersichtskarte von Tirol. Universitätsverlag Wagner, Innsbruck. Also published in Tiroler Heimat, Jahrbuch für Geschichte und Volkskunde, 48./49. Band, 1985.

Bucher, S., Ulardic, C, Bousquet, R., Ceriani, S., Fügenschuh, B., Gouffon, Y. & Schmid, S. M., 2004, Tectonic evolution of the Briançonnais units along a transect (ECORS-CROP) through the Italian-French Western Alps. Eclogae geologicae Helvetiae, 97/3, 321–345.

Burkhard, M., 1990, Aspects of large scale Miocene deformation in the most external part of the Swiss Alps (Subalpine Molasse to Jura fold belt). Eclogae geologicae Helvetiae, 83/3, 559–583.

Burkhard, M. & Sommaruga, A., 1998, Evolution of the western Swiss Molasse basin: structural relations with the Alps and the Jura belt. In: Mascle, A., Puigdefabregas, C., Luterbacher, H. P. & Fernandez, M. (eds), Cenozoic Foreland Basins of Western Europe, Geological Society London, Special Publications, 134, 279–298.

Buxtorf, A., 1907, Geologische Beschreibung des Weissenstein-Tunnels und seiner Umgebung. Beiträge geologische Karte Schweiz, neue Folge 21.

Buxtorf, A., 1916, Prognosen und Befunde beim Hauensteinbasis- und Grenchenbergtunnel und die Bedeutung der letztern für die Geologie des Juragebirges. Verhandlungen der Naturforschenden Gesellschaft in Basel, Band XXVII, 184–254.

Castellarin, A. & Cantelli, L., 2006, Neo-Alpine evolution of he Southern Eastern Alps. Journal of Geodynamics, 30, 251–274.

Castellarin, A., Nicolich, R., Fantoni, R., Cantelli, L., Sella, M. & Selli, L., 2006, Structure of the lithosphere beneath the Eastern Alps (southern sector of the TRANSALP transect). Tectonophysics, 414, 259–282.

Ceriani, S. & Schmid, S. M., 2004, From N–S collision to WNW-directed post-collisional thrusting and folding: Structural study of the Frontal Penninic Units in Savoie (Western Alps, France). Eclogae geologicae Helvetiae, 97/3, 347–369.

Diel, T., Husen, S., Kissling, E. & Deichmann, N., 2009, High-resolution 3-D P-wave model of the Alpine crust. Geophysical Journal International, 179, 1133–1147.

Doglioni, C., 2007, Tectonics of the Dolomites. Bulletin for Applied Geology, 12/2, 11–15.

Eisbacher, G. H., Linzer, H.-G., Meier, L. & Polinski, R., 1990, A depth-extrapolated structural transect across the Northern Calcareous Alps of western Tirol. Eclogae geologicae Helvetiae, 83/3, 711–725.

Engi, M., Berger, A. & Roselle, G. T., 2001, Role of the tectonic accretion channel in collisional orogeny. Geology, 29/12, 1143–1146.

Escher, A., Hunziker, J. C., Marthaler, M., Masson, H., Sartori, M. & Steck A., 1997, Geologic framework and structural evolution of the Western Swiss-Italian Alps. In: Pfiffner, O. A., Lehner, P., Heitzmann, P., Mueller, & Steck, A. (eds), Deep Structure of the Swiss Alps: Results of NRP 20. Birkäuser Verlag Basel, 205–222.

Franks, S., Trümpy, R. & Auf der Maur, J., 2000, Aus der Frühzeit der alpinen Geologie. Johann Gottfried Ebels Versuch einer Synthese (1808). Neujahrsblatt der Naturforschenden Gesellschaft in Zürich, 68 pp.

Funk, Hp., Labhart, T., Milnes, A. G., Pfiffner, O. A., Schaltegger, U., Schindler, C., Schmid S. M. & Trümpy, R., 1983, Bericht über die Jubiläumsexkursion 'Mechanismus der Gebirgsbildung' der Schweizerischen Geologischen Gesellschaft in das ost- und zentralschweizerische Helvetikum und in das nördliche Aarmassiv vom 12. bis 17. September 1982. Eclogae geologicae Helvetiae, 76/1, 91–123.

Guellec, S., Mugnier, J.-L., Tardy, M. & Roure, F., 1990, Neogene evolution of the western Alpine foreland in the light of ECORS data and balanced cross-section. In: Roure, F., Heitzmann, P. & Polino, R. (eds), Deep Structure of the Alps, Mémoires de la Société Géologique de France, Paris, 156, 165–184.

Hamilton, W., Wagner, L. & Wessely, G., 2000, Oil and Gas in Austria. In: Neubauer, F. & Höck, V. (eds), Aspects of Geology in Austria. Österreichische Geologische Gesellschaft, Wien, 235–262. Also published in Mitteilungen der Österreichischen Geologischen Gesellschaft, 92. Band (1999).

Hänni, R. & Pfiffner, O. A., 2001, Evolution and internal structure of the Helvetic nappes in the Bernese Oberland. Eclogae geologicae Helvetiae, 94, 161–171.

Heim, Alb., 1878, Untersuchung über den Mechanismus der Gebirgsbildung im

Anschluss an die Geologische Monographie der Tödi-Windgällen-Gruppe. Schwabe, Basel.

Herwegh, M., Hürzeler, J.-P., Pfiffner, O. A., Schmid, S. M., Abart, R. & Ebert, A., 2008, Excursion guide to the field trip of the Swiss Tectonic Studies Group (Swiss Geological Society) to the Glarus nappe complex (14.–16.09.2006). Swiss Journal of Geosciences, 101/2, doi: 10.1007/s00015-008-1259-z.

Kammer, A., 1985, Bau und Strukturen des nördlichen Aarmassivs und seiner Sedimente. Unveröff. Dissertation Université de Neuchâtel, 103 pp.

Kaufmann, F. J., 1877, Geologische Beschreibung der Kalkstein- und Schiefergebiete der Kantone Schwyz, Zug und des Bürgenstocks bei Stanz. Beiträge geologische Karte Schweiz, 14/2. Abteilung, 180 pp.

Kempf, O. & Pfiffner, O. A., 2004, Early Tertiary evolution of the North Alpine Foreland Basin of the Swiss Alps and adjoining areas. Basin Research, 16, 549–567, doi: 10.1111/j.1365-2117.2004.00246.x.

Lammerer, B. & Weger, M., 1998. Footwall uplift in an orogenic wedge: the Tauern Window in the Eastern Alps of Europe. Tectonophysics, 285, 213–230.

Lammerer, B., Gebrande, H., Lüschen, E. & Vesela, P., 2008, A crustal-scale cross-section through the Tauern Window (eastern Alps) from geophysical and geological data. In: Siegesmund, S., Fügenschuh, B. & Froitzheim, N. (eds), Tectonic Aspects of the Alpine–Dinaride–Carpathian System. Geological Societey, London, Special Publications, 298, 219–229.

Lardeaux, J. M., Schwartz, S., Tricart, P., Paul, A., Guillot, S., Béthoux, N. & Masson, F., 2006, A crustal-scale cross-section of the south-western Alps combining geophysical and geological imagery. Terra Nova, 18, 412–422, doi: 10.1111/j.1365-3121.2006.00706.x.

Laubscher, H. P., 1961, Die Fernschubhypothese der Jurafaltung. Eclogae geologicae Helvetiae, 54/1, 2221–2282.

Laubscher, H. P., 1965, Ein kinematisches Modell der Jurafaltung. Eclogae geologicae Helvetiae, 58/1, 231–318.

Lickorish, W. H. & Ford, M., 1998, Sequential restoration of the external Alpine Digne thrust system, SE France, constrained by kinematic data and synorogenic sediments. In: Mascle, A., Puigdefabregas, C., Luterbacher, H. P. & Fernandez, M. (eds), Cenozoic Foreland Basins of Western Europe, Geological Society, London, Special Publications, 134, 189–211.

Lippitsch, R., 2002, Lithosphere and Upper Mantle P-Wave Velocity Structure Beneath the Alps by High-Resolution Teleseismic Tomography. Dissertation ETH No. 14726, 137 pp.

Lippitsch, R., Kissling, E. & Ansorge, J., 2003, Upper mantle structure beneath the Alpine orogen from high-resolution teleseismic tomography. Journal of Geophysical Research, 108/B8, 2376, doi: 10.1029/2002JB002016.

Lüschen, E., Lammerer, B., Gebrande, H., Millahn, K., Nicolich, R. & TRANSALP Working Group, 2004, Orogenic structure of the Eastern Alps, Europe, from TRANSALP deep seismic reflection profiling. Tectonophysics, 388, 85–102.

Mandl, G. W., 2000, The Alpine sector of the Tethyan shelf – Examples of Triassic to Jurassic sedimentation and deformation from the Northern Calcareous Alps. In: Neubauer, F. & Höck, V. (eds), Aspects of Geology in Austria, 61–77. Also published in den Mitteilungen der Österreichischen Geologischen Gesellschaft, 92 (1999).

Masson, H., 1976, Un siècle de géologie des Préalpes: de la décourverte des nappes à la recherche de leur dynamique. Eclogae geologicae Helvetiae, 69/2, 527–575.

Morag, N., Avigad, D., Harlavan, Y., McWilliams, M. O. & Michard, A., 2008, Rapid exhumation and mountain building in the Western Alps: Petrology and 40Ar/39Ar geochronology of detritus from Tertiary basins of southeastern France. Tectonics, 27, TC2004, doi: 10.1029/2007TC002142, 18 pp.

Mosar, J., Stampfli, G. M. & Girod, F., 1996, Western Préalpes Médianes romandes: timing and structure. A review. Eclogae geologicae Helvetiae, 89/1, 389–425.

Mugnier, J.-L., Guellec, S., Ménard, G., Roure, F., Tardy, M. & Vialon, P., 1990, A crustal scale balanced cross-section through the external Alps deduced from the ECORS profile. In: Roure, F., Heitzmann, P. & Polino, R. (eds), Deep Structure of the Alps, Mémoires de la Société Géologique de France, Paris, 156, 203–216.

Nicolas, A., Polino, R., Hirn, A., Nicolich, R. & Ecors-Crop working group, 1990, Ecors-Crop traverse and deep structure of the western Alps: a synthesis. In: Roure, F., Heitzmann, P. & Polino, R. (eds), Deep Structure of the Alps, Mémoires de la Société Géologique de France, Paris, 156, 15–28.

Ortner, H., Reiter, F. & Brandner, R., 2006, Kinematics of the Inntal shear zone – sub-Tauern ramp fault system and the interpretation of the TRANSALP seismic section, Eastern Alps, Austria. Tectonophysics, 414, 241–258.

Pfiffner, O. A., 1986, Evolution of the north Alpine foreland basin in the Central Alps. In: Allen, P. A. & Homewood, P. (eds), Foreland Basins, Special Publication of the International Association of Sedimentologists, 8, 219–228.

Pfiffner, O. A., 1993b, The structure of the Helvetic nappes and its relation to the mechanical stratigraphy. Journal of Structural Geology, 15/3–5, 511–521.

Pfiffner, O.A., 2011, Explanatory Notes to the Structural Map of he Helvetic Zone of the Swiss Alps, including Vorarlberg (Austria) and Haute Savoie (France). Geological Special Map 128, Text and 10 plates. Swiss Geological Survey, swisstopo.

Pfiffner, O. A. & Hitz, L. (1997): Geologic interpretation of the seismic profiles of the Eastern Traverse (lines E1–E3, E7–E9): eastern Swiss Alps. In: Pfiffner, O. A., Lehner, P., Heitzmann, P., Mueller, S. & Steck, A. (eds), Deep Structure of the

Swiss Alps: Results of NRP 20, Birkäuser Verlag Basel, 73–100.

Pfiffner, O. A., Lehner, P., Heitzmann, P., Mueller, S. & Steck, A. (eds), 1997a, Deep Structure of the Swiss Alps: Results of NRP 20, Birkhäuser, 380 pp.

Pfiffner, O. A., Erard, P.-F. & Stäuble, M., 1997b, Two cross-sections through the Swiss Molasse Basin (lines E4–E6, W1, W7–W10). In: Pfiffner, O. A., Lehner, P., Heitzmann, P., Mueller, S. & Steck, A. (eds), Deep Structure of the Swiss Alps: Results of NRP 20, Birkhäuser, 64–72.

Pfiffner, O. A., Sahli, S. & Stäuble, M., 1997c, Structure and evolution of the external basement uplifts (Aar, Aiguilles Rouges/Mt. Blanc). In: Pfiffner, O. A., Lehner, P., Heitzmann, P., Mueller, S. & Steck, A. (eds), Deep Structure of the Swiss Alps: Results of NRP 20, Birkhäuser, 139–153.

Pfiffner, O. A., Ellis, S. & Beaumont, C., 2000, Collision tectonics in the Swiss Alps: Insight from geodynamic modeling. Tectonics, 19/6, 1065–1094.

Pfiffner, O. A., Burkhard, M., Hänni, R., Kammer, A., Kligfield, R., Mancktelow, N., Menkveld, J.-W., Ramsay, J. G., Schmid, S. M. & Zurbriggen, R., 2010, Structural Map of the Helvetic Zone of the Swiss Alps, including Vorarlberg (Austria) and Haute Savoie (France). Geological Special Map, 128, 1:100 000, 7 Map sheets. Swiss Geological Survey, swisstopo.

Philippe, Y., Deville, E. & Mascle, A., 1998, Thin-skinned inversion tectonics at oblique basin margins: example of the western Vercors and Chartreuse Subalpine massifs (SE France). In: Mascle, A., Puigdefabregas, C., Luterbacher, H. P. & Fernandez, M. (eds), Cenozoic Foreland Basins of Western Europe, Geological Society London, Special Publications, 134, 239–262.

Plancherel, R., 1979, Aspects de la déformation en grand dans les Préalpes médianes plastiques entre Rhône et Aar. Implications cinématiaques et dynamiques. Eclogae geologicae Helvetiae, 72/1, 145–214.

Polinski, R., 1991, Ein Modell der Tektonik der Karawanken, Südkärnten, Österreich. Dissertation Universität Karlsruhe, 143 pp.

Prosser, G., 1998, Strike-slip movements and thrusting along a transpressive fault zone: The North Giudicarie line (Insubric line, northern Italy). Tectonics, 17/6, 921–937.

Ratschbacher, L. & Frisch, W., 1993, Palinspastic reconstruction of the pre-Triassic basement units in the Alps: the Eastern Alps. In: von Raumer, J. & Neubauer, F. (eds), Pre-Mesozoic Geology in the Alps, Springer Verlag, 41–51.

Ratschbacher, L, Frisch, W., Linzer, H. G. & Merle, O., 1991, Lateral extrusion in the eastern Alps, part II: structural analysis. Tectonics, 10, 257–271.

Roure, F., Bergerat, F., Damotte, B., Mugnier, J.-L. & Polino, R. (1996), The Ecors-Crop Alpine seismic traverse, Mémoires de la Société Géologique de France, 170, 113 pp.

Schaad, W., 1995, Beiträge zur Entstehung und Bedeutung alpintektonischer Abscher-horizonte in den Schweizer Alpen. Unveröff. Dissertation University Bern.

Scharf, A., Handy, M.R., Favaro, S., Schmid, S.M., 2013, Modes of orogeny-parallel stretching and extensional exhumation in response to microplate indentation and roll-back subduction (Tauern Window, Eastern Alps). Internationl Journal of Earth Sciences, doi:10.1007/s00531-013-0894-4.

Scheiber, T., Pfiffner, O.A. & Schreurs, G., 2013, Upper crustal deformation in con-tinent-continent collision: case study central Bernhard nappe complex (Valais, Switzerland). Tectonics, 32, 1–23, doi:10.1002/tect.20080.

Schmid, S. M. & Kissling, E., 2000, The arc of the Western Alps in the light of geo-physical data on deep crustal structure. Tectonics, 19/1, 62–85.

Schmid, S. M., Pfiffner, O. A., Froitzheim, N., Schönborn, G. & Kissling, E., 1996, Geogphysical-geological transect and tec-tonic evolution of the Swiss-Italian Alps. Tectonics, 15/5, 1036–1064.

Schmid, S. M., Pfiffner, O. A., Schönborn, G., Froitzheim, N. and Kissling, E., 1997a, Integrated cross section and tectonic evolution of the Alps along the Eastern Traverse. In: Pfiffner, O. A., Lehner, P., Heitzmann, P., Mueller, S. & Steck, A. (eds), Deep Structure of the Swiss Alps: Results of NRP 20, Birkhäuser Verlag Basel, 289–304.

Schmid, S. M., Pfiffner, O. A. & Schreurs, G., 1997b, Rifting and collision in the Penninic Zone of eastern Switzerland. In: Pfiffner, O. A., Lehner, P., Heitzmann, P., Mueller, S. & Steck, A. (eds), Deep Structure of the Swiss Alps: Results of NRP 20, Birkhäuser Verlag Basel, 160–185.

Schumacher, M., 1997, Geological interpretation of the seismic profiles through the Southern Alps (lines S1-S7 and C3-south). In: Pfiffner, O. A., Lehner, P., Heitzmann, P., Mueller, S. & Steck, A. (eds), Deep Structure of the Swiss Alps: Results of NRP 20, Birkhäuser Verlag Basel, 101–114.

Selzer, C. Buiter, S. J. H. & Pfiffner, O. A., 2008, Numerical modeling of frontal and basal accretion at collisional margins. Tectonics, 27, TC3001, doi: 10.1029/2007TC002169, 26 pp.

Steininger, F. F. & Wessely, G., 2000, From the Tethyan Ocean to the Paratethys Sea: Oligocene to Neogene Stratigraphy, Paleogeography and Paleobiogeography of the circum-Mediterranean region and Oligocene to Neogene Basin evolution in Austria. In: Neubauer, F. & Höck, V. (eds), Aspects of Geology in Austria. Österreichische Geologische Gesellschaft, Wien, 95–116. Also published in Mitteilungen der Österreichischen Geologischen Gesellschaft, 92. Band (1999).

Steinmann, M., 1994, Ein Beckenmodell für das Nordpenninikum der Ostschweiz. Jahrbuch der Geologischen Bundesanstalt, Bd 137/4, 675–721.

Trümpy, R., 1975, Penninic-Austroalpine boundary in the Swiss Alps: A presumed former continental margin and its problems. American Journal of Science, 275-A, 209–238.

Trümpy, R. & Westermann, A., 2008, Albert Heim (1849–1937): Weitblick und Verblendung in der alpentektonischen Forschung. Vierteljahrsschrift der Naturforschenden Gesellschaft in Zürich, 153 (3/4), 67–79.

Wagner, M., Kissling, E. & Husen, S., 2012, Combining controlled-source seismology and local earthquake tomography to derive a 3-D crustal model of the western Alpine region. Geophysical Journal International, 191, 789–802.

Waldhauser, F., Lippitsch, R., Kissling, E. & Ansorge, J., 2002, High-resolution teleseismic tomography of upper-mantle structure using an *a priori* three-dimensional crustal model. Geophysical Journal International, 150, 403–414.

Wissing, S. B. & Pfiffner, O. A., 2002, Structure of the eastern Klippen nappe (BE, FR): Implications for Alpine tectonic evolution. Eclogae geologicae Helvetiae, 95, 381–398.

6 Tectonic Evolution of the Alps

The Alpine mountain chain formed due to convergent plate movements between the two large plates of Europe and Africa, but it also must be noted that a series of smaller continental blocks between Europe and Africa played a role in events: in the case of the Alps, these were the Adriatic continent and the Briançon microcontinent. The sea basins between the continental blocks were closed by subduction as a result of the convergent plate movements. The actual growth of the Alpine mountain chain occurred in the final phase of these convergent movements, during the collision between the European and the Adriatic continental margins. The early history of convergence and subduction is revealed in the internal nappe structure of the Alps, as described in the previous chapter. The collision between Europe and Adria led to the disappearance of the sea basins, but convergence remains active and responsible for recent uplifts, even if this is a slow process (see Fig. 1.8 and associated discussion). Two stages must be distinguished in the tectonic evolution of the Alps.

Two orogenies in the Alps

- In the Cretaceous, convergence movements between Europe and Adria were more east–west-directed. The oceanic lithosphere of the Piemont Ocean plunged beneath the Adriatic plate and was gradually subducted. The upper plate, the Adriatic continental margin, was compressed during these plate movements, nappes were formed and parts of the continental margin were uplifted into an early orogen. At this time, the stretched European continental margin drifted more or less passively towards the subduction zone. This first stage is called the Cretaceous orogeny.

- In the Cenozoic, the more east–west-directed plate movements continued initially, but then gave way to north–south-directed convergence. After the Piemont Ocean had closed completely and the Briançon microcontinent had entered into the subduction zone, the European continental margin was then also compressed and dismembered into nappes. Over time, more external parts of the European continental margin were also affected by the deformation. The entrance of the actual European continent, the crustal substratum of the Helvetic and Dauphinois nappe systems, set a strong signal, with subduction becoming a collision after the arrival of a non-subductable continental crust that was approximately 30 kilometres thick. The continuing plate convergence caused further nappe stacking whereby upper crustal pieces moved upward along dipping thrust faults on both plate margins, which led ultimately to the growth of the Alpine mountain range: this second stage corresponds to the Cenozoic orogeny (or the Tertiary orogeny).

The evolutionary history of the two orogenies yields a highly complex picture overall, not least due to the irregular form of the plate boundaries and continental margins, as well as the changes in direction of the plate movements. Dating this evolutionary history

Geology of the Alps: Revised and updated translation of Geologie der Alpen, Second Edition. O. Adrian Pfiffner.
© 2014 John Wiley & Sons, Ltd. Published 2014 by John Wiley & Sons, Ltd.

relies on a variety of independent methods: petrological and geochronological data ('metamorphic age') provide information on the subduction and exhumation history of rocks; structural analyses permit the relative dating of thrusts, folds, cleavages and the growth of metamorphic mineral parageneses; stratigraphical and sedimentological investigations yield maximum and minimum ages, angular unconformities and information about the supraregional environment provided by the sedimentary associations. The application of all these methods is limited by the circumstance that many of the critical rocks are absent due to erosion, have been covered by Quaternary sediments or are simply still deep underground. The available observations therefore yield an incomplete picture of the evolution of the Alps, which is also manifest in the contradictory views that sometimes appear in the published literature.

6.1 Alpine Metamorphism

When rocks are buried due to subduction and nappe formation they are exposed to high pressures and temperatures. The metamorphic overprint that occurs under these circumstances is also preserved – at least in the form of a relic – during the subsequent exhumation, associated with decreasing pressures and temperatures, and is therefore an excellent indicator for the regional and supraregional pattern of the orogenic processes. However, incorporating metamorphic data into the evolution of the Alpine orogeny requires the study of the regional distribution of metamorphic rocks as well as petrological studies that unravel the pressure and temperatures conditions that were prevalent during the metamorphic overprint.

Accordingly this section is organized in the following way:

• first a very general overview is given of the regional distribution of the Alpine metamorphic overprint;
• followed by a more detailed discussion on the pressure-dominated metamorphism, including discussion on the evolution of pressures and temperatures with time (so-called P–T–t paths);
• the section on regional metamorphism focuses on the local distribution of metamorphic overprints and their relation to nappe structures;
• followed by a section on the effect of Cenozoic intrusions on their country rocks.

Regional Distribution of Alpine Metamorphism

The regional distribution of metamorphic facies zones or metamorphic relics is elaborated from a great number of local analyses of very different metamorphic rock types, mineral parageneses, mineral transformations, etc., but comparison of such independent data sets requires careful calibration. Oberhänsli et al. (2004) compiled metamorphic data covering the entire Alps. Although the primary data yield detailed distribution patterns of individual mineral parageneses, which in turn can be converted into temperature and/or pressure distributions, grouping these petrological systems into more general metamorphic facies zones makes the data more comprehensible when presented on a map. The map shown in Fig. 6.1 is a simplified metamorphic map of the Alps, based on the condensed information from a map in Oberhänsli et al. (2004). On the one hand, metamorphic ages (Middle and

▶ **Figure 6.1** Simplified metamorphic map of the Alps – based on Oberhänsli et al. (2004). Regional Barrovian type metamorphism is of Cretaceous age in the Eastern Alps and of Cenozoic age in the Western and Central Alps. The Eastern Alps are characterized by a Cretaceous aged high-pressure metamorphic overprint dated at around 100 million years ago. In the Western Alps a high-pressure metamorphic overprint occurred between 45 and 35 million years ago. Source: Adapted from Oberhänsli et al. (2004) © CCGM-CGMW (2012).

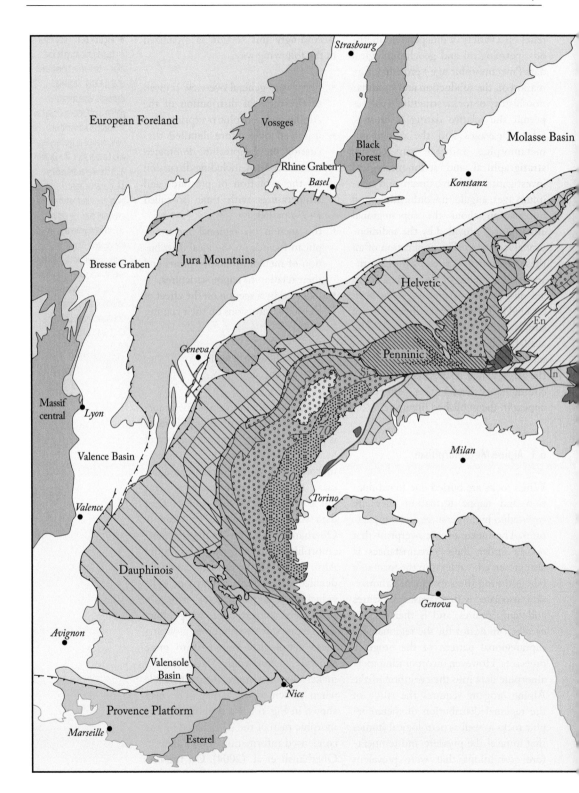

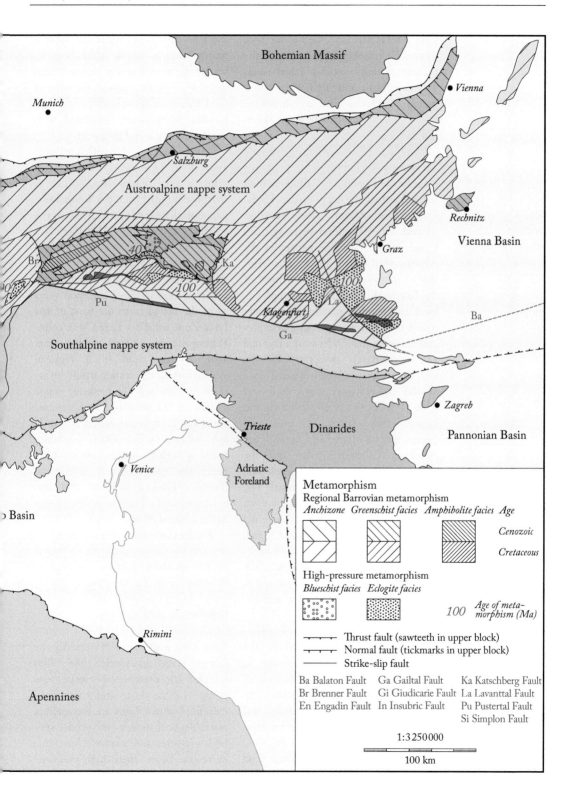

Bohemian Massif

Munich

Vienna

Salzburg

Austroalpine nappe system

Rechnitz

Vienna Basin

Br

40

Ka

Graz

100

Pu

La

Ba

Klagenfurt

Southalpine nappe system

Ga

Zagreb

Dinarides

Pannonian Basin

Trieste

Venice

Adriatic
Foreland

Basin

Rimini

Apennines

Metamorphism

Regional Barrovian metamorphism

Anchizone	*Greenschist facies*	*Amphibolite facies*	*Age*
			Cenozoic
			Cretaceous

High-pressure metamorphism

Blueschist facies   *Eclogite facies*

100   *Age of meta-morphism (Ma)*

Thrust fault (sawteeth in upper block)
Normal fault (tickmarks in upper block)
Strike-slip fault

Ba Balaton Fault   Ga Gailtal Fault   Ka Katschberg Fault
Br Brenner Fault   Gi Giudicarie Fault   La Lavanttal Fault
En Engadin Fault   In Insubric Fault   Pu Pustertal Fault
                                        Si Simplon Fault

1:3 250 000

100 km

Late Cretaceous and Cenozoic) and, on the other, metamorphic types (high-pressure eclogitic and blueschist-facies metamorphism and high-temperature greenschist-facies and amphibolite-facies metamorphism) have been categorized.

We notice immediately that the Cretaceous metamorphism predominates in the Eastern Alps and is limited to the Austroalpine nappe system. High-pressure assemblages, namely eclogites formed 110–90 million years ago, are found in the crystalline nappes in the extreme southeast of the Eastern Alps and to the southeast and southwest of the Tauern Window. Areas of amphibolite-facies regional metamorphism surround the domain containing eclogites and the rocks that underwent a regional greenschist-facies overprint in the Cretaceous form a broad external zone around these highly metamorphosed rocks. Finally, the zone exhibiting the lowest grade, anchizonal metamorphism, is concentrated in a band in the Northern Calcareous Alps and the southeastern corner of the Austroalpine nappe system east of Klagenfurt. The rocks in the Southalpine nappe system and in the Dolomites show no signs of a Cretaceous metamorphism. Overall, this picture shows a core zone of high-grade Cretacous metamorphism in the southern part of the Austroalpine nappe system that becomes broader towards the east and the west.

High-pressure blueschist-facies metamorphism is mostly limited to the Penninic nappe system. Its age is Cenozoic as it overprints sediments of Cretaceous age. Blueschist-facies rocks are found in the Rechnitz, Tauern and Engadine windows in the Eastern Alps, and in the main body of the Penninic nappe system in the Central and Western Alps. Of note is the fact that the blueschist-facies rocks also form a narrow band in the immediate hanging wall of the basal Penninic thrust, which indicates high pressures at the base of the Penninic nappes possibly caused by tectonic overpressure associated with nappe formation and transport.

There is a larger area with a regional amphibolite-facies metamorphic overprint in the Central Alps, which is sharply delimited from the Southalpine nappe stack by the Insubric Fault in the south. In the Western Alps there is a zone with high-pressure rocks (eclogites) to the east of the blueschist zone, the formation of which is dated largely to the Cenozoic (35–59 million years ago); the eclogites to the west of the Ivrea Zone exhibit a Late Cretaceous–Palaeogene age (89–60 million years ago); a thin band of a regional greenschist-facies metamorphic overprint runs through the Penninic nappe system in the west of the blueschist zone; the Dauphinois nappe system is split in two by the 300 °C boundary marking the limit between grennschist-facies (or epizonal) and anchizonal metamorphism, a structure that also continues in the Helvetic nappe system in the Central Alps.

Finally, Cenozoic high-pressure eclogites and blueschist overprints also occur in the southern part of the Tauern Window in the Eastern Alps.

High-Pressure Metamorphism

Pioneering work in the Western Alps by Chopin (1984) and Goffé (1984–1986) revealed the presence of coesite–pure pyrope in bluschists and carpholite–chloritoïd assemblages in metapelites, assemblages that form only under very high pressures and relatively low temperatures. Since then high-pressure–low-temperature assemblages have been

Cretaceous and Cenozoic metamorphism

found in many places throughout the Western, Central and Eastern Alps. Key rocks are eclogites, mafic rocks that contain garnet and omphacite and formed mainly from gabbro or basalt. In the Alps, eclogites formed in some instances from the mafic igneous rocks of the Piemont Ocean. For these and other eclogites the P–T–t paths determined illustrate how individual tectonic units were buried to great depths and later exhumed to finally reach the surface. In the following, a series of typical P–T–t paths chosen from the literature are discussed in more detail. It has to be noted that eclogites and other high-pressure assemblages were overprinted by a subsequent Barovian-type regional metamorphism at lower pressures and higher temperatures and are preserved as relics only. Their preservation was favoured by the fact that they consist of anhydrous phases such as garnet, pyroxenes, quartz, kyanite and olivine (Brouwer et al. 2005).

A word of caution is necessary regarding the conversion of pressure (as determined from metamorphic mineral assemblages) to depth at which these reactions occurred because in many instances, these pressures are taken to represent purely lithostatic pressures. However, as Petrini & Podladchikov (2000) and Mancktelow (2008) argue, tectonic overpressure may contribute to up to 50% of these pressures and hence, when converting to depth, only the remaining 50% of lithostatic pressure should be taken into account. When constructing the palinspastic, retro-deformed cross-sections across the Alps for the various time frames discussed in Section 6.3, palaeodepths were determined accordingly. In addition, as pointed out by Desmons et al. (1999a), experimental data cannot be simply converted to natural processes and in

the rocks analysed, peak conditions may not be recorded.

The high-pressure metamorphism in the Western Alps is summarized in the review paper by Desmons et al. (1999b). Three examples for P–T–t paths are given in Fig. 6.2:

- the Dora Maira, a basement nappe made of continental crust of the Briançon Rise;
- the Zermatt–Saas Fee Zone derived from oceanic crust of the Piemont Ocean;
- the Dent Blanche nappe made of continental crust of the Adriatic margin.

The P–T–t path of Dora Maira is based on Castelli et al. (2007). Maximum pressures at 4 gigapascals corresponding to a burial depth of at least 50 kilometres were attained at around 35 million years ago, and only shortly later, at 32 million years ago, the rocks were decompressed to ca. 0.5 gigapascals and exhumed to a depth of 10–15 kilometres. The samples from the Zermatt–Saas Fee Zone at Valtournanche reached pressures of 3 gigapascals at around 44 million years ago and were subsequently decompressed rapidly by around 38 million years ago (P–T–t path in Fig. 6.2 after Reinecke 1998). Although this P–T–t path resembles that in Dora Maira, tha latter seems to have descended to depth somewhat later. Considering that the Piemont Ocean descended the former subduction zone prior to the arrival of the Briançon Rise, this time lag seems at least plausible. The third example, the Dent Blanche nappe, reached somewthat lower maximum pressures compared with the two other examples (P–T–t path in Fig. 6.2 after Höpfer & Vogler 1994). In the Austroalpine Sesia

P-T-t paths

Figure 6.2 High-pressure metamorphism in the Western Alps. P–T–t paths are from the Penninic Dora Maira basement and the Zermatt–Saas Fee Zone in Valtournache, and from the Austroalpine Dent Blanche nappe.

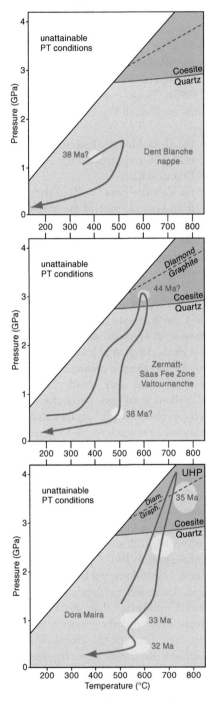

years ago, separated by a low-pressure event at around 73 million years ago: they refer to this as yo-yo subduction, meaning that certain units were subducted, exhumed and subducted again in an oblique subduction system. It could be argued that at least part of the pressure differences was caused by short-lived tectonic overpressures.

In the Central Alps, high-pressure assemblages are preserved in the Lepontine area in a complex stack of slices of oceanic crust and lenses of mantle rocks embedded in continental crust. This mélange has been called a 'tectonic accretion channel' by Engi et al. (2001) to emphasize its process of formation in a subduction zone. The largest unit in this zone is the Adula nappe, but smaller units outcrop to the south near the Insubric Fault (notably Alp de Trescolmen, Cima Lunga, Monte Durio and Alpe Arami). P–T–t paths of these units and the Tambo nappe are summarized in Frey & Ferreiro Mählmann (1999) and are shown in Fig. 6.3A. In the Alpe Arami lens, which is interpreted to represent mantle material expelled along the former subduction zone, maximum pressures of 5 gigapascals were reached at 40–35 million years ago: pressures then dropped to about 1 gigapascals by 33–27 million years ago. In comparison, P–T–t paths of samples from the central part of the Adula nappe indicate a rapid rise of pressures to a maximum of 2.5 gigapascals followed by a rapid drop to around 0.7 gigapascals. The present-day distance of around 20 kilometres between these locations can explain a difference in overburden of around 15 kilometres equivalent to 0.45 gigapascals lithostatic pressure only. Therefore we must conclude that the Adula nappe was a much longer unit when it was subjected to subduction and was subsequently shortened. This

Zone, Rubatto et al. (2011) report two high-pressure events less than 20 million years apart at 79 and 64 million

deformation can be envisaged as intensive shearing in the subduction channel, which could have been accompanied by tectonic overpressure.

For the tectonically higher Tambo nappe a similar path has been proposed by Marquer et al. (1994), with peak pressures of 1.2 megapascals reached at around 38 million years ago and 0.5 megapascals at around 30 million years ago. Obvously the Adula nappe was buried to greater depth compared with the Tambo nappe, but the rapid decompression between 40–35 million years ago and 30 million years ago occurred simultaneously in the two units. The change to a retrograde path at around 30 million years ago coincides with the intrusion of the Bregaglia Pluton in the vicinity of the Insubric Fault, which demarcates the Lepontine area in the south.

Figure 6.3B and C illustrates the morphological expression of the erosion-resistant eclogite lenses in the central part of the Adula nappe and the internal structure of thse lenses at outcrop. The eclogites (with garnets) are preserved in the core of the green lenses while towards the border, towards the surrounding paragneisses, the eclogites were amphibolitized and show the typical dark green colour.

In the Eastern Alps, high-pressures are recorded in the Austroalpine nappe complex west and east of the Tauern window, as well as within the Penninic ophiolite nappes within the Tauern window. The age of the high-pressure eclogitic metamorphism in the Austroalpine units is of Late Cretacous age (Hoinkes et al. 1999), with typical ages around 100 million years ago, as indicated in Fig. 6.1. A much younger age for eclogite foration has been determined in the Penninic nappes: 39 million years ago in the Moderreck unit

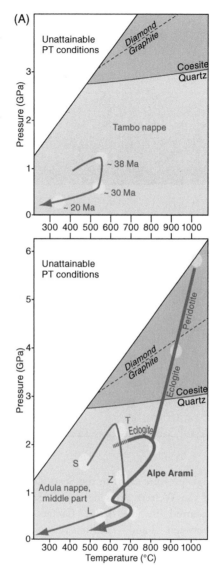

Figure 6.3 High-pressure metamorphism in the Central Alps.

(A) P–T–t paths are from the Penninic Adula nappe and Tambo nappe.

by Kurz et al. (2008) and 45–42 million years ago in the eclogite zone by Ratschbacher et al. (2004). The unmistakable difference between the Late Cretaceous event in the Austroalpine and the Late Eocene to Early Oligocene event in the Penninic nappes bears witness to two distinct orogenies in the (Eastern) Alps. The

(B)

(C)

(B) Morphological expression of eclogite outcrops in the Adula nappe. View is to the south-southeast. Easterly dip of lithologic contacts and foliation of the Adula nappe are clearly visible in the distance. Source: Photograph courtesy of Alfons Berger. Reproduced with kind permission.

(C) Mafic boudins (green) within paragneisses (grey). Eclogites are preserved in the core of the boudins while the outer rims are amphibolitized. Outcrop is at Alp di Trescolmen in the central part of the Adula nappe. Source: Photograph courtesy of Lukas Rohrbach. Reproduced with kind permission.

high-pressure event in the Sesia Zone of the Western Alps described above occurred in the latest Cretaceous (between 79 and 64 million years ago) and seems to mark a transition between the two orogenies.

The $P–T–t$ paths of the Austroalpine Ötztal nappe and the Penninic Eclogite Zone shown in Fig. 6.4 (taken from Hoinkes et al. 1999) show peak pressures of 1.2 gigapascals at 100 million years ago and peak pressures of 2.3 gigapascals at or just after 45 million years ago, respectively. The ensuing decompression in the time span from 42 to 40 million years ago was followed by a thermal heating event at 35 million years ago, which correlates in time with

the regional metamorphism that will be discussed below.

Temperature-Dominated Regional Metamorphism

A regional, temperature-dominated metamorphism of Barrovian type reached amphibolite facies conditions in the Eastern Alps in the Cretacous and in the Lepontine area of the Central Alps in the Cenozoic. The latter has been studied extensively by petrologists and has the status of a classic example showing the interaction of nappe formation and the evolution of pressure and temperature. A more detailed discussion on the history of research can be found in Frey & Ferreiro Mählmann (1999), and for the analysis of pressure and temperaturs conditions, the pioneering work by Engi et al. (1995) and Todd & Engi (1997) must be mentioned. Figure 6.5 gives a compilation of $P–T$ conditions and mineral assemblages based on Engi (2011): 6.5A is a tectonic map showing the nappe structure; 6.5B shows the distribution of index minerals and reaction isograds. The limits of sillimanite and kyanite clearly cross-cut the nappe contacts of the Penninic basement nappes (Antigorio, Maggia, Simano, Adula) and the underlying Leventina basement, and the reaction isograds of the middle-pressure metamorphic sequence also show cross-cutting relationships. Similarly, the boundary between amphibolite and greenschist facies in the north between Olivone and Airolo does not follow nappe contacts. Obviously the metamorphic stages recorded by the limits of index minerals, the reaction isograds and facies boundaries were attained when the nappe stack had already been formed. Figure 6.5C shows the isotherms of maximum temperatures reached during this regional metamorphism, and they cut the nappe contacts as well as the facies boundaries and indicate a temperature high in the south between Bellinzona and Locarno just north of the Insubric Fault. However, the 300 °C boundary has been offset by the basal thrust fault in the Penninic nappes in the extreme east of the Central Alps. The maximum temperatures were reached 28 million years ago in the south and later, 26–21 million years ago, in the north (Engi et al. 2004). Interestingly the pressures attained at the moment of maximum temperatures (see Fig. 6.5D) show a different pattern, with a maximum of 0.7 gigapascals stretching across the Lepontine in a WSW–ENE direction. But the maximum pressures reached during this regional metamorphism were much higher, in the order of 1.2 gigapascals. Isotherms as well as isobars are cut by the Insubric Fault in the south. The Southalpine nappe stack to the south of this fault shows a very weak Alpine metamorphic overprint characterized by the local occurrence of stilpnomelane and prehnite (Colombo & Tunesi 1999).

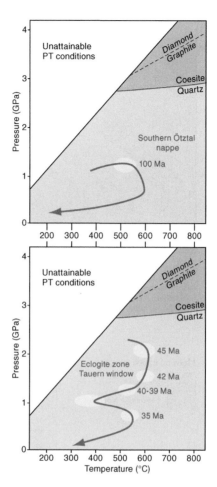

Figure 6.4 High-pressure metamorphism in the Eastern Alps. P–T–t paths are from the Austroalpine Southern Ötztal nappe and the Penninic Eclogite Zone in the Tauern Window. Source: Hoinkes et al. (1999). Reproduced with permission.

Contact Metamorphism

Along the Peri-Adriatic fault system, several plutons intruded in Late Eocene and Early Oligocene times and submitted the country rock to higher temperatures. As a consequence new minerals grew within the contact aureols and highlight the temperature gradients attained in the country rock. As an example the contact metamorphism in the east of the Bregaglia Pluton is shown in Fig. 6.6. The reaction isograds and first appearances of index minerals are based on Trommsdorff & Nievergelt (1983), Puschnig (1996) and Trommsdorff & Connolly (1996). Both sets of isolines

form a concentric pattern around the intrusive contact of the Bregaglia Pluton and crosscut the thrust fault between the Margna nappe and the Malenco–Forno–Lizun nappe. The isograd marking the disappearance of antigorite, however, is offset by the Muretto Fault, a young brittle fault post-dating the intrusion. According to thermal modelling performed by Trommsdorff & Connolly (1996), ambient temperatures in the mafic–ultramafic country rock are estimated to be around 350 °C. In the course of the (modelled) intrusion of the tonalite, which forms an outer shell in

Figure 6.5 Regional metamorphism in the Leventina (Central Alps). Source: Adapted from Engi (2011). See text for discussion.

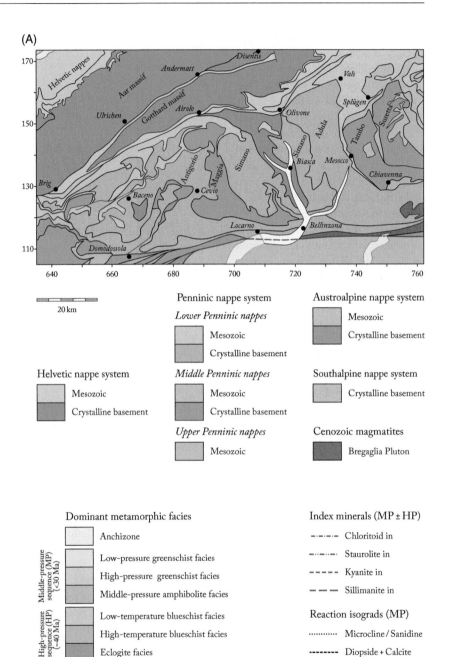

the Bregaglia Pluton, rocks in the vicinity of the intrusive contact were heated to around 570 °C, while rocks 2 kilometres away from this contact reached approximately 470 °C. The tonalite intruded at 32 million years ago and the granodiorite at 30 million years ago.

(B)

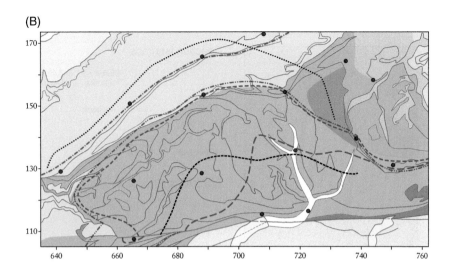

(C)

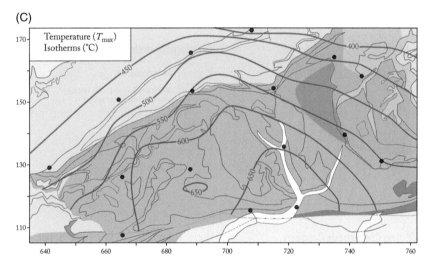

Temperature (T_{max})
Isotherms (°C)

(D)

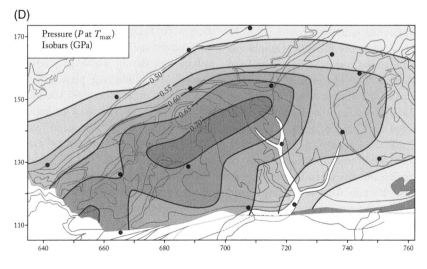

Pressure (P at T_{max})
Isobars (GPa)

Figure 6.6 Contact metamorphism in the contact aureole of the Bregaglia Pluton (Central Alps). See text for discussion.

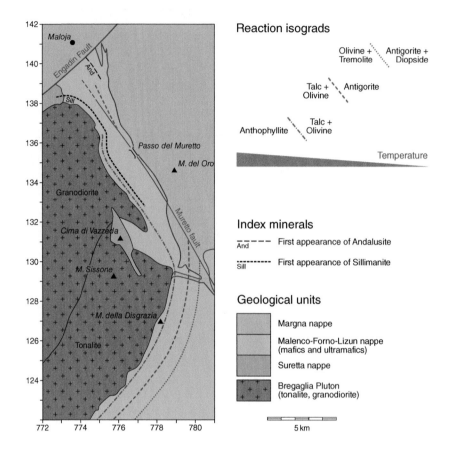

Reaction isograds

Olivine + ⋰⋱ Antigorite +
Tremolite ⋱ Diopside

Talc + ⋱ Antigorite
Olivine ⋱

Anthophyllite ⋱ Talc +
⋱ Olivine

Temperature

Index minerals

‒ ‒ ‒ ‒ ‒ First appearance of Andalusite
And

⋯⋯⋯⋯ First appearance of Sillimanite
Sill

Geological units

Margna nappe

Malenco-Forno-Lizun nappe
(mafics and ultramafics)

Suretta nappe

Bregaglia Pluton
(tonalite, granodiorite)

5 km

6.2 The Cretaceous Orogeny

The Cretaceous orogeny affected the Austroalpine realm in the Eastern Alps, while sediments continued to be deposited in the Helvetic–Dauphinois and Penninic realms in the Central and Western Alps. The palaeogeographical maps of the Early Cretaceous (Aptian, 125 million years ago) in Fig. 3.28 and of the Late Cretaceous (Turonian, 90 million years ago) in Fig. 4.9 show these conditions. The rocks with an eclogite and blueschist facies overprint in the Eastern Alps in the Lower and Middle Penninic crystalline nappes indicate that these units had been deeply buried. Figure 6.1 shows that

these eclogites formed over a period of 110–90 million years ago, that is, at the transition from Early to Late Cretaceous (Albian to Turonian). Berger & Bousquet (2008) interpret the isotopic age as a result of a lengthy subduction, during which fragmentation of the subducting plate can be assumed in order to explain the local differences in age. Structural geological indicators point to west-northwest-directed nappe movements or east-southeast-directed subduction of the Piemont Ocean. However, the Cretaceous age of the nappe movements has also been observed in the Austroalpine sedimentary nappes, in the Northern Calcareous Alps. Ortner (2001) reviews

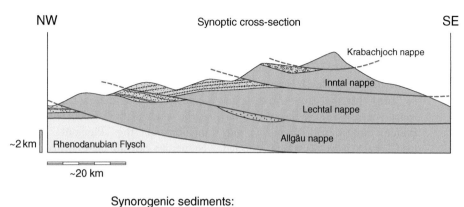

NW   Synoptic cross-section   SE

▲ Figure 6.7 Synoptic cross-section through the Northern Calcareous Alps in the Cretaceous showing synorogenic Gosau sediments – compiled from Ortner (2001). Source: Adapted from Piller & Rasser (2001).

the literature and discusses a kinematic model for the synorogenic sediments. Figure 6.7 shows a synoptic cross-section that summarizes the structural relationships in the synorogenic sediments and the different nappe contacts. At the base, the synorogenic sediments comprise the Lech Formation, which is composed of siliciclastics that also contain ophiolitic components. The Branderfleck Formation above it and the Gosau Group at the top are carbonate–siliciclastic successions. These were deposited under shallow-marine or in continental conditions. In several cases, they lie unconformably on top of the eroded Mesozoic substratum. As is shown in Fig. 6.7, the different nappes both overthrust synorogenic sediments of different ages and such sediments were also deposited on top of them. Ortner (2001) demonstrated that while the positioning of the various thrusts progressed from southeast to northwest, more internal thrust faults were also activated occasionally by

'out-of-sequence thrusting'. The oldest thrust fault, which resulted in the Juvavic nappe complex overlying the Tyrolean nappe complex, originated as early as the end of the Barremian, about 120 million years ago (not shown in Fig. 6.7). In the Aptian–Albian (approximately 100–110 million years ago), the Lechtal thrust was activated for the first time. The Inntal thrust followed next at the end of the Cenomanian (approximately 94 million years ago), even though it is located higher up in the nappe pile and more internally. The Lechtal thrust was then activated for a second time at the end of the Turonian (approximately 89 million years ago). Finally, the Krabachjoch nappe overthrust the Gosau Group at the end of the Cretaceous (approximately 65 million years ago).

Froitzheim et al. (1994, 1997) analysed the structural evolution of the Austroalpine nappe system in Graubünden and compared it with geochronological dates obtained from new mineral formation in shear zones and pseudotachylites. A Cretaceous age was also determined for thrusts and folds in this nappe complex. However, it must be noted that the entire Austroalpine nappe system was later thrust onto the Penninic nappes, in the Cenozoic, with the Cenozoic movements being directed more towards the north. Froitzheim et al. (1994) classified the orogenic movements into phases, each named after a type locality. Two Late Cretaceous phases are included in the Cretaceous orogeny. The older of the two, the Trupchun Phase, comprised nappe stacking in a westerly direction and local strike-slip motion. During the later phase, the Ducan-Ela Phase, east- to southeast-directed normal faults overprinted the higher portion of the nappe pile. Extension, in combination with vertical shortening, produced folds in the deeper portion of the nappe pile. Recumbent folds developed in locations where the strata were already vertical

Cretaceous orogen overprinted by Cenozoic orogeny

prior to the vertical shortening. Some of the Ducan-Ela Phase normal faults reactivated the Trupchun Phase thrust faults. A larger shear zone, the Vinschgau shear zone, delimits the Campo and the Ötztal nappes, two Upper Austroalpine crystalline nappes. Geochronological data indicate an age of about 100 million years for the activation of this strike-slip fault.

Figure 6.8 illustrates the different phases of deformation, sedimentation and metamorphism in a time–space diagram, or an 'orogenic timetable'. This clearly shows the restriction of the Cretaceous orogeny to the Austroalpine units. Conversely, it also makes clear that uplift continued in the Eastern Alps in the Cenozoic.

Figure 6.9 (inspired by Froitzheim et al. 1997) shows a schematic illustration of the situation in the latest

Cretaceous (after the Ducan-Ela Phase, approximately 70 million years ago). The Eastern Alps jutted out of the surrounding sea in the form of a flat mountain chain. There is a thick nappe pile in the extreme east, which is the result of the east-directed subduction of the Piemont Ocean. Numerous east-directed normal faults originated during the Ducan-Ela extension phase. The Briançon Rise ends towards the northeast and pelagic sedimentation was still occurring there. The Piemont Ocean and the Valais Trough merge in the northeast. Strike-slip faults to the north and south of the Eastern Alps permitted the Austroalpine nappe system to move independently from the Southalpine nappe system and the Penninic Ocean. The southern strike-slip fault was a precursor to the Peri-Adriatic fault system.

The later Cenozoic movements in the Eastern Alps resulted in the entire Austroalpine nappe system being moved northwards onto the Penninic and Helvetic nappe systems. At the same time, in the south, the Southalpine nappe system and the Dolomites were compressed and parts of the Adriatic crust were thrust under the Eastern Alps. These processes can easily be compared to those of the Cenozoic orogeny in the Central Alps and will therefore be discussed in the next section.

6.3 The Cenozoic Orogeny

Alpine orogeny occurred in several stages in the Cenozoic. After an initial phase of subduction, the actual collision occurred between the Adriatic and European continental margins. The orogenic timetable in the Eastern, Central and Western Alps does show common characteristics in relative progression, but there is no chronological correlation between similar processes along the Alps, which is why the three segments are discussed separately. The orogeny was accompanied by large-scale thrusts and folds, which were ultimately also responsible for the uplift and relief formation.

The large tectonic units are shown in their original relative positions in the orogenic timetable for the Eastern Alps and Dolomites in Fig. 6.8. A first glance is sufficient to show that, over time, deformation prograded into the external areas and that sedimentation continued for longest in the foreland basin in these areas. This corresponds to an orogen that was growing simultaneously on both sides. On the northern side, the Austroalpine nappes were thrust onto the Penninic realm. During this process, the Penninic sediments were doubled up: the southern Bündnerschiefer Group, which contains ophiolites, was thrust on top of the northern Bündnerschiefer Group, which does not contain ophiolites. Due to continuing convergence, the Penninic nappes, together with the Austroalpine lid on top of them, then ended up on top of the Helvetic realm. The Ultrahelvetic sediments were sheared off and dragged along during this process, such that they eventually ended up overlying the Subalpine Molasse. As is demonstrated by the antiformal structure of the basal thrusts in the Penninic and Austroalpine nappes, the updoming of the Tauern massif occurred only once the Penninic–Austroalpine nappe pile had been emplaced onto the Tauern massif. The uplift is probably linked to a normal fault (see Fig. 5.3-2) that connects with the thrusts in the Subalpine Molasse further to the north. The exact chronological sequence for

the individual deformation events has not been established, but must have started after the end of deposition of the Rhenodanubian Flysch, and the Subalpine Molasse was not overthrust before the Early Miocene (Fig. 6.8). In the case of the Glockner nappe and the Tauern massif, the deformation caused penetrative foliations, and horizontal shortening of the Tauern massif and the associated thickening caused a conspicuous vertical uplift. Normal faults formed (Brenner and Katschberg normal faults in Fig. 6.8) to the west and east of the Tauern massif as a reaction to this uplift. Froitzheim et al. (1994) proposed that in the Austroalpine nappe system itself, the overthrusting onto the Penninic nappe system caused internal folding, which is attributed to the Blaisun Phase. Renewed extension parallel to the orogen corresponds to the Turba Phase. The Turba normal fault affects the Austroalpine–Penninic boundary area, but can probably be attributed to the uplift of the Penninic nappe pile in the Lepontin region in the Central Alps. The final phase, the Domleschg Phase, caused fairly small-scale NNW–SSE-directed shortening.

Deformation also migrated towards the external areas on the southern side, in the Dolomites. The oldest deformation events in the Karawanks and the Dolomites are attributed to tectonic movements in the adjacent Dinarides (Polinski 1991, Castellarin et al. 2006). The later Valsugana thrust is contemporaneous, but spatially separated from the sedimentation in the Po Basin. In the case of the Montello thrust, the most recent Pliocene and Pleistocene deposits were affected by the deformation. The arguments made in support of the age of deformation in the Karawanks and the Dolomites are discussed in more detail in the context of Figs 5.3-8 and 5.3-9.

Orogenic timetable: Eastern Alps

Figure 6.8 Orogenic timetable of nappe formation, sedimentation, metamorphism and magmatism in the Eastern Alps.

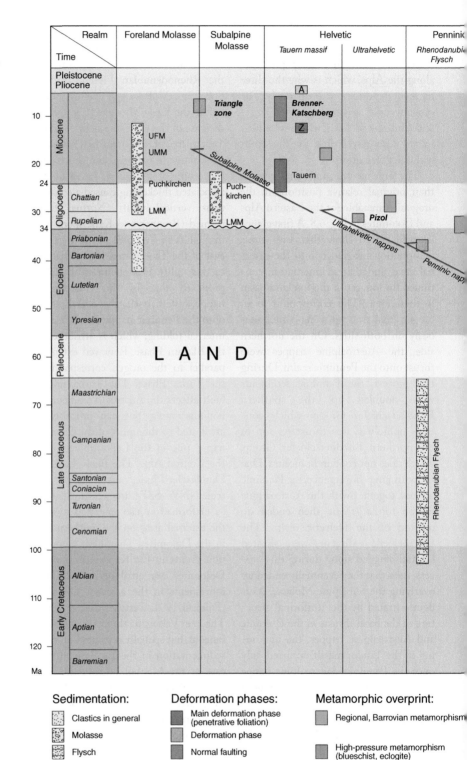

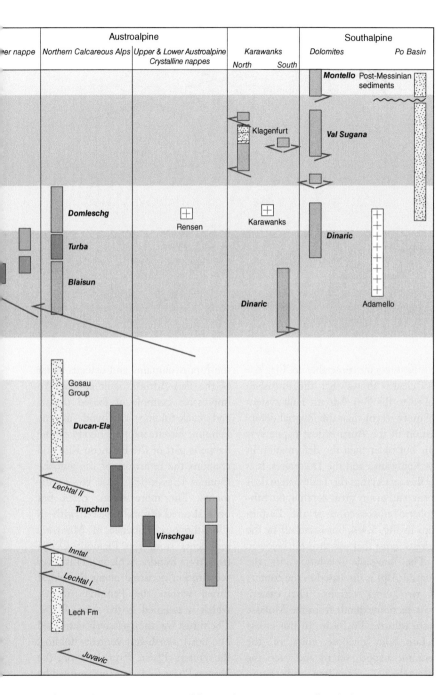

Fission-track ages

| A | Apatite |
| Z | Zircon |

Magmatism:

⊞ Plutons

Symbols:

〜〜〜 Unconformity

＞ Thrusting

Figure 6.9 Schematic block diagram of the Eastern Alps in the Late Cretaceous – inspired after Froitzheim et al. (1997). A west-vergent nappe stack is bordered in the north and south by strike-slip faults. Sedimentation prevails in the future Central and Western Alps. Source: Froitzheim et al. (1997). Reproduced with permission of Elsevier.

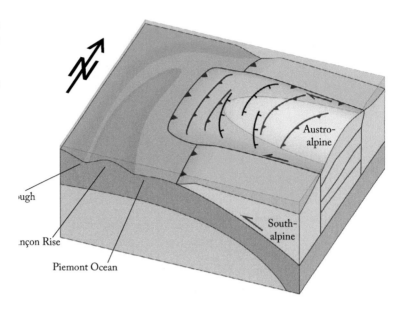

The orogenic timetable in Fig. 6.8 also clearly shows that the intrusives linked to the Peri-Adriatic fault system are more recent than the internal deformation of the Austroalpine nappe system, but older than the deformation in the Karawanks and the Dolomites. It is for this reason that this fault system does not extend to any great depth in the lithospheric cross-section of the Eastern Alps in Fig. 5.4-4, but is cut off by the Adriatic crust that was thrust in below it.

The orogenic evolution in the Central Alps is discussed in the context of two cross-sections. The eastern cross-section extends from the Molasse Basin into the Po Basin. In this cross-section, the deepest units of the Penninic nappe system and even the Ultrahelvetic crystalline substratum below it are accessible for observation in the field. Ductile folded thrusts and folded folds can be observed in the units with up to amphibolite-facies metamorphic overprint. The second, western cross-section also comprises

the Jura Mountains and extends as far as the Peri-Adriatic fault system. An impressive example of folded thrusts and back-folding is found in the Penninic Suretta nappe. This crystalline nappe is part of the Briançon Rise and contains the remnants of the autochthonous Triassic–(?)Jurassic cover sediments. The more recent cover has been sheared off and replaced with an allochthonous sequence of Mesozoic sediments from the Piemont Ocean (the Avers Bündnerschiefer). This cover substitution occurred along the oldest thrust within the Penninic nappes, which is assigned to the Avers Phase. The thrust was then passively deformed by north-northwest-vergent folding, the Ferrera Phase. Even later on, the Ferrera Phase folds were overprinted by south-southeast-vergent back-folding – the Niemet–Beverin Phase.

Figure 6.10A shows a folded thrust at the base of the Avers Bündnerschiefer. The thrust contact and the contact between the Permian Rofna Porphyry

(A)

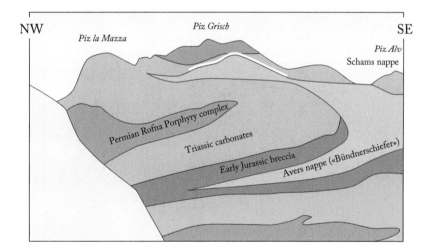

Figure 6.10 Folded folds and thrust faults.

(A) In the Middle Penninic Suretta nappe (Val Ferrera, canton Graubünden, Switzerland), a southeast-vergent fold beneath Piz Grisch folds an older, Avers Phase thrust fault, along which the Southpenninic Avers nappe came to lie on top of Triassic carbonates pertaining to the Suretta nappe. The southeast-vergent fold was initially a north-vergent Ferrera Phase fold which was back-folded by shearing of the Niemet–Beverin Phase.

Folded thrust faults

and the Triassic carbonates depict southeast-vergent folds. The anticline with Rofna Porphyry in the core closes towards the southeast and not towards the northeast, as is usual in the Penninic nappe pile. The folds themselves formed during the Ferrera Phase and were sheared back later on, during the Niemet–Beverin Phase. This back-folding is also visible in the cross-section in Fig. 5.2-19.

Figure 6.10B shows the case of a folded thrust in the Helvetic nappe sys-tem. The sedimentary cover of the Trun submassif has been stripped off except for the lowermost (Triassic) strata and replaced by an already inverted sequence pertaining to the Cavistrau nappe (Käch 1969, Pfiffner 1978). This cover substitution is attributed to the Cavis-trau Phase. Subsequently the thrust fault at the base of the Cavistrau nappe was folded and even overturned. As a consequence, the sedimentary sequence within the Cavistrau nappe seems to be upright, although it had been overturned

(B) A folded and overturned, steeply south-dipping thrust fault marks the contact between the Trun submassif of the Aar massif and the Mesozoic of the Cavistrau nappe. The thrust fault follows a gully to the summit of Piz Dado and continues at distance into the summit area of Piz Dadens. A thin layer of Triassic remained attached to the crystalline basement. The sediments of the Cavistrau nappe young towards the thrust fault (indicated by arrows). A penetrative (Calanda Phase) foliation cuts across and post-dates the (Cavistrau Phase) thrust fault.

(B)

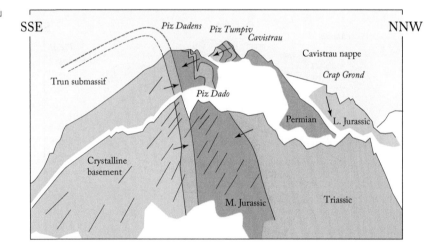

twice. This later folding was accompanied by a pervasive foliation shown in Fig. 6.10B and is attributed to the Calanda Phase (Milnes & Pfiffner 1977, Pfiffner 1978).

The succession of deformation phases in the Penninic and Helvetic nappe systems is discussed in detail in Schmid et al. (1997) and Pfiffner (1986). Figure 6.11 shows the orogenic timetable for the eastern Central Alps. This is based on the publications cited above and is amended by data from Schönborn (1992) for the

Southalpine nappe system. The orogenic timetable for the Austroalpine nappes has also been inserted, which was taken from Fig. 6.8. Deformation started in the Penninic nappe system after cessation of sedimentation of the Prättigau and Arblatsch flyschs, that is, no earlier than the Early Eocene. This also applies to the Cenozoic portion of deformation in the Austroalpine nappe system. Deformation became progressively more recent in the direction of the Helvetic realm and the Molasse Basin.

The first phase in each realm (Avers Phase in the Penninic realm, Pizol Phase in the Helvetic realm) is characterized by detachment of sedimentary units off their crystalline basement and their emplacement on the still undeformed, more external areas. The more ductile deformation, which also affected the pre-Triassic basement (Ferrera and Calanda phases), does not start until after this. The thrust tectonics were associated with the tectonic burial of the affected units, which is manifested in high-pressure metamorphism (blueschist-facies according to Oberhänsli et al. 2004; see Fig. 6.1). According to Berger & Bousquet (2008), the Suretta nappe, which corresponds to part of the Briançon Rise, was overprinted by high-pressure metamorphism 49–42 million years ago, in the Middle Eocene (Lutetian). This is only slightly more recent than the youngest sediments in the areas immediately adjacent to the Briançon Rise (Southpenninic Arblatsch Flysch and Northpenninic Prättigau Flysch), which indicates very rapid burial and overprinting. High-pressure meta-morphism gave way to more temperature-dominated regional meta-morphism, which shifted to more and more external areas over time. Nappe stacking in the Penninic realm was followed by two phases of extension, indicated by two larger normal faults, the Turba and Forcola normal faults. In the external area in the north, the most recent deformation started no earlier than the Early Oligocene (after the Northhelvetic Flysch had been deposited) and lasted into the Miocene. Sedimentation in the Subalpine Molasse lasted only until the very early Miocene: the Upper Marine Molasse and the Upper Fresh-water Molasse are limited to the Plateau Molasse. The sedimentation area of the Subalpine Molasse had probably already been overthrust by the Helvetic

and Penninic nappes in the Early Miocene. The deformation that then resulted in the formation of the most external structure, the triangle zone between the tilted Plateau Molasse and the Subalpine Molasse, must have taken place in the Late Miocene.

In the Oligocene, a steep, south-directed thrust fault that was accompa-nied by a dextral strike-slip became active in the vicinity of the Peri-Adriatic fault system. The Bregaglia intrusion and a regional, mainly temperature-driven metamorphism that reached amphibolite facies higher up, also occurred around this time. The activity along the Insubric Fault in the Miocene was limited to dextral strike-slip motion.

Orogenic movements started in the south of this cross-section before the Adamello intrusion occurred, and Schönborn (1992) even proposes a Cretaceous age for this Orobic Phase. With regard to the Coltignone Phase, Schönborn (1992) states that a subdivi-sion into a pre-Adamello and a syn-Adamello phase can be assumed. During the Oligocene and the Miocene, the orogenic movements progressed further and further southwards into the Po Basin. The most recent Lombardian Phase caused a fold belt in the subsur-face of the region of Milano, which is covered by the most recent Po Basin fill, and the related angular unconformity at the base of the post-Messinian sedi-ments is a known feature of this.

Taking the Alps as a whole, then the cross-section of the eastern Central Alps reveals an orogen that grew simul-taneously towards the north and south, which overthrust the erosional products of the Eocene–Oligocene ancestral Alps in the Miocene. The growth of the Alps in this cross-section is illustrated using a sequence of palaeogeographical cross-sections (Fig. 6.12). The recon-struction of these sections is based on a

Orogenic timetable: eastern Central Alps

Figure 6.11 Orogenic timetable of nappe formation, sedimentation, metamorphism and magmatism in the eastern Central Alps. Source: Compiled and complemented from Schönborn (1992), Schmid et al. (1997b) and Pfiffner et al. (2002).

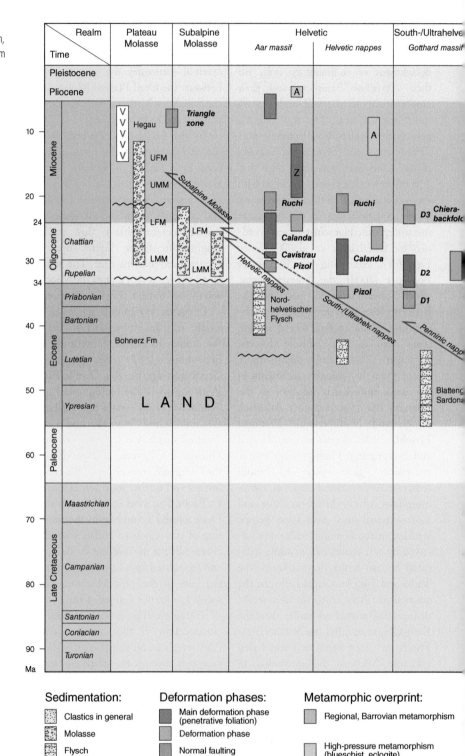

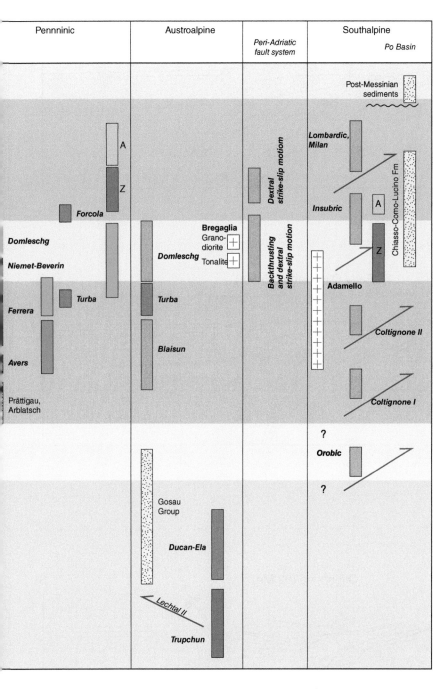

Fission-track ages

A	Apatite
Z	Zircon

Magmatism:

V	Volcanism
+	Plutons

Symbols:

 Unconformity

Thrusting

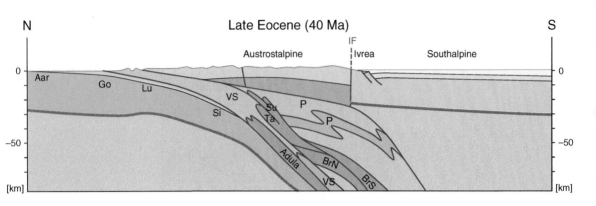

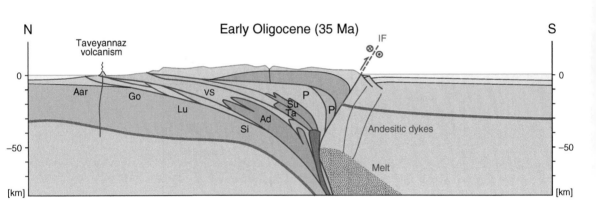

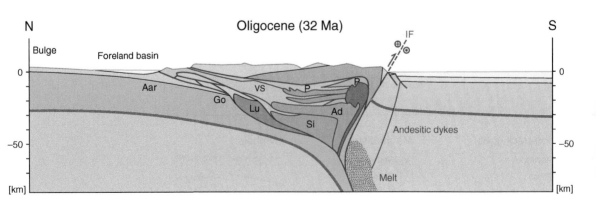

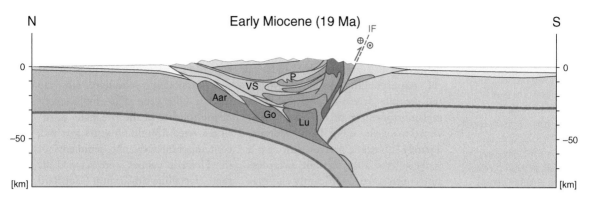

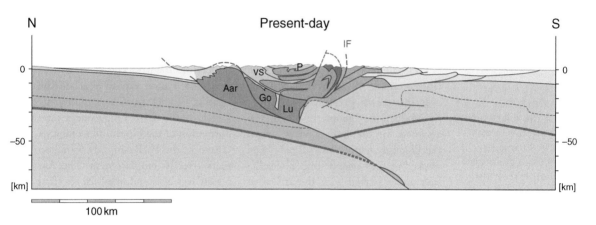

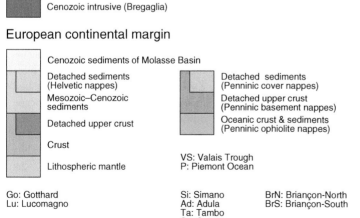

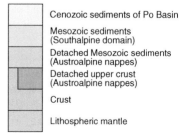

Cenozoic intrusive (Bregaglia)

European continental margin

Cenozoic sediments of Molasse Basin

Detached sediments
(Helvetic nappes)

Mesozoic–Cenozoic
sediments

Detached upper crust

Crust

Lithospheric mantle

Detached sediments
(Penninic cover nappes)

Detached upper crust
(Penninic basement nappes)

Oceanic crust & sediments
(Penninic ophiolite nappes)

VS: Valais Trough
P: Piemont Ocean

Adriatic continental margin

Cenozoic sediments of Po Basin

Mesozoic sediments
(Southalpine domain)

Detached Mesozoic sediments
(Austroalpine nappes)

Detached upper crust
(Austroalpine nappes)

Crust

Lithospheric mantle

Go: Gotthard
Lu: Lucomagno

Si: Simano
Ad: Adula
Ta: Tambo
Su: Suretta

BrN: Briançon-North
BrS: Briançon-South

IF Insubric Fault (segment of the
    Peri-Adriatic fault system)

⊗ motion away from observer
⊙ motion towards observer

◀ **Figure 6.12** Tectonic evolution of the Central Alps of eastern Switzerland depicted as a series of palinspastic cross-sections. Note that because of strike-slip motion along the Insubric Fault, some units of the Southalpine nappe system disappear with time and are replaced by others. Growth of the orogen is accomplished by successive incorporation of detached upper crustal blocks (e.g. Simano, Lucomagno, Gotthard, Aar). The sedimentary cover of these units was stripped off and moved to the north in an earlier stage. The core of the orogen made up of Penninic basement nappes was uplifted by retro-thrusting along the Insubric Fault and basal accretion of Simano, Lucamagno, Gotthard and Aar basement blocks. Uplift was aided by erosional removal of the Austroalpine nappe pile. Source: Adapted from Schönborn (1992).

Tectonic evolution in time frames

reversal of the deformation caused by thrusting and folding. The depth of individual points is more or less defined by the metamorphic grades of the corresponding rocks. However, the pressure estimates obtained from metamorphic data are not simply converted to an equivalent depth assuming a purely lithostatic pressure. Instead, it is considered that tectonic overpressure contributed significantly and may account for up to almost half the value obtained from the metamorphic assemblages (Petrini & Podladchikov 2000, Mancktelow 2007). The chronological markers are deduced from the arguments given in the orogenic timetable in Fig. 6.11. A more detailed discussion can be found in Schmid et al. (1997), where a first version of these cross-sections disregarding tectonic overpressure was presented.

Subduction geometry can be assumed for the Late Eocene, when the Penninic nappes had already been stacked up into a nappe pile. The collision phase started with the arrival of the actual European continental margin with the crust of the Ultrahelvetic nappe complex, which was of normal thickness. Large parts of the Penninic nappes underwent eclogitic and blueschist-facies overprinting under the lid of the Austroalpine nappe system. Rapid exhumation of these units prevented an equilibration of the mineral parageneses, which is why the indicators of the high-pressure metamorphism remain preserved. The European upper crust began to escape upwards from the Early Oligocene onwards along a steep, south-directed thrust fault (the Insubric Fault) and a backfold started to form in its hanging wall. At this point, the Ultrahelvetic nappes were sheared off their crystalline substratum and remained at a shallow

depth, while the crystalline basement moved down to form the crystalline nappe (Lucomagno) under which the crystalline rock of the Gotthard massif was eventually subducted in the Middle Oligocene. The emplacement of the tonalite in the Bregaglia intrusion started 32 million years ago and, contemporaneously, the sediments of the Helvetic nappes were scraped off their crystalline substratum (Gotthard massif). The back-thrust at the Insubric Fault was now accompanied by pervasive back-folding at depth: the nappe contacts of the Penninic nappes were passively folded into an antiform by the upward escape of the upper crustal material (Simano, Adula). This back-folding and back-thrusting continued into the Miocene, such that the Bregaglia Pluton rapidly reached the surface. Pebbles from this pluton are found in the Late Oligocene Como Formation of the Gonfolite Lombarda Group in the Po Basin only 5 million years after its emplacement. The Aar massif was also updomed for the first time in the Early Miocene. This dome also included the Helvetic (and higher-level) nappes that were already overlying it. Now the dextral movements at the Insubric Fault introduced a new portion of the Adriatic continental margin into the cross-section under observation. Up to the Late Oligocene, this was the block containing the Ivrea Zone (lower crust) and its associated mantle. From about the Miocene onwards, it was then a normal crustal sequence that was further shortened by the continuing convergence. The south-vergent nappes of the Southalpine nappe system were formed in the upper crust, while the lower crust was wedged into the European crust that was itself undergoing deformation. The length of this indentor, made up of lower crust, as it appears in the present-day cross-section in Fig. 6.12, is directly proportional to the original length of the upper crust. As was demonstrated by Schönborn (1992), shortening in the Southalpine nappe system increases towards the east in the current cross-section, so the indentor must become longer. The entire Southalpine system is offset at the Giudicarie fault. The Dolomites located to the east of this fault system have been shifted further to the north and, as is illustrated in Fig. 5.4-3, the entire Adriatic plate plunges beneath the Alps towards the north.

The western cross-section through the Central Alps includes the Jura Mountains at its northern end. In the region of the Helvetic nappe system, it cuts through the Aar massif where uplift was greatest (see Fig. 5.4-1). Furthermore, the kinematics of the Helvetic nappes at this location is well demonstrated, as shown in Fig. 6.13 (based on Herwegh & Pfiffner 2005), in which the sequence in the nappe pile at the western end of the Aar massif is greatly simplified. In the earliest Oligocene, the Ultrahelvetic (and Penninic) nappes had already been thrust onto the future Wildhorn nappe. The Wildhorn nappe was placed onto the Gellihorn nappe, and both of these on the future Doldenhorn nappe, as early as the Early Oligocene. The sediments of the Doldenhorn nappe were deposited in an asymmetric basin with a thick Early and Middle Jurassic sequence. This semi-graben fill was then squeezed out during the formation of the Doldenhorn nappe in the Late Oligocene and a large recumbent fold was formed (Kiental Phase). During this process, the overlying nappes were passively folded as well. The Aar massif then underwent

further uplift (Grindelwald Phase). The basal thrust in the Helvetic nappes at the foot of the north face of the Eiger, to the south of Grindelwald, was rendered vertical by this and even overturned at some locations (see Figs 5.2-8 and 5.2-9).

We see almost isoclinal large-scale folds at some locations in the Helvetic nappes in central Switzerland, the limbs of which have been cross-cut by normal faults. For example, the Axen nappe in Fig. 6.14 contains a series of tightly folded recumbent anticlines and synclines – some of the anticlines are even plunging folds. The Widderfeldstock anticline (Wf) was transported to the north on a large thrust, which with other thrusts produced an imbricate structure in the limestones of the Late Jurassic (Quinten Limestone). The subsequent deformation caused by the uplift of the Aar massif passively rotated the thrust faults, such that they are now actually present in the form of normal faults (the hanging wall moved down relative to the footwall). The normal limbs of the large-scale folds are cross-cut by numerous normal faults, and recumbent folds developed from upright folds that originally exhibited a steep axial surface, but were subsequently highly sheared. The cores of the folds were filled with the argillaceous successions of the Early and Middle Jurassic. These folds were then first sheared in a northerly direction, such that the axial surface was dipping towards the south, but passive rotation on the northern margin of the Aar massif then resulted in the present-day recumbent to plunging configuration. Both the shearing of the folds and the normal faults on the fold limbs, mentioned above, indicate horizontal extension and vertical shortening of the nappe, which can be

Orogenic collapse by gravity spreading

Figure 6.13 Kinematic
evolution of the Helvetic
nappes near the western end
of the Aar massif. Source:
Herwegh & Pfiffner (2005).
Reproduced with permission
of Elsevier.

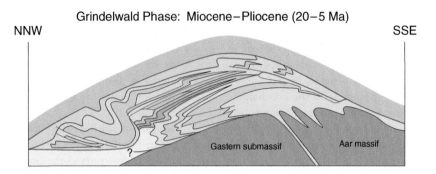

Grindelwald Phase: Miocene−Pliocene (20−5 Ma)

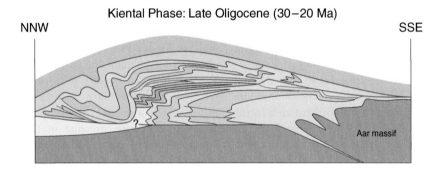

Kiental Phase: Late Oligocene (30−20 Ma)

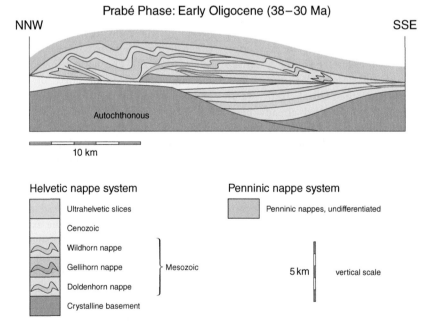

Prabé Phase: Early Oligocene (38−30 Ma)

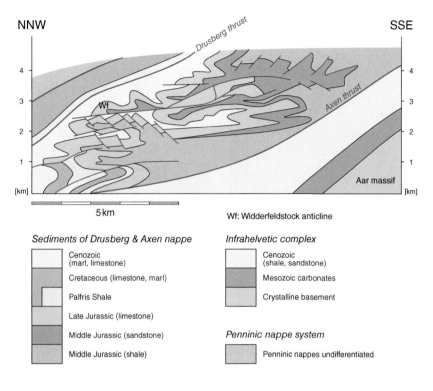

NNW   SSE

Drusberg thrust

Axen thrust

Wf

Aar massif

[km]   [km]

5 km

Wf: Widderfeldstock anticline

Sediments of Drusberg & Axen nappe

  Cenozoic (marl, limestone)

  Cretaceous (limestone, marl)

  Palfris Shale

  Late Jurassic (limestone)

  Middle Jurassic (sandstone)

  Middle Jurassic (shale)

Infrahelvetic complex

  Cenozoic (shale, sandstone)

  Mesozoic carbonates

  Crystalline basement

Penninic nappe system

  Penninic nappes undifferentiated

Figure 6.14
Geological cross-section through the Axen nappe of central Switzerland – redrawn after Menkveld (1995). The plunging isoclinal anticlines are evidence of orogenic collapse above the updoming Aar massif. Source: Adapted from Menkveld (1995).

explained by gravitational spreading whereby the nappe pile was drifting apart under its own weight – a process that is also called gravitational or orogenic collapse.

The Jura Mountains were the last mountain range to form during the Cenozoic orogeny. The growth of this mountain range by folding and thrusting has been dated using different methods. For example, the disposition of the Ajoie, Laufen and Delémont basins, which contain sandstones and conglomerates sourced from the north, is often cited, because anticlinal ridges now partially cut off these basins from the source area of these conglomerates. The sediments are of Middle and Late Miocene age (Kälin 1997), thus indicating a maximum age of 10–11 million years for these anticlines. Unfolded fissure fills in the core of the Vue-des-Alpes anticline provide more

information, as they indicate a minimum age for the folding of the Jura Mountains. Dating based on mammalian teeth resulted in a Middle Pliocene age of 3 million years for the fissure fills (Bolliger et al. 1993), thus the main folding of the Jura Mountains can be placed between 10 and 3 million years ago.

However, older unconformities are also observed, for example, there is an angular unconformity between the sandstones of the Early Miocene Upper Marine Molasse (UMM) and the karstified Early Cretaceous limestones (Marbre bâtard) near Les Verrières (see Fig. 6.15). The Cretaceous limestones had clearly been tilted to an angle of over 25° before the UMM strata were deposited. This tilting can be explained only through an early, pre-Miocene phase of folding. The originally horizontal UMM strata were then folded

Early folding in the Jura Mountains

Figure 6.15 Early phase
of folding in the Jura
Mountains indicated by
angular unconformities.
Source: Adapted from
Aubert (1975).

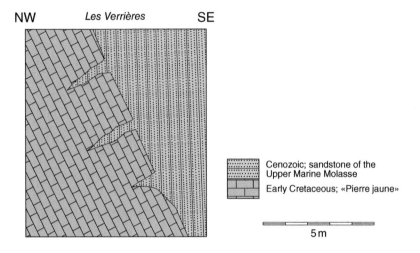

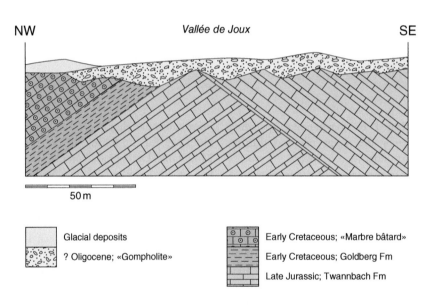

Figure 6.15 Early phase of folding in the Jura Mountains indicated by angular unconformities. Source: Adapted from Aubert (1975).

and rendered vertical, as observed at outcrop now. It is conceivable that parts of the future folded Jura Mountains had already developed into a high zone during this pre-Miocene folding phase and formed a ridge that was uplifted between the Molasse Basin in the southeast and an oceanic arm that ran lengthways through the Jura Mountains from Delémont, via Les Verrières to Bellegarde (see Aubert 1975).

Aubert (1975) demonstrated that a limestone conglomerate formation (called 'gompholites' locally) lies unconformably on top of an anticline in the Vallée de Joux (see Fig. 6.15). In its core, this anticline exhibits a thrust that is also unconformably overlain. Aubert (1975) proposes that the conglomerate, which was subsequently tilted by folding, dates from the Oligocene and correlates the rock with the Lower Freshwater Molasse (LFM).

In addition to these older deformation events, there are also events that are younger than the main folding. For example, to the north of the Jura Mountains, in the southern Rhine Graben, there are the Pliocene Sundgau Gravels, which are approximately 3 million years old and have been tilted by more recent movements (Giamboni et al. 2004).

Figure 6.16 gives the orogenic timetable for nappe formation and backfolding in the western Central Alps compiled based on Escher et al. (1997), Pfiffner et al. (2002) and Scheiber et al. (2013). Once again, the deformation front migrates outward from the core of the orogen, the Penninic nappes, towards the Jura Mountains in this cross-section. This progression is undelined by the successively younger foreland deposits migrating from the Penninic Niesen Flysch northward to the Northhelvetic Flysch and the Upper Freshwater Molasse (UFM). Likewise, nappe stacking progressed outward with time. Within the Penninic nappes three phases of back-folding follow in time (Mischabel, Vanone, Berisal) and are interspersed by phases of normal faulting (Simplon, Aosta-Ranzola).

In the Helvetic realm, the most internal Mesozoic–Cenozoic sediments of the Ultrahelvetic realm were sheared off, first by emplacement of the Penninic sedimentary nappes in the Plaine-Morte Phase, and transported to the north. Helvetic nappe stacking then occurred subsequently, initially during the Prabé Phase (see Fig. 6.13), and then later the sediments on the crest of the Aar massif and the Aar massif itself were overthrust and folded during the Kiental Phase. The final uplift of the Aar massif during the Grindelwald

Phase was contemporaneous with the Berisal back-folding in the south, which was also responsible for the steepening of the southern edge of the Aar massif. Steck (1984), Burkhard (1988), Escher et al. (1997), Pfiffner et al. (1997) and Scheiber et al. (2013) provide detailed discussions of the deformation processes in the Penninic and Helvetic realms.

In the Molasse Basin, sedimentation in the Subalpine Molasse ended with the LFM, probably also due to the foreland basin being overthrust by the Helvetic (east of the Aare) and Penninic (west of the Aare) nappes. Initial folding in the Jura Mountains also occurred at about the time of thrusting in the Subalpine Molasse. At this time, the Molasse Basin clearly transferred tectonic stresses far into the foreland, which was then followed by the formation of the triangle zone at the contact between Subalpine and Plateau Molasse and the main phase of folding in the Jura Mountains. However, both events are also chronologically correlated with the squeezing out of the Aar massif (Grindelwald Phase in the north and Berisal back-folding in the south), the Berisal back-folding in the Penninic nappe system and the most recent activity of the Simplon Fault.

To show the growth of the Alpine nappe stack along this transect, a series of palaeogeographical cross-sections were constructed (Fig. 6.17). The cross-sections are based on Scheiber et al. (2013) for the southern more internal part and on Wissing & Pfiffner (2002) for the external part. They highlight the successive incorporation of the European margin into the orogen. At around 65 million years ago, the Simme nappe of the Piemont Ocean had been subducted, but sedimentation continued in the Gurnigel Flysch and the adjacent

Orogenic timetable: western Central Alps

Tectonic evolution in time frames

Figure 6.16 Orogenic timetable of nappe formation, sedimentation, metamorphism and magmatism in the western Central Alps. Source: Compiled and complemented from Escher et al. (1997), Pfiffner et al. (2002) and Scheiber et al. in press.

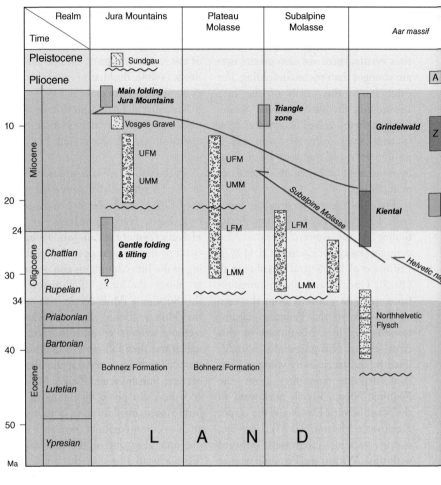

European margin (Briançon Rise, Valais Trough and Helvetic realm).

At 45 million years ago the Briançon Rise had entered the subduction zone. Its cover sediments, the Klippen nappe were sheared off and remained stuck near the surface while the crystalline basement was subducted. From the Eocene onwards, after the Niesen Flysch had been deposited, the Mesozoic sediments of the Valais Trough were also sheared off their substratum and stacked up in a nappe pile, and once again the (thinned) crust of the Valais Trough was subducted. This process is attributed to the Evolène Phase. The cover nappes involved in this are the Gurnigel nappe (which originated in the Piemont Ocean), the Klippen and Breccia nappes (Briançon Rise) and the Niesen nappe

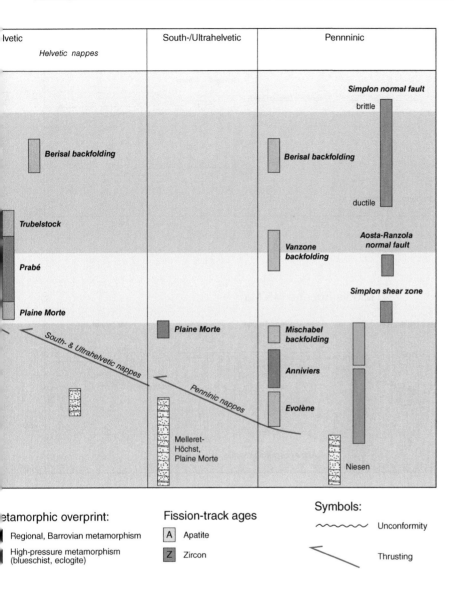

lvetic	South-/Ultrahelvetic	Pennninic

Helvetic nappes

Simplon normal fault

brittle

Berisal backfolding

Berisal backfolding

ductile

Trubelstock

Aosta-Ranzola
normal fault

Vanzone
backfolding

Prabé

Simplon shear zone

Plaine Morte

Plaine Morte

Mischabel
backfolding

South- & Ultrahelvetic nappes

Anniviers

Penninic nappes

Evolène

Melleret-
Höchst,
Plaine Morte

Niesen

Symbols:

etamorphic overprint:	Fission-track ages	
Regional, Barrovian metamorphism	A Apatite	∿∿∿∿ Unconformity
High-pressure metamorphism (blueschist, eclogite)	Z Zircon	⟵ Thrusting

(Valais Trough), that were all stacked on top of each other by frontal accretion such that they are now all exposed in the northern edge of the Alps. As a rule in Alpine nappe stacks, higher units derive from palaeogeographically more internal parts. However, irregularities occurred in the sequence of nappe stacking in this transect. So-called 'out-of-sequence thrusting' led to the Gurnigel nappe

(Piemont Ocean) ending up in a structural position underneath the Klippen nappe (Briançon Rise). The Tsaté nappe (Piemont Ocean) and the 'Bündnerschiefer' Group (Valais Trough) were subducted and metamorphosed at depth in the internal region of the Alps. The blueschist-facies overprint close to the base of the Penninic sedimentary nappes (see Fig. 6.1) in the internal

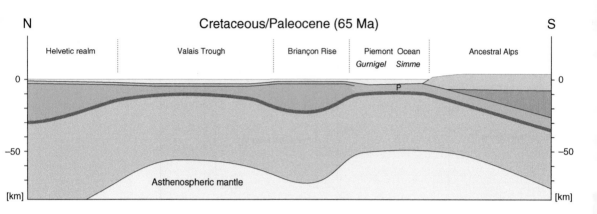

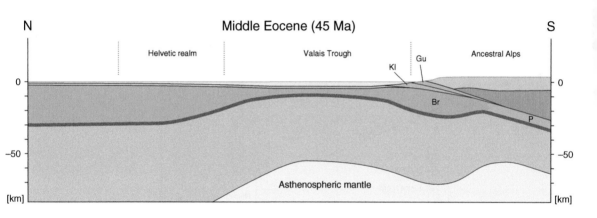

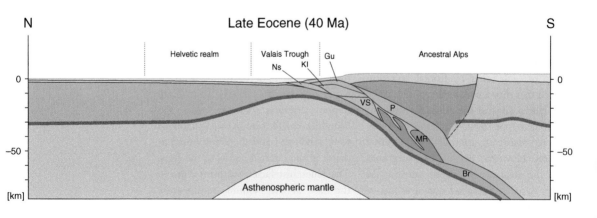

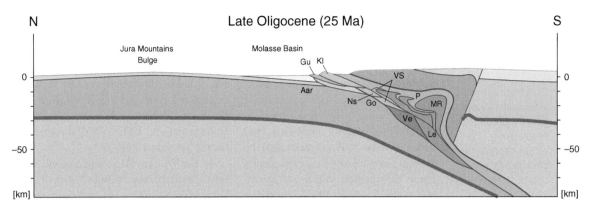

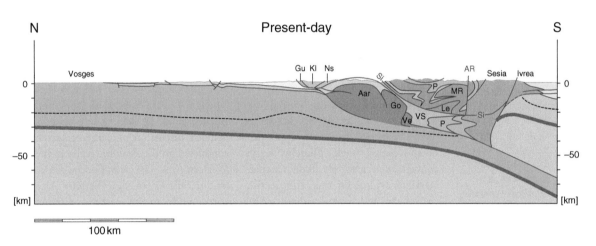

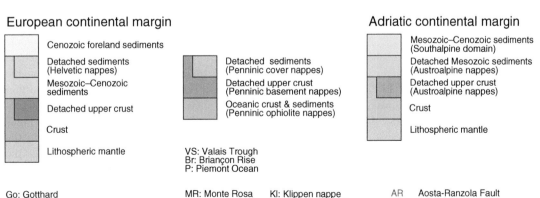

European continental margin

- Cenozoic foreland sediments
- Detached sediments (Helvetic nappes)
- Mesozoic–Cenozoic sediments
- Detached upper crust
- Crust
- Lithospheric mantle

- Detached sediments (Penninic cover nappes)
- Detached upper crust (Penninic basement nappes)
- Oceanic crust & sediments (Penninic ophiolite nappes)

Adriatic continental margin

- Mesozoic–Cenozoic sediments (Southalpine domain)
- Detached Mesozoic sediments (Austroalpine nappes)
- Detached upper crust (Austroalpine nappes)
- Crust
- Lithospheric mantle

VS: Valais Trough
Br: Briançon Rise
P: Piemont Ocean

Go: Gotthard
Ve: Verampio
Le: Lebendun

MR: Monte Rosa

Kl: Klippen nappe
Gu: Gurnigel nappe
Ns: Niesen nappe

AR   Aosta-Ranzola Fault
Si   Simplon Fault

◀ **Figure 6.17**
Kinematic evolution of the Central Alps in western Switzerland. Bernhard nappe complex after Scheiber et al. (2013), Prealps of western Switzerland after Wissing & Pfiffner (2002). Growth of the orogen was accomplished by successive basal accretion of upper crustal blocks (e.g. Verampio, Gotthard, Aar). The sedimentary cover of the upper crustal units was stripped off in an earlier phase and piled up at the front of the orogen by frontal accretion. Out-of-sequence thrusting during frontal accretion was responsible for the Southpenninic, Piemont derived Gurnigel nappe now being beneath the Middle Penninic, Briançon derived Klippen nappe. In the core of the orogen, the Penninic basement nappes (Monte Rosa, Bernhard nappe complex) were intensively squashed, stacked and back-sheared. Combined with retro-thrusting along a steeply dipping south-vergent thrust fault and aided by erosional removal of the overlying Austroalpine nappes, these units were exhumed. Source: Adapted from Scheiber et al. in press and Wissing & Pfiffner (2002).

region of the Alps indicates relatively high pressures for the Mesozoic sediments, which resulted most probably from the combination of lithostatic overburden and tectonic overpressure.

The crystalline basement, namely the upper crust of the Briançon Rise, was compressed only somewhat later (at around 40 million years ago) by thrusting and folding at greater depth. These deformation events are attributed to the Anniviers Phase. The internal structure of the Bernhard nappe complex and the Monte Rosa nappe (see Fig. 5.2-17) are good examples exhibiting the ductile behaviour under elevated temperatures.

By 25 million years ago, the upper crustal slabs of the Lebendun and Verampio nappes were plucked off their lower crust and incorporated in the nappe stack, a process referred to as basal accretion, which was accompanied by back-folding in the upper part of the nappe stack. Thereby, nappe contacts were passively folded, a characteristic typical for this cross-section (Mischabel and Vanzone phases). This late folding (also called 'post-nappe folding') gives the false impression of ubiquitous ductile deformation, however, it can be assumed that the earliest deformation associated with thrusting processes was more brittle in nature, which occurred before the rocks had been heated to temperatures enabling a ductile behaviour.

In the course of the Miocene (after 20 million years ago), detached upper crust of the European margin led to the formation of the Gotthard and Aar massifs. These basement uplifts formed a dome whereby on the southside nappe contacts were steepened and overturned (Berisal Phase). A further important point is the activity of the Aosta–Ranzola and the Simplon normal faults.

Both post-date the nappe formation of the Evolène and Anniviers phases, as is evident from the cross-cutting relationship in cross-sectional view (see Fig. 5.2-17). Both fault systems were extensional, oriented north–south in the case of the Aosta–Ranzola normal fault system, and ESE–WNW directed for the Simplon Fault.

In the Western Alps, orogenic evolution started in the Eocene (Fig. 6.18), but an earlier phase of deformation took place in the Provence. This is linked to the formation of the Pyrenees and produced east–west-trending folds in the region of Dévoluy, which were then overprinted by the subsequent north–south-oriented Alpine structures.

The orogenic evolution in the Penninic realm illustrated in Fig. 6.18 is based on Ceriani & Schmid (2004) and Bucher et al. (2004), who provide a detailed discussion of earlier research conducted on the Penninic realm in the Western Alps. The early, Eocene deformation history of the Western Alps corresponds to an easterly dipping subduction, implicating the Piemont Ocean, the microcontinent of the Briançon Rise and the Valais Trough. A clear indication for subduction is provided by the high-pressure metamorphism that produced eclogites in the internal portion of the Penninic nappe system and led to the blueschist-facies overprint at the base of the Penninic nappes, in the external portion (see Fig. 6.1). Berger & Bousquet (2008) determined the age of the high-pressure metamorphism linked to the subduction, which occurred in the internal portions of the Penninic nappes during the interval between 50 and 35 million years ago (Eocene): a slightly greater age (60–50 million years ago, i.e., Paleocene to Early Eocene) has been assumed for the Penninic nappe system to the south of the Western Alps (including Dora Maira and Monviso). Even older ages are indicated in the Austroalpine Sesia Zone where Rubatto et al. (2011) report two eclogitic overprints in the time interval from 79 to 64 million years ago, which were separated by a low-pressure metamorphic overprint: they report pressures of up to 2.2 gigapascals for the high-pressure event and argue that they correspond to mantle depths of roughly 60 kilometres. Even if significant tectonic overpressure caused by the eo-alpine collision is taken into account, an assumption that the Sesia Zone had been deeply buried in the Late Cretaceous seems inevitable.

Bucher et al. (2004) propose a first phase of isoclinal folding in the Middle (Briançon Rise) and Upper (Piemont Ocean) Penninic nappes that is linked to subduction and high-pressure metamorphism, and a second phase of isoclinal folding that is associated with exhumation of the same units. During these phases, the direction of thrusting was oriented to the northwest to northnorthwest and the formation of the nappe pile can be seen as a growing accretionary prism in the subduction zone. During a third phase, the nappes, including their basal thrusts, were folded into large-scale folds and in a fourth and finalphase, the nappe pile was subjected to shallow, dome-like uplift. Deformation in the adjacent Lower Penninic nappes (Ceriani & Schmid 2004) started slightly later, but also with a first phase of isoclinal folding and overthrusting in a more northerly direction. In a subsequent phase, the Lower Penninic nappes were thrust in a west-northwest direction onto the Ultradauphinois realm. The contemporaneous deformation events in the

Penninic nappe system in the Central Alps included north-directed overthrusts, in contrast to the Western Alps. Ceriani & Schmid (2004) proposed decoupling along the, also contemporaneous, Peri-Adriatic fault system and the Simplon Fault to explain this circumstance: the southern or upper block in both fault systems was shifted towards the west.

Orogenic movements in the Dauphinois realm began after cessation of flysch sedimentation in the internal region (Villabella) in the Priabonian and in the external region (Barrème) in the Chattian. Analogous to the Eastern Alps (Tauern massif) and the Central Alps (Aar massif), the Dauphinois sedimentary cover was thrust over the top of the future external massifs (Argentera and Belledonne massifs). Due to subsequent shortening of the crust, the crystalline basement bulged upwards to form the external massifs and folded the overlying nappes in the process.

The uplift and exhumation of the external massifs can be dated using fission-track cooling ages. According to the compilation by Fügenschuh & Schmid (2003), cooling progressed systematically from the internal regions towards the west, that is, to the external regions. Cooling started 32 million years ago in the internal region and approximately 10 million years ago in the external region (see Fig. 6.18). In contrast to the situation in the Eastern Alps and the Central Alps, there are no clear indications of extension after nappe stacking in the Western Alps, although the Subbriançonnais Fault at the western end of the Middle Penninic nappes provides an exception to this. The fission-track data show that this normal fault has been active during the

Orogenic timetable: Western Alps

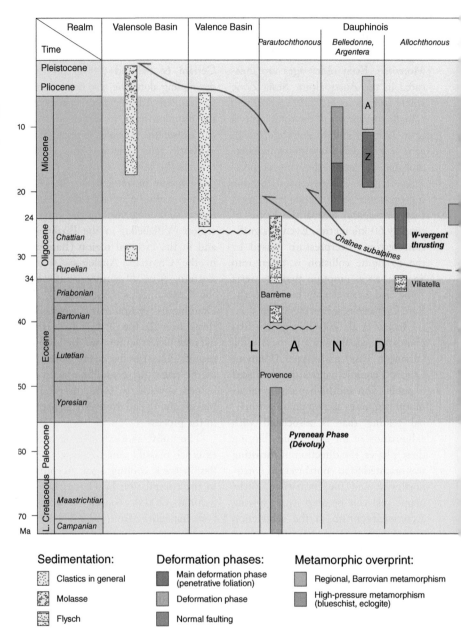

Figure 6.18 Orogenic timetable of nappe formation, sedimentation, metamorphism and magmatism in the Western Alps. Source: Adapted from Ceriani & Schmid (2004) and Bucher et al. (2004). Reproduced with permission of Swiss Topo.

past 5 million years, although this activity may well continue to this day (Fügenschuh & Schmid 2003; Ceriani & Schmid 2004).

In the extreme south, the parautochthonous units of the Subalpine chains in the Vercors area have been thrust onto the Late Miocene sediments of the Valence Basin (see Fig. 5.1-6) and the Digne thrust resulted in the Mesozoic sediments being placed on top of the Pliocene fill

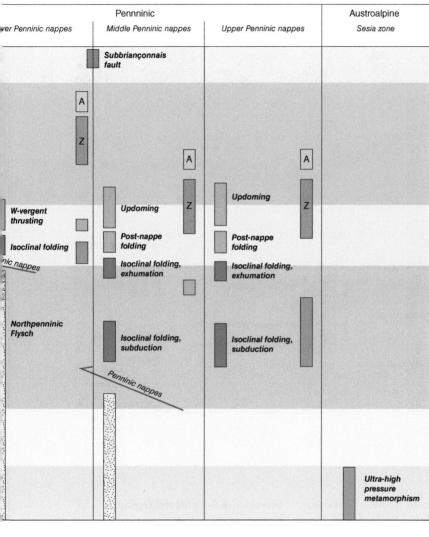

| | Pennninic | | Austroalpine |
| Lower Penninic nappes | Middle Penninic nappes | Upper Penninic nappes | Sesia zone |

Fission-track ages:

A Apatite
Z Zircon

Symbols:

Thrusting
Unconformity

of the Valensole Basin (see Fig. 5.1-7). The Digne thrust bulges upwards over the Argentera massif but the Alpine metamorphic overprint and heating were insufficient to allow successful fission-track dating of the uplift.

Ceriani & Schmid (2004) detected folded folds in the Lower and Middle Penninic cover nappes. These formed when the sediments had been scraped off their crystalline substrate and were telescoped together into a nappe pile

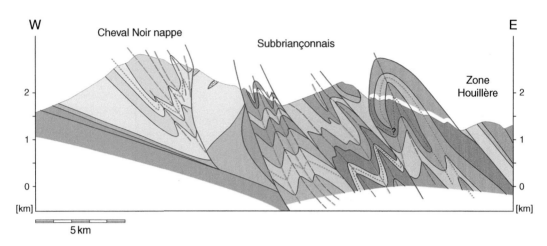

Penninic nappe system

Dauphinois nappe system

Middle Penninic nappes/Briançon Rise

	Lower Malm — Late Jurassic
	Upper Dogger
	Lower Dogger — Middle Jurassic
	Upper Lias
	Middle Lias — Early Jurassic
	Lower Lias
	Late Triassic
	Early Triassic
	Permian
	Carboniferous

Lower Penninic nappes/Valais Trough

Late Eocene

	Flysch with olistoliths
	Breccia
	«Wildflysch» with olistoliths
	Permian – Late Cretaceous

	Mesozoic

Axial surfaces:

2nd phase
- – – – Antiform
- – – – Synform

1st phase
- ·········· Anticline
- ·········· Syncline

▲ **Figure 6.19** Folded folds in the Penninic nappes of the Western Alps. Isoclinal folds of the first phase can be recognized from the symmetric pattern of the ages of strata with either young or old strata in the fold cores. These folds were refolded by a second west-vergent shearing. Source: Adapted from Ceriani & Schmid (2004) and Bucher et al. (2004). Reproduced with permission of Swiss Topo.

during the accretionary process. Figure 6.19, adapted from Ceriani & Schmid (2004), shows the complex geometry that was created through polyphase folding of the sediments. The fact that early folding in the Middle Penninic nappes is linked to a transpressive movement must be taken into consideration. Several sinistral strike-slip faults, where each eastern block is offset to the north, cross-cut the folds as they formed. Only the Lower Penninic Cheval Noir nappe exhibits a west-vergent thrust fault at its base, with the most recent folding shortening the older folds within the nappe in an east–west direction.

6.4 Uplift and Erosion

Orogenesis does not simply refer to the production of topographic relief, but also to the formation of the internal structure of a mountain chain. This structure is characterized by folds, nappes and thrusts, as well as by normal faults and strike-slip faults. The processes that produce these structures are the result of the convergent movements between tectonic plates. When considered in this way, the production of relief is essentially caused by deformation and associated movement patterns within the mountain chain. Parts of the Earth's crust are uplifted through folding and thrusting, which then also raises the level of the land surface. The orogen that thereby develops is, however, immediately exposed to surface processes: weathering disintergrates the rocks near the land surface; water, ice and gravity then transport this loose material away. Erosion therefore counteracts uplift, but numerical models reveal that erosion can also result in local acceleration of uplift, by facilitating the upwards escape of deeply buried rocks through erosion at the surface (Pfiffner et al. 2000). Furthermore, it must be noted that the eroded material is not fully removed from the mountain system, but rather is stored on a temporary basis, and the foreland basins that are linked to mountain chains represent one such temporary storage location. Foreland basins form in depressions that result from the plate being pushed downwards in the foreland by the weight of the mountain chain – the orogen, as it were, creates the catch basin for the erosional debris.

It is not possible to adequately illustrate the highly complex interplay between surface processes (erosion and temporary storage) and processes taking place at depth (folding, overthrusting, etc.) using only two-dimensional cross-sections and orogenic timetables. Palaeogeographical maps are, however, a useful tool for gaining an insight into the creation of the Alpine orogen and understanding the spatial and temporal interactions between mountain-building, uplift and erosion, sedimentation in the foreland basins, etc. Based on this principle: Fig. 4.10 illustrates the Early Eocene (55 million years ago) palaeogeographical conditions during the sedimentation of the Northpenninic Flysch together with the transpressive movements in the future Western Alps; Fig. 4.11 shows the same thing for the

Northhelvetic Flysch in the Early Oligocene (35 million years ago), and, in particular, the juxtaposition of sedimentation in the foreland, volcanism, bivergent thrusts and uplift in the developing Central Alps; finally, Fig. 4.12 illustrated the early stages of the Molasse Basin in the Late Oligocene, 25 million years ago. Large alluvial fans were built up by the ancestral rivers flowing from the growing mountain chain in the north of the Central and Eastern Alps. Submarine alluvial fans developed in the Po Basin to the south of the Alps. Bivergent thrusts shape the northern and southern edges of the Alps, while back-folding started within the orogen itself, and passively folded the already existing nappe structures. We can assume that the Alps were a formidable mountain chain at this time and that they exhibited substantial local relief, given the thick, coarse clastic alluvial fans in the north and south. Interestingly, the first deposits of conglomerates started earlier in the eastern alluvial fans than in the western ones (32 and 25 million years ago, respectively). Pfiffner et al. (2002) propose that the orogen developed into a high-altitude mountain range earlier in the eastern Central Alps than in the western Central Alps, which can be explained by dextral strike-slip along the Peri-Adriatic fault system. First, the Ivrea Body migrated in a western direction relative to the mountain chain due to this strike-slip motion. Second, during convergence of the two plates, this heavy and rigid body acted like a kind of buffer that deformed the European crust, which lay to its north, particularly strongly during the collision and thus forced uplift to occur. This is a good example of an interaction between surface processes and endogenous events.

Figure 6.20 shows the situation 22 million years ago, towards the end of the

The Alps in Early Eocene times

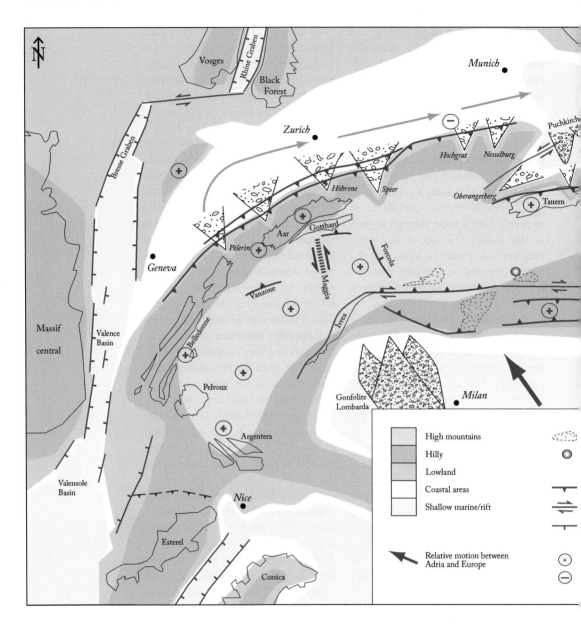

Oligocene, that is, slightly later than in Fig. 4.12. At the end of the Oligocene, rapid uplift was underway within the orogen, as is indicated, for example, by the cooling ages yielded by zircon fission-track data. These uplifts were caused, not least, by the uplift of the domes of the external massifs (Tauern and Aar massifs) and the stacking up of the basement nappes in the Penninic nappe system. In the region of the Lepontine dome (to the east of the Maggia transverse zone), in particular, these uplifts caused extension parallel to the orogen, as indicated by the Forcola normal fault. This extension could be interpreted as an orogenic collapse: local uplift produced steep gradients in the top level of the mountain

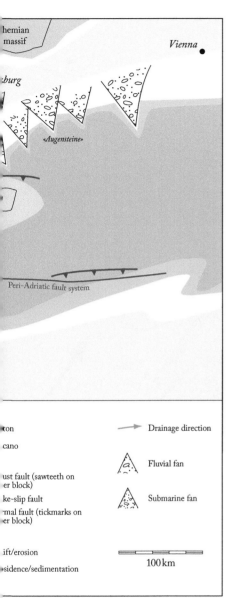

◀ **Figure 6.20**
Palaeogeographical map of
the Alps towards the end of
the Oligocene (22 Ma). Source:
Adapted from Frisch et al.
(1999), Frisch et al. (2000a)
and Frisch et al. (2000b).

barely changed at all if we make comparisons with the situation 25 million years ago (Fig. 4.12). The large ancestral rivers drained an area in which substantial uplift occurred in their lower reaches, but despite this uplift the rivers did not deviate from their courses, and instead cut down further into the substratum more or less in the same locations. It is possible that the persistent uplift preserved steep channel gradients and raised or maintained the rivers' erosion potential. The northwest-directed motion of the Adriatic plate led to progressively more external areas being involved in thrust tectonics in the Western Alps, which resulted in the Western Alps becoming higher and broader. In the south, Corsica started to separate from the European margin (Esterel).

At this time, the Molasse Basin was still draining in an easterly direction, into the Para-Tethys Basin, which extended across Vienna and into the Salzburg region. Thrust faults developed at the southern margin of the Molasse Basin, which moved thrust sheets of Subalpine Molasse northwards. In the western Molasse Basin, compressive stress was transferred as far as the Jura Mountains, causing a first phase of folding (see Fig. 6.15).

An enormous change then occurred in the Early Miocene. Figure 6.21 shows the palaeogeographical situation 15 million years ago (in the Langhian or Badenian) at the time of the deposition of the Upper Freshwater Molasse, with the palaeogeography of the Eastern Alps being based on Frisch et al. (1999, 2000b). Due to continuing convergence, the Alps had become narrower, but higher. Intensive uplift was underway in a variety of areas (external massifs, Southern Alps–Dolomites, southern Karawanks). In the case of the Tauern massif, normal faults were active in the

The Alps in Miocene times

chain, which triggered the onset of gravitational spreading. At this time, the mountain chain was growing towards the south, which is evident from the new thrust faults that developed in the Southalpine nappe system (to the south of the Adamello intrusion), in the Dolomites and in the Karawanks. The large alluvial fans in the Molasse Basin

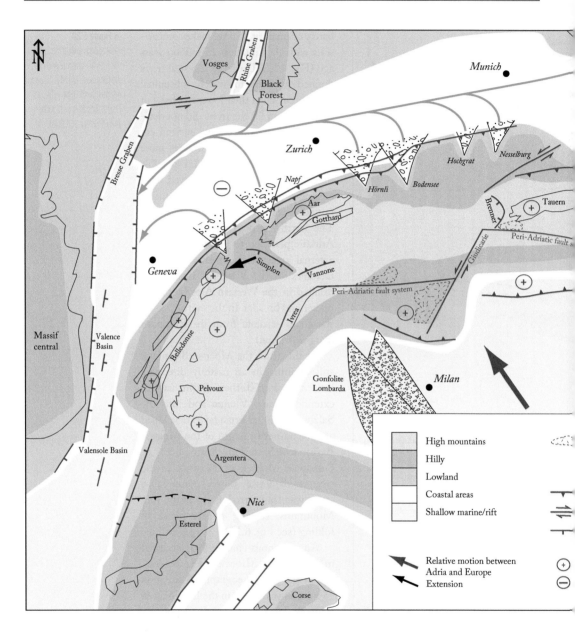

east and the west (Brenner and Katschberg normal faults) and the Tauern massif was subjected to horizontal shortening and squeezing out. The shedding of the 'Augensteine' conglomerates ceased and gave way to new new alluvial fans in the eastern (Hausruck, Wachtberg) and western Molasse Basin (Bodensee, Hörnli). However, the greatest change is the fact that drainage now occurred in a westerly direction, into the Rhone system. Associated with this was significant subsidence in the westernmost Molasse Basin at this time, which is clearly reflected in the thickness of the sediments assigned to this period. Pfiffner et al. (2002) analysed the situation and reached the conclusion that

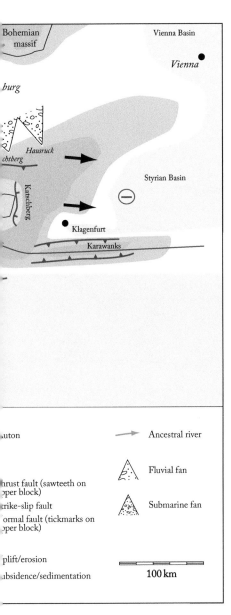

Bohemian massif

Vienna Basin

Vienna

_burg

Hausruck
_chtberg

Styrian Basin

Katschberg

Klagenfurt

Karawanks

_uton — Ancestral river

Fluvial fan

Submarine fan

_hrust fault (sawteeth on _pper block)
_trike-slip fault
_ormal fault (tickmarks on _pper block)

_plift/erosion
_ubsidence/sedimentation

100 km

◄ **Figure 6.21**
Palaeogeographical map of the Alps in the Middle Miocene (15 Ma). Source: Adapted from Frisch et al. (1999), Frisch et al. (2000a) and Frisch et al. (2000b).

wedged into the European crust has a greater density than the normal crustal rocks) caused greater down-flexing of the European plate and thus made the foreland basin deeper. The west-directed drainage is also revealed in a channel filled with sandstones containing micas (the so-called 'Glimmersandrinne') that can be followed from the Bohemian massif towards the west, and into the Jura Mountains. In the extreme east, the intramontane Styrian Basin and, an appendage of this, the Klagenfurt Basin, subsided at this time. In the Miocene, the entire Eastern Alps were extended, parallel to the Alpine chain, by 150 kilometres (Ratschbacher & Frisch 1993) to 300 kilometres (Frisch et al. 2000a). This extension is interpreted as the lateral escape of parts of the Eastern Alps during the north–south collision. The east-directed extension continued in the adjacent Pannonian Basin.

Erosion and uplift are clearly discerned by the mapped pattern of zircon and apatite fission-track ages. Zircon ages record the age a rock sample cooled down below ca. 230 °C, whereas apatite ages record cooling below ca. 110 °C. These cooling ages correspond to exhumation, that is, simultaneous rock uplift and erosion. Figure 6.22 shows the regional occurrence of fission-track ages that are less than 20 million years ago and less than 8 million years ago. The data are from Vernon et al. (2008) for the Western and Central Alps and from Luth & Willingsdhofer (2008) for the Eastern Alps. Outside the areas highlighted in Fig. 6.22, fission-track ages are older than 20 million years ago or the rocks have never attained the closure temperatures in the course of the Alpine orogeny. Zircon ages younger than 20 million years ago cover the area of the Tauern Window in the Eastern Alps. In the Western and Central Alps

Uplift and erosion from cooling ages

this increase in subsidence was caused by the additional load placed on the European lithosphere by the Ivrea Body, which attained a more westerly location due to the dextral strike-slip motion at the Peri-Adriatic fault system (more specifically, at the segment of the Insubric Fault). The associated additional intracrustal load (the Ivrea Body

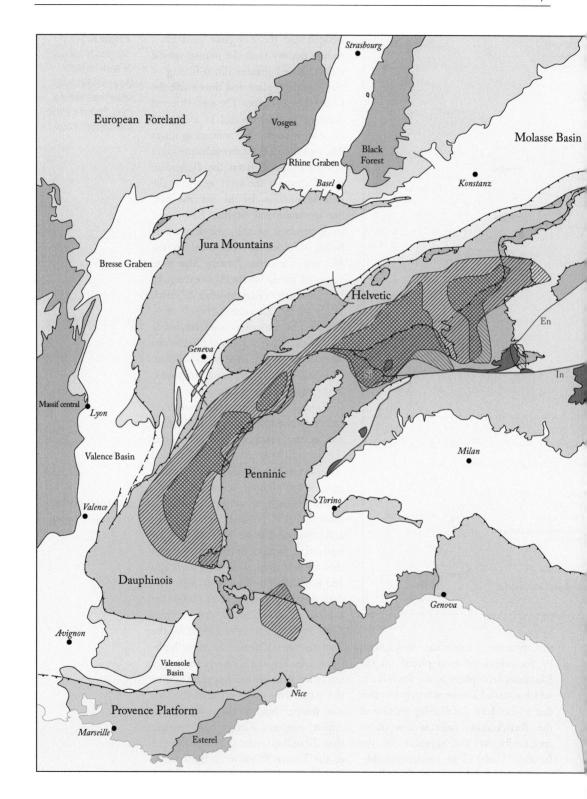

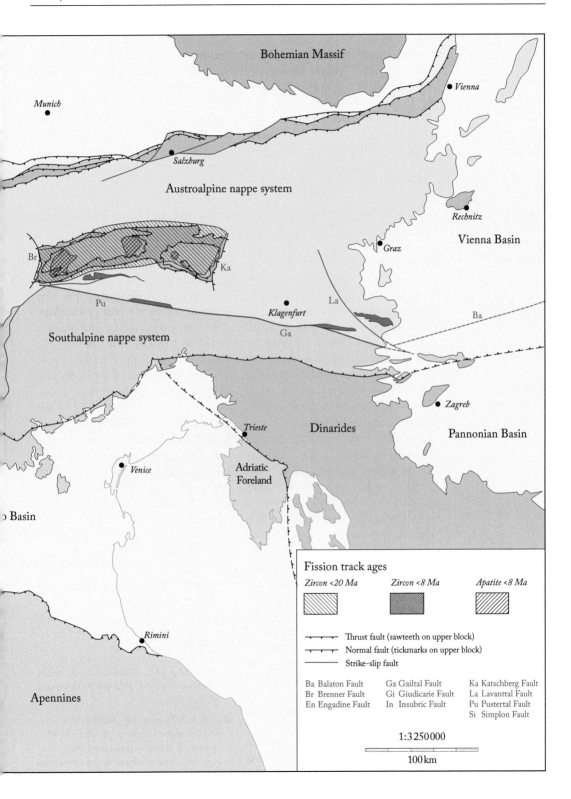

◀ **Figure 6.22** Map
showing the late Neogene
cooling history of the Alps.
Zircon fission-track ages
indicate the time when the
rock sample cooled below
about 230 °C. For Apatite this
temperature corresponds to
about 110 °C. Data in the
Western and Central Alps are
from Vernon et al. (2008)
and data in the Eastern Alps
from Luth & Willingshofer
(2008). Source: Data from
Vernon et al. (2008) and Luth
& Willingshofer (2008).

Exhumation rates:
0.4–1.5 mm/year

they correspond to the regions of the external massifs (Aar, Gotthard, Mont Blanc, Aiguilles Rouges, Belledonne and Pelvoux), and in the Central Alps the Penninic nappes between these massifs and the Insubric Fault. Zircon fission-track ages younger than 8 million years ago are restricted to the core of the Toce dome located in the Penninic nappes east of the Simplon fault. For apatite, Fig. 6.22 displays only fission-track ages younger than 8 million years ago. They spread over much the same areas as zircon ages younger than 20 million years ago. In the Tauern Window of the Eastern Alps they mark three smaller domains, in the Western and Central Alps they cover a somewhat larger area, including also the Argentera massif (north of Nice). Generally speaking this means that the external massifs (including Tauern Window) and the Penninic nappes north of the Insubric Fault experienced a late phase of exhumation in the Miocene and Pliocene.

In many instances, rock samples cooled down below ca. 230 °C at less than 20 million years ago and to below ca. 110 °C at less than 8 million years ago, which corresponds to an average cooling by ca. 120 °C within 12 million years or, for an average geothermal gradient of 25 °C per kilometre, an erosional removal of a ca. 5-kilometre-thick pile – an equivalent erosion rate of around 0.4 millimetres per year. Considering the zircon fission-track ages of less than 8 million years ago in the Penninic nappes east of the Simplon Fault, the time span of cooling is only a few million years and thus the local erosion rate must be much higher. Regarding the apatite fission-track ages, an average regional erosion rate of less than 0.55 millimetres per year is obtained for cooling of 110 °C in less than 8 million years. Locally, an even

higher erosion rate of more than 1 millimetres per year is obtained for those areas where apatite fission-track ages are as young as 3 million years ago. Interestingly these high erosion rates coincide with areas that show exceptionally high uplift rates of 1.5 millimetres per year, as determined from geodetic levelling (discussed in Chapter 7). Examples of such local 'hot spots' are the eastern and western end of the Aar massif and the eastern margin of the Tauern Window.

Summarizing it seems that uplift and erosion since Miocene times followed a pattern of continued exhumation at ca. 0.4 millimetres per year concentrated in the core of the Alps. This core covers the external massifs, including Tauern window in the Eastern Alps and the Penninic nappes north of the Insubric Fault in the case of the Central Alps. In the area outside, exhumation started earlier and/or the rocks at the surface now were never buried deep enough to reach the closure temperatures for resetting fission-track ages. However, in the core of the Alps, much higher exhumation rates affected small areas over short time spans. The formation of the Alpine mountain range must be seen as a regional uplift interspersed by local bursts of very rapid uplift, a pattern that is still going on, as will be discussed in Chapter 7.

References

Aubert, D., 1975, L'évolution du relief jurassien. Eclogae geologicae Helvetiae, 68/1, 1–64.

Berger, A. & Bousquet, R., 2008, Subduction-related metamorphism in the Alps: a review of isotopic ages based on petrology and their geodynamic consequences. In: Siegesmund, S., Fügenschuh, B. &

Froitzheim, N. (eds), Tectonic Aspects of the Alpine-Dinaride-Carpathian System. Geological Society London, Special Publications, 298, 117–144.

Bolliger, Th., Engesser, B. & Weidmann, M., 1993, Première découverte de mammifères pliocènes dans le Jura neuchâtelois. Eclogae geologicae Helvetiae, 86/3, 1031–1068.

Brouwer, F. M., Burri, T., Engi, M. & Berger, A., 2005, Eclogite relics in the Central Alps: PT-evolution, Lu-Hf ages and implications for formation of tectonic mélange zones. Schweizerische Mineralogische und Petrographische Mitteilungen, 85, 147–174.

Bucher, S., Ulardic, C, Bousquet, R., Ceriani, S., Fügenschuh, B., Gouffon, Y. & Schmid, S. M., 2004, Tectonic evolution of the Briançonnais units along a transect (ECORS-CROP) through the Italian–French Western Alps. Eclogae geologicae Helvetiae, 97/3, 321–345.

Burkhard, M., 1988, L'Helvétique de la bordure occidentale du massif de l'Aar (évolution tectonique et métamorphique). Eclogae geologicae Helvetiae, 81/1, 63–114.

Castellarin, A., Nicolich, R., Fantoni, R., Cantelli, L., Sella, M. & Selli, L., 2006, Structure of the lithosphere beneath the Eastern Alps (southern sector of the TRANSALP transect). Tectonophysics, 414, 259–282.

Castelli, D., Rolfo, F., Groppo, C. & Compagnoni, R., 2007, Impure marbles from the UHP Brosasco-Isasca Unit (Dora Maira Massif, western Alps): Evidence for Alpine equilibration in the diamond stability field and evaluation of the $X(CO_2)$ fluid evolution. Journal of Metamorphic Geology, 25/6, 587–603.

Ceriani, S. & Schmid, S. M., 2004, From N-S collision to WNW-directed post-collisional thrusting and folding: Structural study of the Frontal Penninic Units in Savoie (Western Alps, France). Eclogae geologicae Helvetiae, 97/3, 347–369.

Chopin, C., 1984, Coesite and pure pyrope in high-grade blueschists of the Western Alps: a first record and some consequences. Contributions to Mineralogy and Petrology, 86, 107–118.

Colombo, A. & Tunesi, A., 1999, Alpine metamorphism of the Southern Alps west of the Giudicarie line. Schweizerische Mineralogische und Petrographische Mitteilungen, 79, 183–189.

Desmons, J., Aprahamian, J., Compagnoni, R., Cortesogno, L. & Frey, M., with collaboration of Gaggero, L., Dallagiovanna, G. & Seno, S., 1999a, Alpine metamorphism of the Western Alps: I Middle to high T/P metamorphism. Schweizerische Mineralogische und Petrographische Mitteilungen, 79, 89–110.

Desmons, J., Compagnoni, R. & Cortesogno, L., with collaboration of Frey, M., Gaggero, L., Dallagiovanna, G., Seno, S. & Radelli L., 1999b, Alpine metamorphism of the Western Alps: II. High-P/T and related pre-greenschist metamorphism. Schweizerische Mineralogische und Petrographische Mitteilungen, 79, 111–134.

Engi, M., 2011, Structure et evolution métamorphique des Alpes centrales. Géochronique, 117, 22–26.

Engi, M., Todd, C. S. & Schmatz, D. R., 1995, Tertiary metamorphic conditions in the eastern Lepontine Alps. Schweizerische Mineralogische und Petrographische Mitteilungen, 75, 347–369.

Engi, M., Berger, A. & Roselle, G. T., 2001, Role of the tectonic accretion channel in collisional orogeny. Geology, 29/12, 1143–1146.

Engi, M., Bousquet, R. & Berger, A., 2004, Explanatory notes to the map: Metamorphic structure of the Alps, Cental Alps. Mitteilungen der Osterreichischen Mineralogische Gesellschaft, 149, 157–173.

Escher, A., Hunziker, J. C., Marthaler, M., Masson, H., Sartori, M. & Steck A., 1997, Geologic framework and structural evolution of the Western Swiss-Italian Alps. In: Pfiffner, O. A., Lehner, P., Heitzmann, P.,

Mueller, S. & Steck, A. (eds), Deep Structure of the Swiss Alps: Results of NRP 20. Birkäuser Verlag Basel, 205–222.

Frey, M. & Ferreiro Mählmann, R., 1999, Alpine metamorphism of the Central Alps. Schweizerische Mineralogische und Petrographische Mitteilungen, 79, 135–154.

Frisch, W., Brügel, A., Dunkl, I., Kuhlemann, J. & Satir, M., 1999, Post-collisional large-scale extension and mountain uplift in the Eastern Alps. Memorie di scienze geologiche, 51/1, 3–23

Frisch, W., Dunkl, I. & Kuhlemann, J., 2000a, Post-collisional orogen-parallel large-scale extension in the Eastern Alps. Tectonophysics, 327, 239–265.

Frisch, W., Székely, B., Kuhlemann, J. & Dunkl. I., 2000b, Geomorphological evolution of the Eastern Alps in response to Miocene tectonics. Zeitschrift für Geomorphologie, Neue Folge 44/1, S. 103–138.

Froitzheim, N., Schmid, S. M. & Conti, P., 1994, Repeated change from crustal shortening to orogen-parallel extension in the Austroalpine units of Graubünden. Eclogae geologicae Helvetiae, 87/2, 559–612.

Froitzheim, N., Conti, P. & van Daalen, M., 1997, Late Cretaceous, synorogenic, low-angle normal faulting along the Schlinig fault (Switzerland, Italy, Austria) and its significance for the tectonics of the Eastern Alps. Tectonophysics, 280, 267–293.

Fügenschuh, B. & Schmid, S. M., 2003, Late stages of deformation and exhumation of an orogen constrained by fisstion-track data: A case study in the Western Alps. Geological Society of America Bulletin, 115/11, 1425–1440.

Giamboni, M., Ustaszewski, K., Schmid, S. M., Schumacher, M. E. & Wetzel, A., 2004, Plio-Pleistocene transpressional reactivation of Paleozoic and Paleogene structures in the Rhine–Bresse transform zone (northern Switzerland and eastern France). International Journal of Earth Sciences, 93, 207–223.

Goffé, B., 1984–86, Le faciès à carpholite-chloritoïde dans la couverture briançonnaise des Alpes ligures: un témoin de l'historie tectono-métamorphique régionale. Memorie della Societa Geologica Italia, 28, 461–479.

Herwegh, M. & Pfiffner, O. A., 2005, Tectono-metamorphic evolution of a nappe stack: A case study of the Swiss Alps. Tectonophysics, 404, 55–76.

Höpfer, N. & Vogler, W.S., 1994, Alpiner Druck-Temperatur-Deformationspfad in der südlichen Dent Blanche-Decke. Bericht der DMG, European Journal of Mineralogy, 6/1, 117.

Hoinkes, G., Koller, F., Rantitsch, G., Dachs, E., Höck, V., Neubauer, F. & Schuster, R., 1999, Alpine Metamorphism of the Eastern Alps. Schweizerische Mineralogische und Petrographische Mitteilungen, 79, 155–181.

Käch, P., 1969, Zur Tektonik der Brigelserhörner. Eclogae geologicae Helvetiae, 62/1, 173–183.

Kälin, D., 1997, Litho- und Biostratigraphie der mittel- bis obermiozänen Bois de Raube-Formation (Nordwestschweiz). Eclogae geologicae Helvetiae, 90/1, 97–114.

Kurz, W., Handler, R. & Bertoldi, C., 2008, Tracing the exhumation of the Eclogite Zone (Tauern Window, Eastern Alps) by $^{40}Ar/^{39}Ar$ dating of white mica in eclogites. Swiss Journal of Geosciences, Supplement 1, 101, 191–206.

Luth, S.W. & Willingshofer, E., 2008, Mapping of the post-collisional cooling history of the Eastern Alps. Swiss Journal of Geosciences, Supplement 1, 101, 207–223.

Mancktelow, N., 2007, Tectonic pressure: Theoretical concepts and modelled examples. Lithos, 103, 149–177, doi: 10.1016/ j.lithos.2007.09.013.

Marquer, D. Baudin, T., Peucat, J.-J. & Persoz, F., 1994, Rb-Sr mica ages in the Alpine shear zones of the Truzzo granite: Timing of the Tertiary alpine P-T-deformations in the Tambo nappe (Cental Alps, Switzerland). Eclogae geologicae Helvetiae, 87, 225–239.

Menkveld, J.-W., 1995, Der geologische Bau des Helvetikums der Innerschweiz. Unpublished Dissertation Universität Bern, 165 pp and Profile.

Milnes, A. G. & Pfiffner, O. A., 1977, Structural evolution of the Infrahelvetic complex, eastern Switzerland. Eclogae geologicae Helvetiae, 70/1, 83–95.

Oberhänsli, R., Bousquet, R., Engi, M., Goffé, B., Gosso, G., Handy, M., Höck, V., Koller, F., Lardeaux, J.-M., Polino, R., Rossi, Ph., Schuster, R., Schwartz, S. & Spalla, I., 2004, Metamorphic structure of the Alps (Karte 1:1 000 000). Commission for the Geological Map of the World, SGMW, Paris.

Ortner, H., 2001, Cretaceous thrusting in the western part of the Northeren Calcareous Alps (Austria) evidences from synorogenic sedimentation and structural data. Mitteilungen der Österreichischen Geologischen Gesellschaft, 94, 66–77.

Petrini, K., Podladchikov, Y., 2000, Lithospheric pressure-depth relationships in compressive regions of thickened crust. Journal of Metamorphic Geology, 18, 67–77.

Pfiffner, O. A., 1978, Der Falten- und Kleindeckenbau im Infrahelvetikum der Ostschweiz. Eclogae geologicae Helvetiae, 71/1, 61–84.

Pfiffner, O. A., 1986, Evolution of the north Alpine foreland basin in the Central Alps. In: Allen, P. A. & Homewood, P. (eds), Foreland Basins, Special Publication of the International Association of Sedimentologists, 8, 219–228.

Pfiffner, O. A., Sahli, S. & Stäuble, M., 1997, Structure and evolution of the external basement uplifts (Aar, Aiguilles Rouges/ Mt. Blanc). In: Pfiffner, O. A., Lehner, P., Heitzmann, P., Mueller, S. & Steck, A. (eds), Deep Structure of the Swiss Alps: Results of NRP 20, Birkhäuser, 139–153.

Pfiffner, O. A., Ellis, S. & Beaumont, C., 2000, Collision tectonics in the Swiss Alps: Insight from geodynamic modeling. Tectonics, 19/6, 1065–1094.

Pfiffner, O. A., Schlunegger, F. & Buiter, S. J. H., 2002, The Swiss Alps and their peripheral foreland basin: Stratigraphic response to deep crustal processes. Tectonics, 21/2, 3-1–3-16, doi: 10.1029/2000TC900030.

Polinski, R., 1991, Ein Modell der Tektonik der Karawanken, Südkärnten, Österreich. Dissertation Universität Karlsruhe, 143 pp.

Puschnig, A. R., 1996, Regional and emplacement-related structures at the northeastern border of the Bergell intrusion (Monte del Forno, Rhetic Alps). Schweizerische Mineralogische und Petrographische Mitteilungen, 76, 399–420.

Ratschbacher, L. & Frisch, W., 1993, Palinspastic reconstruction of the pre-Triassic basement units in the Alps: the Eastern Alps. In: von Raumer, J. & Neubauer, F. (eds), Pre-Mesozoic Geology in the Alps, Springer Verlag, 41–51.

Ratschbacher, L., Dingeldey, C., Miller, C., Hacker, B.R. & McWilliams, M.O., 2004, Formation, subduction, and exhumation of Penninic oceanic crust in the Eastern Alps: time contstraints from ^{40}Ar/^{39}Ar geochronology. Tectonophysics, 394/3–4, 155–170.

Reinecke, T., 1998, Prograde high- to ultrahigh-pressure metamorphism and exhumation of oceanic sediments at Lago di Cignana, Zermat-Saas Zone, western Alps. Lithos, 42, 147–189.

Rubatto, D., Regis, D., Hermann, J., Boston, K., Engi, M., Beltrando, M., McAlpine, S. R. B., 2011, Yo-yo subduction recorded by accessory minerals in the Italian Western Alps. Nature/Geoscience, 4, 338–342, doi: 10.1038/NGEO1124.

Scheiber, T., Pfiffner, O.A. & Schreurs, G., 2013, Upper crustal deformation in continent-continent collision: case study central Bernhard nappe complex (Valais, Switzerland). Tectonics,.

Schmid, S.M., Pfiffner, O. A. & Schreurs, G., 1997, Rifting and collision in the Penninic Zone of eastern Switzerland. In: Pfiffner, O. A., Lehner, P., Heitzmann, P., Mueller, S. & Steck, A. (eds), Deep Structure of the Swiss Alps: Results of NRP 20, Birkhäuser Verlag Basel, 160–185.

Schönborn, G., 1992, Alpine tectonics and kinematic models of the central Southern Alps. Memorie di scienze geologiche (Memorie degli Istituti di Geologia e Mineralogia dell'Università di Padova), XLIV, 229–393.

Steck, A., 1984, Structures et déformations tertiaires dans les Alpes centrales. Eclogae geologicae Helvetiae, 77/1, 55–100.

Todd, C. S. & Engi, M., 1997, Metamorphic field gradients in the Central Alps. Journal of Metamorphic Geology, 15, 513–530.

Trommsdorff, V. & Connolly, J. A. D., 1996, The ultramafic contact aureole about the Bregaglia (Bergell) tonalite: isograds and a thermal model. Schweizerische Mineralogische und Petrographische Mitteilungen, 76, 537–547.

Trommsdorff, V. & Nievergelt, P., 1983, The Bregaglia (Bergell) Iorio intrusive and its field relations. Memorie della Societa Geologica Italia, 26, 55–86.

Vernon, A. J., van der Beek, P. A., Sinclair, H. D. & Rahn, M. K., 2008, Increase in late Neogene denudation of the European Alps confirmed by analysis of a fission-track thermochronology database. Earth and Planetary Science Letters, 270, 316–329.

Wissing, S. B. & Pfiffner, O. A., 2002, Structure of the eastern Klippen nappe (BE, FR): Implications for ist Alpine tectonic evolution. Eclogae geologicae Helvetiae, 95, 381–398.

7 The Latest Steps in the Evolution of the Alps

When using the expression 'mountain building', we may intuitively first think of the development of local relief and not of the crustal processes, such as nappe stacking and folding, that occur as a result of convergent plate movement at the margins of the plates that are involved. The development of relief thus appears to be the sole product of deepening of valleys and valley formation and therefore synonymous with mountain building. However, the endogenous processes have a very direct effect on the altitude of the land surface and thereby also control the surface processes, such as weathering and erosion. Erosion results in the removal of rock from mountain slopes, but only some of that material is deposited during transport to the foreland and in the foreland itself. When studying the chronological evolution of the relief, we are therefore put in the difficult position that there is no complete archive that records erosion. Exceptions to this are, for example, the sediments in the foreland that provide information on where the components originated in the mountain chain. Within the mountain chain itself, cooling ages indicate the exhumation rates for the rocks as they moved from deep down to the present-day surface. The precision of both these methods is limited as the processes that are investigated happened several million or tens of million years ago. In particular, it must be noted that the erosion that has taken place over such time intervals has removed up to more than one or even several kilometres of material. The height of the Alps and the floors of the valleys above sea level 5–15 million years ago was probably similar to today: rivers cut into the bedrock as it was being uplifted and more or less kept pace with the uplift. Numerical models suggest that the rivers are not readily forced into changing their courses due to local differences in the rate of uplift (Kühni & Pfiffner 2001a), instead, many of the valleys that run parallel to the mountain ranges within the Alps (Rhône, Vorderrhein, Inn, Adda, Enns, Mürz, to mention just a few) also run parallel to tectonic contacts over longer distances. It is apparent that rocks along tectonic contacts that are easily eroded do play an important role in shaping the course of a rivers. However, as such tectonic contacts can also be offset, such that if rivers follow the erodible rock their course will be displaced laterally along the dipping contact in the course of incision. Furthermore, tectonic contacts are often limited to individual nappe units, and during erosion and incision of the rivers, the topography will thus constantly adjust to the rock types that are at the surface at a given point in time. If zones of easily erodible rocks appear from the subsurface in the course of exhumation, major adjustments of the river courses are to be expected. For all these reasons, it is not an easy task to reconstruct the course of the 'ancestral rivers' in the inner part of the Alps.

In spite of these reservations, it is possible to make some inferences on the most recent development of relief in the Alps and its foreland.

- Sediments in the foreland – even if they are sparse – and the minimal erosion in the foreland in comparison to the mountains in the hinterland provide information on the dynamics in the lower reaches of Pliocene and Pleistocene rivers.

- Although the Pleistocene glaciations repeatedly destroyed evidence of preceding glaciations, evidence of the last ice age, the Last Glacial Maximum (LGM), and for some older glaciations shows the effects of glaciers on the Alpine landscape.
- Recent vertical and horizontal movements, in combination with the present-day seismicity, and cooling ages from fission-track data allow us to envisage what the earlier uplift scenarios might have looked like and what rates of uplift can be assumed.
- Finally, we can use the ongoing, present-day erosional processes, such as mass wasting (debris flows, rockfall, landslides) and rockslides, to improve our understanding of the evolution of relief in the past.

These four approaches will be adopted below in order to understand the present-day relief in the Alps and its formation in more detail, based on individual examples. Although a full discussion is not possible as the processes that shape the surface are usually very localized, we do, at least, have a full analysis of the present-day morphology at our disposal. In the Eastern Alps, Frisch et al. (2000a) analysed the regional distribution of altitude and slope angles and distinguished between five main domains, each with unique characteristics. High and rough relief correlates with those areas that have experienced significant uplift and erosion since the Miocene, in connection with the Cenozoic orogeny. In contrast, the easternmost Eastern Alps exhibit the effects of Miocene extensional tectonics to a greater extent and display a moderate relief, even though there are rocks at the surface that were exhumed from great depths: clearly, signs of the uplift and erosion linked to the

Cretaceous orogeny are no longer depicted in the relief. Kühni & Pfiffner (2001b) carried out a similar analysis for the Central Alps and demonstrated the effects of erodibility of the rocks on the mean altitude: the granitic regions of Mont Blanc, Dent Blanche, Aletsch-Gotthard and Bernina-Bregaglia exhibit the highest mean altitude above sea level and the greatest local relief. Kühni & Pfiffner (2001b) concluded that, 20 million years after an intensive phase of exhumation and over 20 kilometres of erosion (as is the case, e.g., in the Lepontine), the mean height is reduced to a moderate level due to erosion and that only the steep slope angles remain as evidence of the exhumation.

7.1 Miocene and Pliocene Drainage Patterns

In the Messinian (6 million years ago), the entire region in the north of the Central and Eastern Alps clearly drained towards the east. This drastic change, compared with the west-directed drainage in the Middle Miocene (15 million years ago; see Fig. 6.16), is probably linked to the main folding of the Jura Mountains that occurred during the time interval of 10–3 million years ago. The folding of the Jura Mountains and the associated northwest-directed displacement of the western Molasse Basin on a southwest-dipping detachment horizon caused substantial uplift of the basin fill. The effect of this process is revealed in the fact that the Upper Marine Molasse (UMM) now lies at over 1400 metres above sea level in the summit of the Guggershorn in the south and that the majority of the Upper Freshwater and Marine Molasse (UFM and UMM) sediments

The Alps at the end of the Miocene

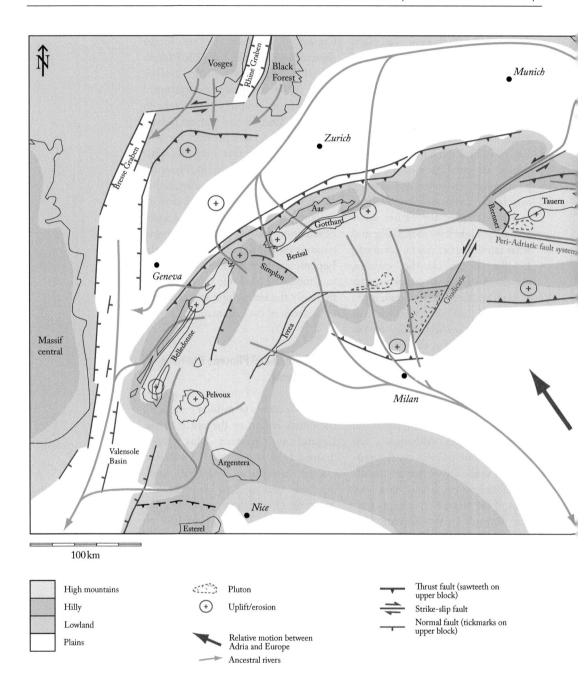

100 km

	High mountains		Pluton		Thrust fault (sawteeth on upper block)
	Hilly	⊕	Uplift/erosion		Strike-slip fault
	Lowland				Normal fault (tickmarks on upper block)
	Plains		Relative motion between Adria and Europe		
			Ancestral rivers		

▲ Figure 7.1
Palaeogeographical map
of the Alps towards the
end of the Miocene
(Messinian, 6 Ma).

were eroded at this time (this is referred to as 'Resthebung' in the local literature; see Chapter 6). The western Molasse Basin was home to the large watershed between the tributaries draining into the Danube system and the Rhône system. The latter consisted of a large north–south oriented depression extending from the Bresse Graben in the north via the Valence and Valensole basins to the Mediterranean.

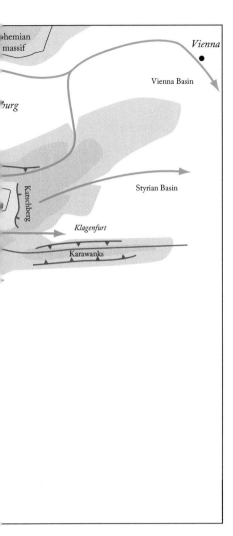

(1974) and Villinger (2003) detected the presence of an Aare–Danube system based on high-altitude gravel deposits containing Alpine components on the Eichberg near Donaueschingen and the Tannberg (see Fig. 7.1). The Alpine Rhine emerging from the Alps was flowing in a notherly direction into the Danube. The Alpine Rhône was redirected to the northeast as it left the Alps, had the Aare as a tributary and unltimately rejoined the Danube north of Zürich. As will be discussed later, the Alpine Rhine and Alpine Rhône were captured by the Rhine and Rhône systems, which drained into the North Sea and the Mediterranean only about 3 million years later, in the Pliocene.

The Western Alps drained towards the west into the Rhone System (Fig. 7.1) and larger valleys cab be assumed to have existed in the region of the Durance and Isère. The Rhône also carved a deep valley into the floors of the Valence and Valensole basins, which were then subsequently filled with Pliocene sediments.

In the Eastern Alps, drainage occurred towards the east via rivers that followed the large faults (Inn, Enns and Gail) and were already present in the Late Oligocene, according to Frisch et al. (2000b). Frisch et al. (2000b) also discuss relics of ancient tilted land surfaces or peneplains (so-called 'Nockfläche') in the eastern Eastern Alps, which formed due to Miocene extensional tectonics in the metamorphic crystalline rocks that are resistant to weathering, and Dachstein surfaces in carbonate-dominated areas that were shaped by karstification.

The history of the ancestral rivers and their lower reaches in the foreland can be reconstructed somewhat more accurately for the end of the Miocene (Messinian) and the Pliocene. For the north side of the Central Alps, Liniger (1966), Hofmann

An important event during this period pertains to the south side of the orogen. During the Messinian, the Mediterranean dried out completely due to insufficient inflow from the Atlantic and thick layers of evaporites

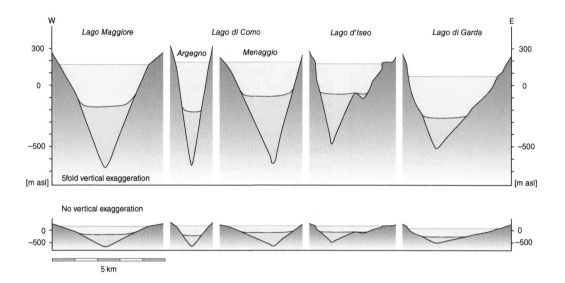

▲ **Figure 7.2** Geological
cross-sections through
valleys carved and
overdeepened by fluvial
erosion. The valleys are to
the south of the Central Alps
and the cross-sections based
on seismic reflection lines.
Source: Finkh (1978).
Reproduced with permission
of Elsevier.

*Deep canyons
carved by rivers*

accumulated on the floor of the basin.
This drying out of the Mediterranean
is also called the 'Messinian Salinity
Crisis'. As a result of this, the erosion
level of the rivers that fed into
Mediterranean reduced substantially,
and due to headward erosion, the rivers
carved deep gorges into the southern
flank of the Alps and into the Valensole
and Valence basins in the Rhone
System. Figure 7.2 shows cross-
sections through the Southalpine lakes
from Lago Maggiore to Lago di Garda,
which are based on seismic investigations (Finckh 1978). The bedrock
surface of the thalwegs lies deeper than
500 metres below sea level (in the case
of Lago Maggiore, almost 700 metres
below sea level). The seismic cross-
sections lie in the region of the end
moraines from the LGM and the
cross-sections of the valleys are
V-shaped, even when depicted without
vertical exaggeration, which favours
fluviatile overdeepening of these valleys rather than glacial. The onset of
the Pliocene starts with the final flooding of the dried out Mediterranean
Basin. The seawater that flowed in

from the Atlantic also flooded the
gorges that had formed during the
Messinian, even into the Alps, which is
revealed, for example, in the marine
shales near Balerna (to the south of
Lago di Lugano). It is unclear to what
extent the Late Miocene deepening of
the gorges extended into the Alps, but
some of these canyons are, indeed,
continued towards the north (Lago
Maggiore in the direction of the
Gotthard pass, Lago di Como in the
direction of the Splügen pass; but Lago
d'Iseo and Lago di Garda have no
major valleys that form a northern
continuation).

There were accentuated uplifts in
the region of the external massifs
(Tauern to Belledonne massifs) within
the Alps. Areas with a tendency to
subside were probably present in the
region next to the Brenner and
Katschberg normal faults in the
Eastern Alps, the Simplon normal
fault in the Central Alps and the
'Subbriançonnais' Fault in the Middle
Penninic realm in the Western Alps
(see Figs 5.1-8 and 6.14). These differential uplifts are underlined by the

colour scheme in Fig. 7.1. The formation of a riverine network must have been particularly dynamic in the source area of the main rivers in the Central Alps. Here, fluvial incision progressed towards the region of the Gotthard massif through headward incision. Kühni & Pfiffner (2001a,b) suggest that the shift of the main watershed between the Po and the Rhine systems towards the north, from its original location further south, occurred at this time. This could have been caused by the fall in base level due to the drying out of the Mediterranean.

The riverine network was fundamentally reorganized in the region to the south of the Rhine Graben as a result of the folding of the Jura Mountains in the Pliocene (see Fig. 7.3). Due to subsidence of the Bresse Graben and the resultant headward incision, the ancestral Aare river was captured and now flowed around the uplifting Jura Mountains, via Sundgau, into the Bresse Graben and into the Mediterranean (a detailed discussion on this can be found in Giamboni et al. 2004): evidence for this is provided by the Sundgau Gravels, which contain components from the Aare catchment area. The Bresse Graben was uplifted and the Rhine Graben subsided immediately afterwards. Headward incision originating in the Rhine Graben caused a further redirection of the Aare, this time into the Rhine draining into the North Sea. This capture included ancestral rivers emerging from the Alps south and southeast of Zürich. The easternmost branch had its source area adjacent to the source area of the Alpine Rhine. The lowered base level caused an important erosional removal of the youngest strata in the Molasse Basin of the Central Alps.

The Alpine Rhine continued flowing in a northerly direction into the Danube (Villinger 2003), an important tributary of which was the ancestral Ill, which drained a part of the Eastern Alps and rejoined the Alpine Rhine to the south of Lake Constance.

On the south and west sides of the Alps, the Mediterranean flooded the different valleys that had deepened in the Late Miocene, for which there is evidence in the form of marine Pliocene sediments (for example, the Balerna Shale at the southern end of Lago di Lugano, mentioned above). These valleys and the entire Po Basin were subsequently progressively filled with the erosional products from the Alps. Erosion continued also on the north side of the Alps, in particular in the western Molasse Basin between Zürich and Geneva. Headward inciscion of the Rhône in the area around Geneva led to the capture of the Alpine Rhône and, as discussed above, of the ancestral Aare River.

The Alps in Pliocene times

Uplift in the Alps probably continued in a similar fashion to during the Late Miocene. According to Apatite fission-track data, uplift and erosion was particularly active in the region of the external massifs (see Vernon et al. 2008). Folding and thrusting in the outermost portion of the Dolomites (e.g. the Montello thrust in Figs 5.2-21, 5.3-9 and 6.3), which involved Pliocene and Pleistocene sediments, created local uplift. Because in the Southalpine nappe system to the west of the Giudicarie Fault, the youngest thrust faults are sealed by post-Messinian sediments (see Fig. 5.2-21), the Giudicarie fault must have been active as a sinistral strike-slip fault in the Pliocene.

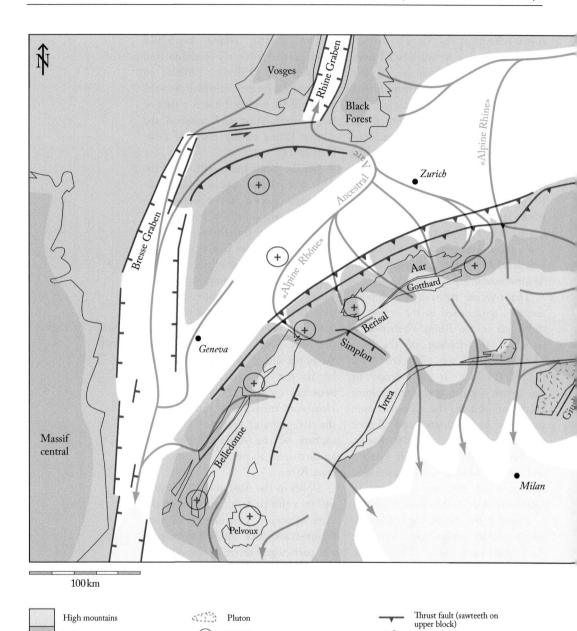

High mountains

Hilly

Lowland

Coastal areas/plains

Shallow marine/rift

Pluton

Uplift/erosion

Relative motion between
Adria and Europe

Ancestral rivers

Thrust fault (sawteeth on
upper block)

Strike-slip fault

Normal fault (tickmarks on
upper block)

7.2 Pleistocene Glaciations

Major climatic changes occurred over
the past, just under, two million years,
in the Pleistocene. Mean annual temperatures varied by about 15 °C
between cold and warm periods.
During the cold periods, the glaciers

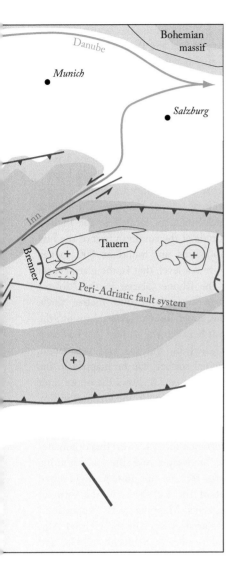

◀ **Figure 7.3**
Palaeogeographical map of the Alps in the Pliocene (3 Ma).

ages, caused enormous changes in the landscape, both in the Alps and in their foreland. The most important changes were the polishing and erosion of surfaces in the Alps by the glaciers, and the (over)deepening and broadening of valleys. Further down the valleys and in the Alpine Foreland, important changes include the development of landforms such as drumlins, roches moutonnées, deposition of glacial till (including moraines) and glaciofluvial outwash gravels close to glacier tongues and farther downstream. Accurate records and dates for the numerous ice ages are difficult to obtain as the most recent advances often removed the older deposits – there is no continuous archive, as there is for the oceans. Even so, careful analysis has resulted in distinguishing evidence for 15 cold periods, far more than the classic four to six ice ages. Figure 7.4 shows the current status for the continental Alpine ice age stratigraphy, based on a summary produced by Schlüchter & Kelley (2000).

It must be noted that these ice ages were also experienced by humans – the earlier ones by *Homo erectus*, the more recent ones by *Homo neandertalensis* and *Homo sapiens* – our early ancestors in the Alps and their foreland. Figure 7.4 therefore also shows some of the stages in human evolutionary history over the past 2 million years. For example, the warm period of 60 000 to 28 000 years ago, between the last ice ages, included the Cro-Magnon culture with the cave paintings in Chauvet (in the foreland of the Western Alps) and the cave bear cult in the Drachenloch near Vättis at the eastern end of the Aar massif.

In the Pleistocene, an important change in the drainage pattern affected

advanced far into the Molasse Basin and reached the Po Basin, overall, at least 15 times. These advances by the glaciers in the cold periods, or ice

the Alpine Rhine river. Headward incision from the southern end of the Rhine Graben captured the Alpine Rhine and deviated it to the west. In the Early Pleistocene, it deposited the sediments that make up the 'Höhere Deckenschotter' (High-level Gravel) on Irchel (to the northwest of Zürich), which contain reworked material from the Upper Freshwater Molasse of the Hörnli Fan (east of Zürich; Graf 1993).

The maximum extension of the glaciers during the last glaciation can be mapped in the foreland based on the end moraines. The advances of the larger ice sheets can also be reconstructed based on the petrographic composition of the glacial till (in particular so-called 'Leitgesteine', i.e. rock types of unambiguous source area within the Alps). The map in Fig. 7.5 is based on data from van Husen (2000) for the Eastern Alps, from Florineth & Schlüchter (1998) for the Central Alps and from De Beaulieu et al. (1991) for the Western Alps. This shows the extent of the Alpine ice sheet and glaciers at the LGM, around 20 000 years ago. The huge advances of the ice streams into the Molasse Basin to the north of Lake Constance and to near Munich are clearly visible. In the western Swiss Molasse Basin, the ice sheet flowed as far as the Jura Mountains from the Rhone Valley and, in the case of the Val de Travers (Neuchâtel), the ice actually flowed up the valley, deep into the interior of the Jura Mountains. Local glaciers were present in the remainder of the Jura Mountains, as well as in the Black Forest and the Vosges. The contiguous ice sheet did not extend as far as Vienna in the extreme east of the Alps, and the southernmost portion of the Western Alps also remained

ice-free. Towards the south, glaciers just reached the Po Basin. This asymmetric distribution of ice between the north and south can be explained by climatic differences on the two sides of the mountain chain.

In addition to the extension of the ice sheets, its thickness and surface are also of interest. Information on these aspects is provided by the erratic blocks located at the highest altitudes and the upper limit of glacial erosion. The erratic blocks that are located at the highest altitude at the southern foot of the Jura Mountains are found at 800 metres above sea level, that is, the glacial sheet of the Rhone Glacier in the Molasse Basin was about 300 metres thick. The upper limit of glacial erosion, the limit between the surface that has been polished by the ice and the rough, original surface, can be mapped in the Alps. Figure 7.6 shows the upper limit of glacial erosion in the Grimsel region. The roundish shapes in the foreground have been polished by the ice, while the Finsteraarhorn (4274 m) that dominates the landscape, and the neighbouring horns that are not quite as high, jutted out of the ice sheets in the form of nunataks. Mapping of the upper limit of glacial erosion in the Central Alps by Florineth & Schlüchter (1998) and Kelly et al. (2004) revealed the presence of ice domes in the interior of the Alps. These domes were built up by the local precipitation and, based on their geographical location to the north of the present-day weather divide, we must conclude that precipitation was mainly caused by southwesterly winds, that is, corresponding to föhn conditions. With reference to large-scale circulation systems, these weather conditions favour a southern location of the Atlantic oscillatory system (Florineth & Schlüchter 2000).

Periods			Human history	Time	Glaciations
Geol.	Archaeology		(main events)	years before present	(with glaciers reaching the Alpine foreland)
Holocene	Iron Age		La Tène, Rome, Homer	3 000	Postglacial stages
	Bronze Age		Alphabet, pyramids	5 000	
	Copper Age		Cuneiform script, wheel	7 500	
	Neolithic		Domestic animals, towns		
	Mesolithic			11 700	
Pleistocene	**Palaeolithic**	Late	(Lascaux) Cro Magnon (Chauvet)	28 000	Last Glacial Maximum II («Würm»)
		Middle	Cave bear cult (Drachenloch/Vättis)	60 000	? Last Glacial Maximum I («Würm») ?
				100 000	
				115 000 130 000	Eem Interglacial
				150 000	Large Glaciation III («Riss»)
				190 000 280 000	Large Glaciation II («Mindel»?)
				330 000	
				500 000	
		Early		600 000	Large Glaciation I («Mindel»?)
				700 000	Greatest Glaciation («Günz»)
				800 000	
				1 Ma	Swiss
					High-level Gravel
					Glaciations
				2 Ma 2.58 Ma	
Pliocene				4.1 5.2	«Wanderblock» Formation (relics of fluvial gravel)
				5.3 Ma	

Vertical labels within the Human history column: *H. sapiens*, *Neanderthals*, *Homo erectus*, *H. habilis*, *Paranthropus*, *Ardipithecus ramadus*, *Australo-pithecus*

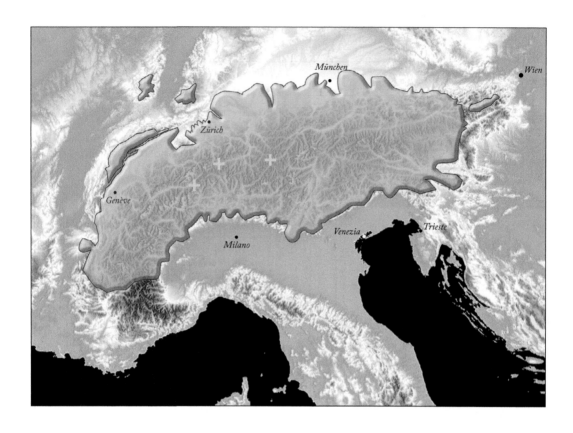

▲ **Figure 7.5** The extent of the Alpine ice sheet and glaciers during the Last Glacial Maximum (20 000 to 15 000 years ago). On the northern side of the Alps, glaciers spread way out onto the Alpine foreland, whereas the Po Basin to the South of the Alps remained almost completely ice free. Three ice domes in the interior of the Alps can be distinguished (marked with +). Source: Adapted from van Husen (2000), Florineth & Schlüchter (1998) and De Beaulieu et al. (1991).

During the advance of the ice sheets through them the Alpine valleys were overdeepened. In several valleys, the bedrock surface lies at sea level in the axis of the valley (the thalweg), or even below sea level. Wildi (1985) and van Husen (2000) provide a compilation of the overdeepened valleys on the north side of the Central and Eastern Alps. Kissling & Schwendener (1990) determined the thickness of the Quaternary valley fills using gravity data. Within the framework of NFP 20, an entire series of seismic cross-sections were recorded through valleys on the northern and southern sides of the Alps. In case of the Rhine Valley upstream of Lake Constance, the thalweg in the bedrock is slightly

above sea level (see Preusser et al. 2010), while in the Rhone Valley it is about 300 metres below sea level. As an example, Fig. 7.7 shows a cross-section through the Rhone Valley near Martigny (adapted from Pfiffner et al. 1997). The unconsolidated rock fill is up to 750 metres thick in total. The uppermost sediments usually correspond to deposits of alluvial fans shed by lateral tributaries. The build-up of sediments in these alluvial fans often resulted in the main river course being shifted to the other side of the valley. Fine-grained, silty to sandy unconsolidated rock is usually found underneath the alluvial fans, which are interpreted as lacustrine and glacial lacustrine sediments. However, it is conceivable that these

Figure 7.6 Limit of glacial polishing related to the Last Glacial Maximum in the area of Finsteraarhorn–Grimselpass (canton Bern, Switzerland).

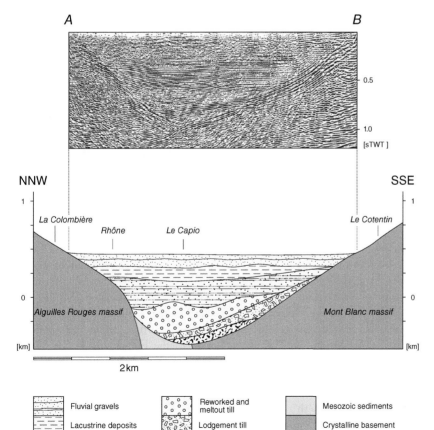

Figure 7.7 Geological cross-section through the glacially overdeepened Rhone Valley near Martigny (canton Valais, Switzerland). Source: Pfiffner et al. (1997d).

fine-grained sediments were deposited underneath a floating sheet of dead ice. Dead ice, the remnant of past glaciers that is covered in debris, protected the subglacial lake from the input of coarse clastic material that was being transported from the sides of the valley that, as yet, had no vegetation cover, and by the meltwater streams on or alongside the glacier. Subglacial till and outwash are found below in the overdeepened valleys, overlying deltaic deposits found right at the base. These latter deposits were probably deposited in the subglacial channels by meltwater, produced by the increase in pressure caused by the load exerted by the ice on the rock–ice contact and was responsible for the actual incision at the glacier bed (Röthlisberger 1972).

7.3 Recent Movements and Seismicity

Convergence between the Adriatic and European plates is still ongoing at the present day, as was discussed in connection with Fig. 1.8. These active movements also manifest themselves in the vertical uplift that has been revealed by the precision levelling surveys undertaken by the national institutions of the Alpine Nations (Federal Office of Metrology and Surveying in Austria, Federal Office of Topography swisstopo in Switzerland, the Istituto Geografico Militare of Italy and the National Institute of Geographic and Forest Information in France). Figure 7.8 shows a simplified compilation of these data on uplift, with those regions that exhibit a tendency to subside and those with a strong tendency to uplift (over 1 millimetre per year) being distinguished. In addition, locations with maximum subsidence or

uplift are indicated. The compilations are based on data provided by Gubler (1991), Kahle et al. (1997), Arca & Beretta (1985) and Jouanne et al. (1995). It must be noted that the vertical movements at each levelling point are related to a reference point (Horn in the Bohemian massif, Aarburg on an anticline of the Subjurassic Zone in the Molasse Basin, Mareografo of Genova at the Alps–Apennines transition, and Pougny in the western Molasse Basin). As these reference points can also move, these data are to be taken as relative uplift and subsidence rates. Furthermore, isolines for equivalent uplift and subsidence for the four surveying networks cannot be drawn.

In spite of these reservations, the map in Fig. 7.8 reveals some well-defined first-order characteristics. The Molasse Basin shows a clear tendency towards subsidence in the east, but this applies only to isolated regions in the west. This difference must be interpreted with caution due to the independent nature of the surveying networks. However, we can point out that the western Molasse Basin has been uplifted since the Miocene in connection with the folding of the Jura Mountains and that this tendency may still be continuing. The Rhine and Bresse grabens at the external margin of the Jura Mountains also exhibit a tendency toward subsidence. Within the Jura Mountains, the internal part in the west is uplifting, while the external parts in the French and in the Swiss survey exhibit a tendency towards subsidence. Subsidence of the eastern Molasse Basin continues in the adjacent Vienna Basin in the east. In the case of the Po Basin south of the Alps, a marked subsidence is observed in its eastern part.

Maximum uplift is noted in the region of the external massifs. This includes the two maxima at the eastern and western edges of the Tauern Window, the maxima at the eastern and western ends of the Aar massif and the maximum at the northeastern end of the Belledonne massif. All these maxima have also been confirmed by very young Apatite fission-track ages, which means that the present-day uplifts were already concentrated in the same regions several million years ago. In the case of the Aar massif, these maxima are also underpinned by the increase in seismicity (a more detailed discussion on this can be found in Persaud & Pfiffner (2004) and Ustaszewski & Pfiffner (2008)). It therefore appears that uplift of the external massifs is still ongoing.

Further maxima for uplift are noted in the northern Western Alps in the Subalpine chains of the Haute Savoie and in the southern Jura Mountains. Jouanne et al. (1995) suggest that these maxima are a manifestation of the continuing folding of the Jura Mountains and the frontalmost anticlines of the Subalpine chains. The fact that this pattern does not continue towards the east (the eastern Jura Mountains exhibit a tendency towards subsidence) and that no maxima for uplift have been detected in the Aiguilles Rouges and Mont Blanc massifs, casts some doubt on this interpretation.

Finally, two maxima for uplift are present within the Penninic nappe system of the Western Alps (west of Turin) and near Domodossola near the northwestern termination of the Ivrea Zone (northwest of Milan). In both places, the uplift rate is more than 2 millimetres per year and even attains 3 millimetres

per year (Noquet et al. 2011, Arca & Beretta 1985).

The cross-section through the western Aar massif in Fig. 7.9 shows the crustal structure, active vertical uplift and seismicity, in order to demonstrate the connectivity between all these parameters. The cross-section shows a simple crustal structure and the hypocentres of observed earthquakes along a strip 70 kilometres wide are projected onto the cross-sections. The uplift rates plotted in the cross-section were determined by intersecting the uplift contours of uplift maps with the plane of the cross-section. A steep uplift gradient is clearly visible on the northern margin of the Aar massif and an uplift maximum is located above the Aar massif, but high rates of more than 1 millimetre per year prevail in the southern part of the Alps and the Po Basin. A steep gradient is observed north-northeast of Ivrea, which suggests that the reverse fault associated with the Adriatic mantle wedge is still active. Seemingly this core of the Alpine orogen is being expelled upward along ramps of opposite dip.

The seismicity in the foreland encompasses the entire crust, whereas within the core of the Alpine orogen only the upper 12–15 kilometres of the crust are seismically active. Some of the observed seismicity is due to strike-slip faults (Fribourg Fault to the northern edge of the Alps). The pronounced seismicity in the Valais on both sides of the Rhone is close to, and to the south of, the Aar massif and exhibits mainly horizontal extension. The hypocentres in the region of the basal thrust fault of the Aar massif, and those over and to the south of the Aar massif with horizontal

▶ **Figure 7.8** Recent uplift rates determined from precision leveling. The map was compiled from various sources: B.E.V. (1991) for the Eastern Alps, Gubler (1991) and Kahle et al. (1997) for the Central Alps, Jouanne et al. (1995) and Nocquet et al. (2011) for the Western Alps, and Arca & Beretta (1985) for the Southern Alps and Po Basin. The easternmost Eastern Alps are subsiding while the remainder of the Alps is uplifting at varying rates.

Miocene uplift maxima continue into the present

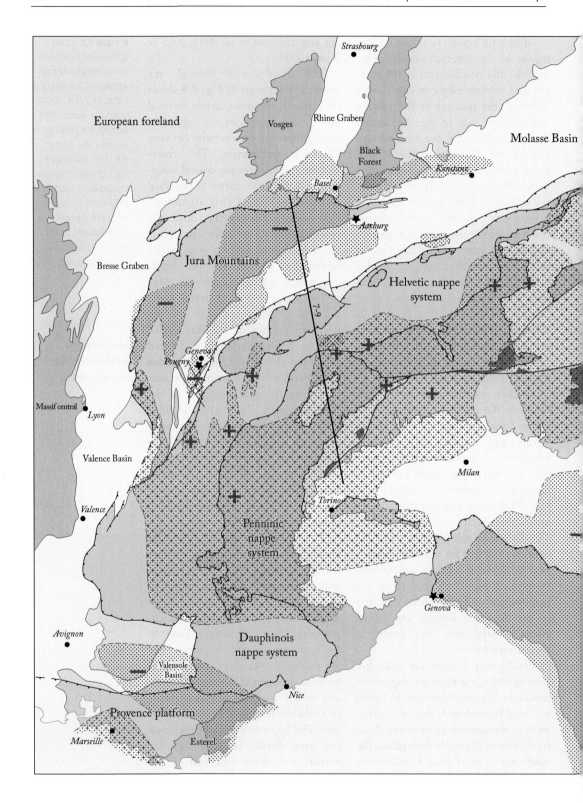

European foreland

Strasbourg

Vosges

Rhine Graben

Black
Forest

Konstanz

Molasse Basin

Bresse Graben

Jura Mountains

Basel

Aarburg

Helvetic nappe
system

Massif central

Lyon

Geneva

Pougny

Valence Basin

Valence

Milan

Torino

Penninic
nappe
system

Genova

Avignon

Dauphinois
nappe system

Valensole
Basin

Nice

Provence platform

Marseille

Esterel

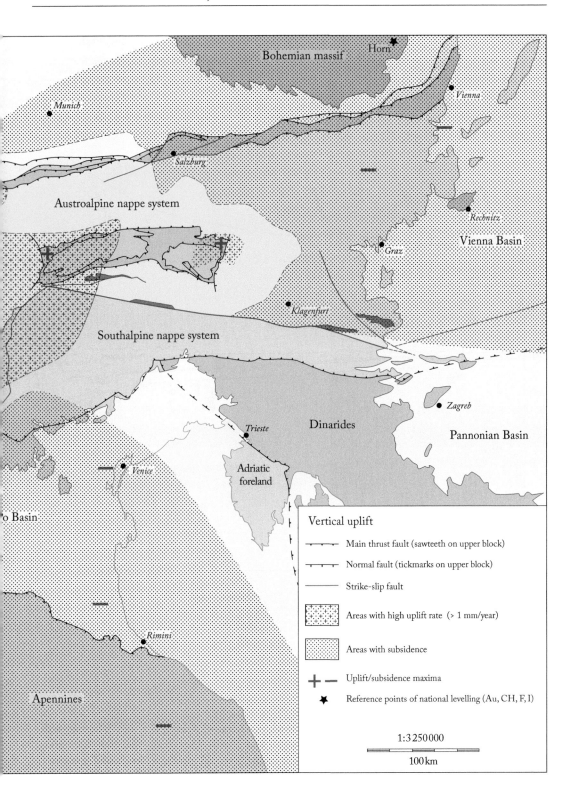

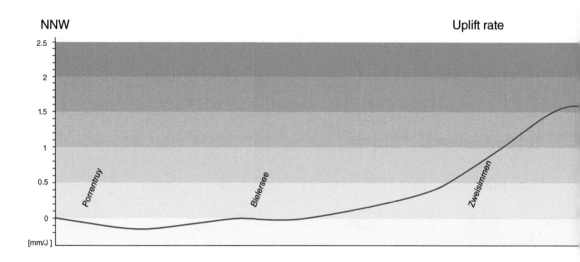

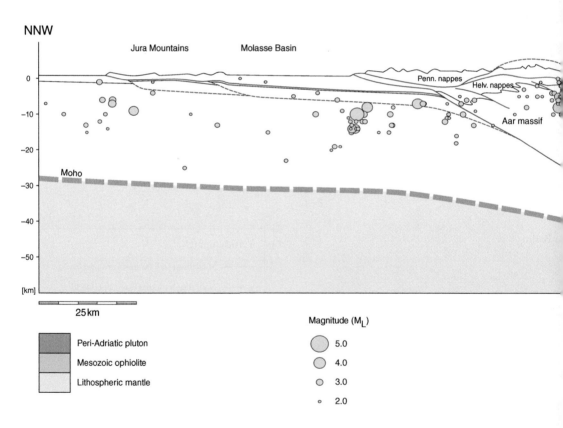

extension, could all be explained by a continuing uplift and bulging of the Aar massif. A basal thrust fault of the Aar massif that was still active would cause folding in the hanging wall, with the result that the updoming would produce horizontal extension in the outer arc of the dome.

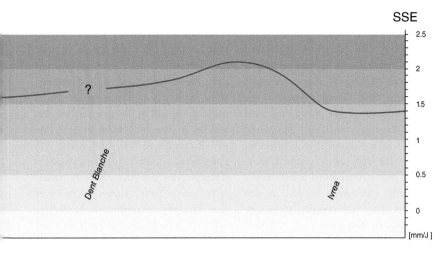

SSE

2.5

2

1.5

1

0.5

0

[mm/J]

? Dent Blanche Ivrea

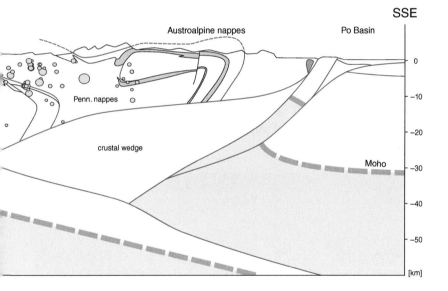

SSE

Austroalpine nappes Po Basin

0

−10

Penn. nappes

−20

crustal wedge

Moho

−30

−40

−50

[km]

Figure 7.9 Geological cross-section through the western Central Alps showing a simple crustal structure along with hypocentres of earthquakes and uplift rates. Note the conspicuous steep gradient of the uplift rates on the northern flank of the Aar massif (near Zweisimmen) and above the Adriatic mantle wedge (north of Ivrea) suggesting that the core of the orogen is being expelled upward. Seismicity is restricted to the upper 12–15 km in the core of the orogen.

However, there is also another explanation for the occurrence of earthquakes with a mechanism of horizontal extension in the core of an orogen. In numerical experiments, the model surface is initially uplifted in the core of the orogen when crustal fragments are sheared off and stacked on top of each other in the core of a collisional orogen by the process of basal accretion (Selzer et al. 2008). Even if erosion is allowed in the models, a positive relief is preserved that then causes a mass movement that diverges towards the foreland due to gravity spreading. This horizontal gravitational spreading results in the apparent contradiction that, in the case of convergent plate motion, that is, horizontal compression, horizontal extension occurs in the direction of the convergence in the core of the orogen. The term used in the literature for this is 'orogenic collapse' and in the case of the Alps, Persaud & Pfiffner (2004) and Delacou et al. (2004) have interpreted the spatial distribution of earthquake mechanisms in the Penninic realm within the meaning of this term.

Recent movements have also caused faults that offset the present-day land surface. A systematic study of such Holocene faults in the Central Alps

by Persaud & Pfiffner (2004) and Ustaszewski & Pfiffner (2008) revealed that many of the faults that are visible are not purely tectonic in origin, but that the main cause is gravitational slope movement. These gravitational movements will be discussed in the next section. Figure 7.10 shows as an example of a tectonic fault that offsets the postglacial landsurface in the western Central Alps (Aiguilles Rouges massif near Lac de Fully). Large blocks cover the glacially overprinted landscape, some of which are erratics that were transported by glaciers, others stem from local rock falls. The orientation of the linear cut through the topography made by the fault plane indicates that an explanation by purely gravitational movement would be highly improbable, but far more likely is a recent tectonic fault (Ustaszewski & Pfiffner 2008). If the frequency of earthquakes of higher magnitudes is considered, we must assume that this (and other tectonic

Figure 7.10 Holocene tectonic fault in the Aiguilles Rouges massif near Lac de Fully (canton Valais, Switzerland). The offset of the glacially scoured suface by the fault is clearly visible. Source: Photograph courtesy of Michaela Ustaszewski. Reproduced with permission.

faults) are more likely to be characterized by aseismic creep and are not the
result of single, large earthquakes.

7.4 Rockslides, Creeping Slopes, Erosion by Modern Rivers

Metamorphic rocks cropping out at the
present-day surface in the Alps demonstrate that a rock unit that was several
kilometres thick has apparently been
eroded. Much of this erosion occurred
during the Cretaceous and Cenozoic
orogenies, although erosion continues
to this day. The growth of the deltas in
Lakes Constance and Geneva due to
deposits from the Rhine and the Rhone
is measured regularly and amounts to
around 3 million cubic metres per year.
This results in a mean erosion rate of
just under 0.5 millimetres per year if
this volume is distributed evenly across
the catchment area (Jäckli 1958). This
rate not only indicates the magnitude of
erosion, but is also comparable to values
determined for other regions in a global
comparison (Summerfield & Hulton
1994). Furthermore, it must also be
taken into consideration that rates of
erosion are substantially influenced by
tectonic processes, and are thus subject
to considerable variation over time: a
detailed discussion of the erosional
history of the Alps since the Oligocene
can be found in Kuhlemann et al.
(2002). Finally, local conditions also
affect erosion. Isolated occurrences of
heavy precipitation, resulting in mass
flow and flooding, are typically restricted
to relatively small areas. However, different types of landscapes indicate that
such isolated events have recurred in
the same place since the glaciers of the
last ice age melted and that they have
contributed towards the build-up of
large and small alluvial fans.

Recent erosional history is also
manifest in local one-off gravitational
slope movements, and rockslides are
a particularly impressive example of
this, with the entire Alpine region
being shaped by numerous smaller
rock falls and multiple rockslides:
Abele (1974) gives an overview of
Alpine rockslides. The dynamics of
these rockslides in shaping the surface
are described in greater detail here
based on the largest recorded rockslide
in the Alps, the Flims rockslide. The
debris avalanche covers an area of
at least 30 square kilometres (see
Fig. 7.11) and the thickness of the
debris avalanche naturally varies: it
is probably just under 1 kilometre
thick at its maximum in the centre.
An average thickness of 500 metres
yields a total volume of about 15 cubic
kilometres for the debris avalanche.
The rockslide broke off from the
Flimserstein (see Fig. 7.11) and flowed
into the valley of the Vorderrhein/
Rein anteriur in a southerly direction.
The debris avalanche crashed into the
slope on the opposite side and forced
its way up the side valley opposite, the
Safien Valley. The debris avalanche
split in the main valley and flowed
both towards the west (in the direction of Glion/Ilanz) and towards the
east (direction of Reichenau). The
mass of the rockslide is composed
mainly of Jurassic, also with some
Cretaceous, limestone from the
Infrahelvetic nappe complex, which
corresponds to the composition of the
Flimserstein. We can also see the edge
where the rockslide broke off in the
landscape, a prominent cliff that is up
to 500 metres high. The basal slip
plane is directly visible in some locations near Flims. It follows bedding
planes in the Quinten Limestone, but
forms steps in some locations, such

The Flims rockslide: the largest in the Alps

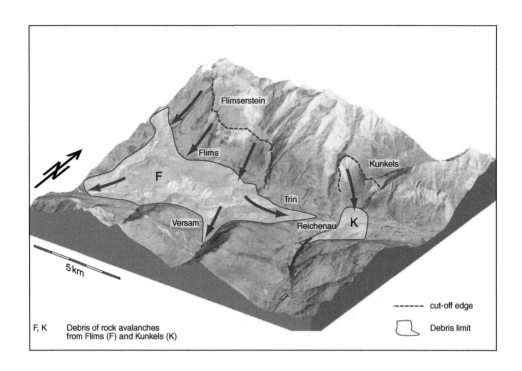

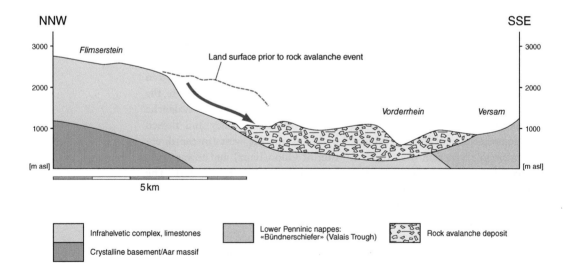

▲ **Figure 7.11** The Flims rock avalanche, the largest one in the Alps: shown as a block diagram (constructed with AdS2) and in cross-sectional view.

that the slip plane becomes successively deeper towards the south.

The lower edge of the debris avalanche is mostly in the subsurface. One exception is found at the exit to the Safien Valley, in the Rabiusa Gorge. Tree trunks

from a forest that had been destroyed by the debris avalanche were found here, which permitted dating of the event (see von Poschinger 2006). According to this, the Flims rockslide occurred about 9300 years ago. As a result, the Vorderrhein/Rein anteriur was dammed upstream and formed Lake Ilanz. Once this lake overflowed, the water broke through the dam and cut a notch into the debris mass and into the landscape downstream (von Poschinger 2006). The lakeburst floodwater flowed into Lake Constance, recording an anomaly in the deposits there. The Vorderrhein/Rein anteriur subsequently cut, step-wise, through the debris avalanche and, during this process, created the magnificent landscape of the Ruinaulta. Today, the river exhibits a longitudinal profile that points to equilibrium with the base level of Lake Constance. The incision into the debris mass created a 300-metre-deep gorge over a period of only about 9000 years, corresponding to an average incision rate of 3.3 centimetres per year. The presence of ancestral gravel terraces above the current level of the Vorderrhein/Rein anteriur (see Fig. 7.12) provides evidence for the fact that this incision probably occurred step-wise. Erosion by the Vorderrhein/Rein anteriur is still ongoing. We can see this by the lateral cutting of the river into the undercut banks of the meanders. Natural forest cover cannot keep pace with erosion in these locations.

The basal height of the main debris mass is difficult to estimate. We know from boreholes and seismic reflection investigations that the bottom of the Rhine Valley lies slightly above sea level further down the valley. An extrapolation of this glacial overdeepening in the direction of Flims would yield a height that is slightly above sea level (see cross-section in Fig. 7.11). However, given the postglacial age of the rockslide, we can also assume that the overdeepened valley had already been filled to a certain extent. Furthermore, it must be assumed that the moving debris avalanche scraped off soft sediments from underneath and although the basal height assumed in Fig. 7.11 allows for these considerations, it is associated with uncertainty. The surface of the debris mass is characterized by numerous funnel-shaped pits and elongate ridges. In the past, isolated erratic blocks of crystalline rocks were interpreted as evidence for subsequent overflowing of the debris mass by a late glacial advance. However, the absence of any moraine material in the depressions and the preservation of the small-scale morphology in the ridges do not support this. The erratic blocks probably lay on the past surface at an altitude of around 2000 metres before the rockslide descended. Erratic blocks from the last glaciation are also found at such altitudes in the surrounding area.

The debris mass comprise shattered limestones, with the degree of shattering subject to substantial local differences. Total shattering produced a whitish appearance for the limestones and if hit with a hammer will break down into sand and dust. At other locations, from a distance, we can still make out a clear internal structure with bedding and folds. Those rocks that were not shattered to any great extent can make up blocks that are several hundred metres in diameter. Clearly, during its descent, the debris avalanche did not reach a state of general turbulent flow, but rather we can assume that the debris moved as a mass of large blocks and mainly broke up when it crashed into the slope on the opposite side of the valley. Greater flow movement at the surface seems evident from the presence of ridges and the fact that large blocks of rock several metres in diameter are often found in the most superficial portion of the debris mass, suggesting that they floated to the surface under flow conditions.

Figure 7.12 Ruinaulta, a gorge that the Vorderrhein River cut through the rock mass displaced by the Flims rock avalanche.

(A) Remains of a gravel terrace representing an ancestral river bed located roughly 50 m above the modern river bed.

(A)

(B) Ancestral valley bottom filled with gravel and boulders located above the modern river bed.

(B)

The question then arises as to what caused this and other rock avalanches. The descent of the Flims rockslide long after the ice of the LGM had melted does not favour an immediate reaction in the flanks of the valley due to the absence of support by the glacier. Labile equilibrium can be disturbed by two factors: earthquakes and groundwater. Earthquakes are known to have triggered historical rockslides, but it is difficult to find evidence for a corresponding causal link for more ancient events. There are also historically proven examples for cases of a reduction in friction due to an increase in the subsurface circulation of water. Palaeoclimatic data could provide indications for the reasons underlying more ancient rockslides. Gruner (2006) has compared the prehistoric rockslides in the Central and Eastern Alps with palaeoclimatic conditions. Apparently, several large rockslides occurred in the Central Alps around the time of 9800 and 9200 years ago (Flims, Kunkels, Kander valley and Köfels in the neighbouring Eastern Alps), while there is a cluster around 3700–4200 years ago in the Eastern Alps. The older of these two time intervals lies in the Boreal, the more recent one in the Subboreal, both periods during which the climate was variable, with corresponding cold phases characterized by high precipitation. However, the longest cold phase in the Holocene that was registered across the globe occurred around 8200 years ago and lasted for about 200 years (Rohling & Pälike 2005). The only rockslide that correlates with this cold period is the Totalp rockslide (near Davos, Graubünden). All we can say about the Flims rockslide, is that it occurred slightly later than the Kunkels rockslide, which dammed the unified Vorder- and Hinterrhein near Reichenau and gravel

and sand were deposited behind the barrier. The Flims rockslide was driven into these unconsolidated sediments, which liquefied them and destroyed all sedimentary structures. Even though the relative sequence of events for the two rockslides appears clear and there is little age difference, any arguments in support of a causal link are absent.

In addition to rockslides, gravitational slope processes also cause slip movements that progress slowly. These are dependent on the composition of the rock substratum (type of rock and orientation of the strata and foliation) and thus vary from one valley to the next. The examples discussed below are to be regarded as a representative selection.

Figure 7.13A shows the landscape in the upper section of the sackung of Cari (Leventina, Central Alps). Antithetic normal faults dipping at a shallow angle parallel to the foliation in the gneisses into the mountain flank are visible in the foreground of the photograph. The resulting uphill-facing scarps can be followed over many kilometres in the landscape in the form of ridges or bands running parallel to the slope. Gravel aprons that are visible from a great distance accumulate behind the uphill facing scarps (visible in the background of the photograph). The normal faults indicate that the entire mass that is slipping towards the valley is being stretched internally, with the lowest, frontal portion of the sackung mass slipping down further than the uppermost portion. The fractured sackung mass offers little resistance to erosion by flowing water.

The photographs of Fig. 7.13B and C were taken in the Val Bedretto on the southern flank of the Gotthard massif. In this case, the main foliation in the gneisses runs almost vertically down to

Figure 7.13 Gravitational slope movements as seen in the field.

(A) Sackung of Cari (Leventina valley, canton Ticino, Switzerland). Motion occurred on surfaces dipping into the mountain flank and being parallel to the foliation in the gneisses.

(A)

(B) Steep faults in Val Bedretto (canton Ticino, Switzerland) forming ridges and uphill-facing scarps. Small ponds and lakes form behind the ridges.

(B)

(C)

(C) Steep fault in Val Bedretto (canton Ticino, Switzerland) offsetting a moraine ridge it intersects at a right angle.

depth. A large number of asymmetrical ridges running parallel to the slope is easily recognizable in the landscape. The flank facing the mountain, an uphill facing scarp, is steeper and often marked by a cliff, which is probably the main reason why these ridges have also been interpreted as tectonic faults. If we follow individual ridges along, then we note that the offset or the height of the ridges can change very rapidly. For example, in Fig. 7.13B, we can clearly see that the height of the ridges on the crest in the background is far greater than in the foreground. The asymmetry of these ridges is also revealed in the distribution of small and largee ponds in depressions along the ridges facing the mountain. This observation of the rapid changes and variability in the relief of the ridges – from several metres to total disappearance over stretches of less than 100 metres – is not indicative of a purely tectonic origin for these faults. In the foreground, Fig. 7.13C shows the steeply inclined foliation in the gneiss and a cliff that demarcates the ridge and indicates a fault scarp. Towards the background, the relief of the ridge becomes smaller, but still offsets the moraine ridge in the background. Clearly, these ridges are post-glacial. Careful mapping of these phenomena in the Central Alps revealed that such steep faults are encountered particularly frequently in those areas where the foliation and/or stratification in the rocky substratum is very steeply inclined (Persaud & Pfiffner 2004, Ustaszewski & Pfiffner 2008): examples of such areas are the southern edge of the Aar and the Gotthard massifs.

The schematic overview in Fig. 7.14 summarizes the phenomena associated with gravitational slope movements.

Figure 7.14 Two types of deep-seated gravitational instabilities.

(A) Antithetic normal faults merging with a basal detachment result in ridges and uphill facing scarps.

(A)

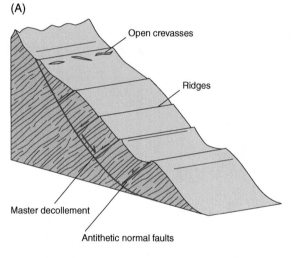

(B) Steep faults delimiting tilted blocks also result in ridges and uphill facing scarps.

(B)

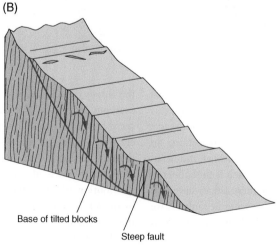

Antithetic normal faults can produce flat uphill-facing scarps where foliation or stratification dips at shallow angle and under the effect of gravity, blocks between faults can be affected by slight, domino-like tilting movements (comparable to toppling). Steep uphill-facing scarps can develop in the case of steeply inclined foliation or stratification and in this case a basal decollement horizon does not have to be present.

An additional process, however, does need to be considered in the case of steep faults, which is the effect of the release of the ice load on valley sides by melting of the glaciers. The Alps sank down under the load of the ice in each of the multiple glaciations and then rose up again after each. A large-scale model in which the European lithosphere is loaded with an ice sheet of a known thickness yields a sinking value for the Alps of slightly over 150 metres beneath the ice domes (Persaud & Pfiffner 2004). At the same time, dome-like uplift by a few metres will probably have occurred far away in the foreland, due to the elastic properties of the plate. The wavelength of these vertical movements was far greater than that for the present-day vertical movements determined geodetically (Fig. 7.8). Furthermore, we must assume that the compensatory movement or uplift induced by the melting of the glaciers started immediately, because an uplift of over 150 metres in 15 000 years yields a mean rate of uplift of 1 centimetre per year – ten times that measured today. In addition to large-scale observations, small-scale observations are also required. In the Alps, the load exerted by a glacier would be far greater along the axis of a valley than on the flanks, given the steep flanks of

the valleys. If we consider the fact that there were already weaknesses in the bedrock in cases of steeply inclined foliation (exacerbated by the multiple upwards and downwards movements over the course of all the ice ages), then we could expect local differences in uplift. A corresponding calculation based on modelling does, indeed, indicate that the valley floor would be uplifted to a greater extent than the flanks of the valley (Ustaszewski et al. 2008). Although this process is linked to gravitational compensatory movements, it is not directly connected to gravitational slip on the flanks of the valleys.

Glacial rebound

Figure 7.15 shows the interaction between the different erosional processes on the southern flank of the Val Bedretto. This southern flank was a rock surface that had been highly polished by the ice 15 000 years ago. Since then, under the high cliff in the shade, a thick layer of debris has built up by rockfall. A gulley on the right-hand side of the picture can be clearly seen, which in recent times numerous mudflows have travelled down, deepening the channel and depositing the material they were transporting in an alluvial fan, and material at the base, or bypassing, the alluvial fan is then transported away by the main river. The overgrown gulleys to the left and right of the currently active gulley are earlier features. Multiple, slightly inclined flattened surfaces transverse to the slope indicate that the slope is moving as a sackung. If we extrapolate the changes in the landscape that have occurred over the past 15 000 years, as visible in this picture, to a geological timescale of many millions of years, then we can better imagine how the rocks in the cliff at the top of the picture moved to the Earth's surface from a depth of 20 kilometres in 30 million years.

Landscapes change rapidly

Figure 7.15 Erosional processes on the southern flank of Val Bedretto (canton Ticino, Switzerland): mass movements, rock falls and debris flows.

References

Abele, G., 1974, Bergstürze in den Alpen. Wissenschaftliche Alpenvereinshefte, 25, 1–230.

Arca, S. & Beretta, G.P., 1985, Prima sintesi geodetico-geologica si movimenti verticali del suolo nell'Italia Settentrionale (1987–1957). Bolletino di geodesia e scienze affini, No. 2, Anno XLIV, 125–156.

B.E.V., 1991, Recent crustal movements in Austria. Map 1: 2 000 000, Bundesamt für Eich- und Vermessungswesen (B.E.V.), Wien.

De Beaulieu, J.-L., Montjuvent, G. & Nicoud, G., 1991, Chronology of the Würmian glaciation in the French Alps: A survey and new hypotheses. In: Frenzel, B. (ed.), Klimagechichtliche Probleme der letzten 130 000 Jahre, Gustav Fischer Verlag, Stuttgart, 435–448.

Delacou, B., Sue, Ch., Champagnac, J.-D. & Burkhard, M., 2004, Present-day geodynamics in the bend of the western and central Alps as constrained by earthquake analysis. Geophysical Journal International, 158, 753–774.

Florineth, D. & Schlüchter, C., 1998, Reconstructin the Last Glacial Maximum (LGM) ice surface geometry and flowlines in he Central Swiss Alps. Eclogae geologicae Helvetiae, 91/3, 391–407.

Florineth, D. & Schlüchter, C., 2000, Alpine evidence for atmospheric circulation patterns in Europe during the Last Glacial Maximum. Quaternary Research, 54, 295–308.

Frisch, W., Dunkl, I. & Kuhlemann, J., 2000a, Post-collisional orogen-parallel large-scale extension in the Eastern Alps. Tectonophysics, 327, 239–265.

Frisch, W., Székely, B., Kuhlemann, J. & Dunkl. I., 2000b, Geomorphological evolution of the Eastern Alps in response to Miocene tectonics. Zeitschrift für Geomorphologie, Neue Folge 44/1, 103–138.

Giamboni, M., Ustaszewski, K., Schmid, S. M., Schumacher, M. E. & Wetzel, A., 2004, Plio-Pleistocene transpressional reactivation of Paleozoic and Paleogene structures in the Rhine-Bresse transform zone (northern Switzerland and eastern France). International Journal of Earth Sciences, 93, 207–223.

Graf, H. R., 1993, Die Deckenschotter der zentralen Nordschweiz. Dissertation ETH Nr. 10205, 151 pp.

Gruner, U., 2006, Bergstürze und Klima in den Alpen gibt es Zusammenhänge? Bulletin für Angewandte Geologie, 11/2, 25–34.

Gubler, E., 1991, The UELN and the Swiss National Leveling Net. In: Report on the geodetic activities in the years 1987–1991. XX. General Assembly of the International Union of Geodesy and Geophysics, Wien 1991. Zürich 1991.

Hofmann, F., 1974, Geologische Geschichte des Bodenseegebietes. Schriftenreihe des Vereins für Geschichte des Bodensees und seiner Umgebung, 92, 251–273.

Jäckli, H., 1958, Der rezente Abtrag der Alpen im Spiegel der Vorlandsedimentation. Eclogae geologicae Helvetiae, 5, 354–365.

Jouanne, F., Ménard, G. & Darmendrail, X., 1995, Present-day vertical displacements in the north-western Alps and southern Jura Mountains: Data from leveling comparisons. Tectonics, 14/3, 606–616.

Kahle, H.-G., Geiger, A., Bürki, E., Gubler, U., Marti, B., Wirth, M., Rothacher, W., Gurtner, W., Beutler, G., Bauersima, I. & Pfiffner, O. A., 1997, Recent crustal movements, geoid and density distribution: Contribution from integrated satellite and terrestrial measurements. In: Pfiffner, O. A., Lehner, P., Heitzmann, P., Mueller, St. & Steck, A. (eds), Deep structure of the Swiss Alps: Results of NRP 20, Birkhäuser, 251–259.

Kelly, M. A., Buoncristiani, J.-F. & Schlüchter, C., 2004, A recontstruction of the last glacial maximum (LGM) ice-surface geometry in the western Swiss Alps and contiguous Alpine regions in Italy and France. Eclogae geologicae Helvetiae, 97/1, 57–75.

Kissling, E., Schwendener, H., 1990, The Quaternary sedimentary fill of some Alpine valleys by gravity modelling. Eclogae geologicae Helvetiae, 82, 311–321, doi.org/10.5169/seals-166589.

Kuhlemann, J., Frisch, W., Székely, B., Dunkl, I. & Kazmér, M., 2002, Post-collisional sediment budget history of the Alps: tectonic versus climatic control. International Journal of Earth Science, 91, 818–837.

Kühni, A. & Pfiffner, O. A., 2001a, Drainage pattern and tectonic forcing: A model study for the Swiss Alps. Basin Research 13, 169–197.

Kühni, A. & Pfiffner, O. A., 2001b, The relief of the Swiss Alps and adjacent areas and its relation to lithology and structure: topographic analysis from a 250-m DEM. Geomorphology, 41, 285–307.

Liniger, H., 1966, Das plio-altpleistozäne Flussnetz der Nordschweiz. Regio Basiliensis, 72, 158–177.

Nocquet, J.-M., Walpersdorf, A., Jouanne, F., Masson, F., Chéry, J., Vernant, Ph. & RENAG team, 2011, Uplift in the Western Alps from a Decade of Conti-nuous GPS Measurements. Colloque G2 – ENSG – Marne La Vallée – 7–9 Novembre 2011.

Persaud, M. & Pfiffner, O. A. (2004): Active tectonics in the eastern Swiss Alps: post-glacial faults, seismicity and surface uplift. Tectonophysics, 385/1–4, 59–84.

Pfiffner, O. A., Heitzmann, P., Lehner, P., Frei, W., Pugin, A. & Felber, M., 1997, Incision and backfilling of Alpine valleys: Pliocene, Pleistocene and Holocene processes. In: Pfiffner, O. A., Lehner, P., Heitzmann, P., Mueller, St. & Steck, A. (eds), Deep structure of the Swiss Alps: Results of NRP 20, Birkhäuser, 265–288.

Preusser, F., Reitner, J.M., Schlüchter, Ch., 2010, Distribution, geometry, age and origin of overdeepened valleys and basins in the Alps and their foreland. Swiss Journal of Geosciences, 103, 407–426, DOI 10.1007/s00015-010-0044-y.

Rohling, E. J. & Pälike, H., 2005, Centennial-scale climate cooling with a sudden cold event around 8200 years ago. Nature, 434/21, 975–979.

Röthlisberger, H., 1972, Water pressure in intra- and subglacial channels. Journal of Glaciology, 11, 177–203.

Schlüchter, C. & Kelley, M., 2000, Das Eiszeitalter in der Schweiz: Eine schema-tische Zusammenfassung. Publikation des Geologischen Institutes der Universität Bern und IGCP-378, 4 pp.

Selzer, C. Buiter, S. J. H. & Pfiffner, O. A., 2008, Numerical modeling of frontal and basal accretion at collisional margins. Tectonics, 27, TC3001, doi: 10.1029/2007TC002169, 26 pp.

Summerfield, M. A. & Hulton, N. J., 1994, Natural controls of fluvial denudation rates in major world drainage basins. Journal of Geophysical Research, 99/87, 13 871–13 883.

Ustaszewski, M. & Pfiffner, O. A., 2008, Neotectonic faulting, uplift and seismicity in the western Swiss Alps. Geological Society of London, Special Publication, 298, 231–249.

Ustaszewski, M. E., Hampel, A. & Pfiffner, O. A., 2008, Composite faults in the Swiss Alps formed by the interplay of tectonics, gravitation and postglacial rebound: an integrated field and modelling study. Swiss Journal of Geosciences, 101, 223–235.

Van Husen, D., 2000, Geological processes during the Quaternary. In: Neubauer, F. & Höck, V. (eds), Aspects of Geology in Austria, Österreichische Geologische Gesellschaft, Wien, 135–256. Also published in Mitteilungen der Österreichischen Geologischen Gesellschaft, 92. Band (1999).

Villinger, E., 2003, Zur Paläogeographie von Alpenrhein und oberer Donau. Zeitschrift der deutschen geologischen Gesellschaft, 154, 193–253.

Von Poschinger, A., 2006, Weitere Erkenntnisse und weitere Fragen zum Flimser Bergsturz. Bulletin für Angewandte Geologie, 11/2, 35–43.

Wildi, W., 1985, Heavy mineral distribution and dispersal pattern in penninic and ligurian flysch basins (Alps, northern Apennines). Giornale di Geologia, seria 3a, 47/1–2, 77–99.

Stratigraphic timetable

after Haq and the "International Commission on Stratigraphy"

Era	Period	Epoch	Stage*		Ma
Cenozoic	Quaternary	Holocene			0.0117
		Pleistocene			
			Gelasian		2.58
	Neogene	Pliocene	Piacenzian		
			Zanclean		5.3
		Miocene	Messinian		
			Tortonian		
			Serravallian		
			Langhian		
			Burdigalian		
			Aquitanian		23
	Paleogene	Oligocene	Chattian		
			Rupelian		
			Priabonian		33.9
		Eocene	Bartonian		
			Lutetian		
			Ypresian		55.8
		Paleocene	Thanetian		
			Selandian		
			Danian		65.5
Mesozoic	Cretaceous	Late	Maastrichtian		
			Campanian	«Senonian»	
			Santonian		
			Coniacian		
			Turonian		
			Cenomanian		99.6
		Early	Albian		
			Aptian		
			Barremian		
			Hauterivian	«Neocomian»	
			Valanginian		
			Berriasian		145.5
	Jurassic	Late / Malm	Tithonian		
			Kimmeridgian		
			Oxfordian		161.2
		Middle / Dogger	Callovian		
			Bathonian		
			Bajocian		
			Aalenian		175.6
		Early / Lias	Toarcian		
			Pliensbachian		
			Sinemurian		
			Hettangian		199.6
	Triassic	Late	Rhätian		
			Norian		
			Carnian		228
		Middle	Ladinian		
			Anisian		245
		Early	Olenekian		
			Induan		251
Palaeozoic	Permian	Late	Thuringian		270.6
		Early	Saxonian		
			Autunian		299
	Carboniferous	Late	Stephanian		
			Westphalian		318.1
		Early	Namurian		
			Visean		
			Tournaisian		359.2
	Devonian				416
	Silurian				443.7
	Ordovician				488.3
	Cambrian				542
Proterozoic					2500
Archean					3800

* Stages as commonly used in the literature on Alpine geology.

Index

Figure numbers are given in bold.